Crop Science: Progress and Prospects

Crop Science: Progress and Prospects

Edited by

J. Nösberger
ETH Zürich, Switzerland

H.H. Geiger
University of Hohenheim, Germany

and

P.C. Struik
Wageningen University, The Netherlands

CABI *Publishing*

CABI *Publishing* is a division of CAB *International*

CABI Publishing
CAB International
Wallingford
Oxon OX10 8DE
UK

CABI Publishing
10 E 40th Street
Suite 3203
New York, NY 10016
USA

Tel: +44 (0)1491 832111
Fax: +44 (0)1491 833508
Email: cabi@cabi.org
Web site: www.cabi.org

Tel: +1 212 481 7018
Fax: +1 212 686 7993
Email: cabi-nao@cabi.org

A catalogue record for this book is available from the British Library, London, UK.

Library of Congress Cataloging-in-Publication Data
Crop science: progress and prospects / edited by J. Nösberger, H.H. Geiger and P.C.
   Struik.
      p. cm.
   Papers presented at the Third International Crop Science Congress in
   Hamburg, Germany.
   Includes bibliographical references (p.).
   ISBN 0-85199-530-6 (hard cover: alk. paper)
     1. Crop science–Congresses. I. Nösberger, J. (Josef) II. Geiger, H.H. (Hartwig
   H.) III.
Struik, P.C. (Paul Christiaan), 1954– IV. International Crop Science Congress
(3rd: 2000: Hamburg, Germany)

SB16.A27 2001
631– –dc21   2001025486

ISBN 0 85199 530 6

Typeset by Wyvern 21 Ltd, Bristol, UK.
Printed and bound in the UK by Cromwell Press, Trowbridge.

Contents

Contributors   ix

Preface   xiii

Foreword   xv

Congress Sponsors   xvii

Part 1: Facing the Growing Needs of Mankind

1   Food Security? We Are Losing Ground Fast!   1
   F.W.T. Penning de Vries

2   The Future of World, National and Household Food Security   15
   F. Heidhues

3   Crop Science Research to Assure Food Security   33
   K.G. Cassman

4   Modifying the Composition of Plant Foods for Better Human   53
   Health
   R.F. Hurrell

5   Facing the Growing Needs of Mankind – Grasslands and   65
   Rangelands
   R.J. Wilkins

Part 2: Stress in Crops and Cropping Systems

6 Abiotic Stresses, Plant Reaction and New Approaches Towards   81
   Understanding Stress Tolerance
   H.J. Bohnert and R.A. Bressan

7 Plant Stress Factors: Their Impact on Productivity of Cropping   101
   Systems
   U.R. Sangakkara

8 Optimizing Water Use   119
   N.C. Turner

9 Abiotic Stresses and Staple Crops   137
   G.O. Edmeades, M. Cooper, R. Lafitte, C. Zinselmeier, J.-M. Ribaut,
   J.E. Habben, C. Löffler and M. Bänziger

10 Biotic Stresses in Crops   155
   R. Nelson

11 Management of Complex Interactions for Growth Resources and   175
   of Biotic Stresses in Agroforestry
   C.K. Ong and M.R. Rao

Part 3: Diversity in Agroecosystems

12 Optimizing Crop Diversification   191
   D.J. Connor

13 Biodiversity of Agroecosystems: Past, Present and Uncertain   213
   Future
   P.J. Edwards and A. Hilbeck

14 Conservation and Utilization of Biodiversity in the Andean Eco-   231
   region
   W.W. Collins

15 The Role of Landscape Heterogeneity in the Sustainability of   243
   Cropping Systems
   J. Baudry and F. Papy

Part 4: Designing Crops and Cropping Systems for the Future

16 Cropping Systems for the Future   261
   J. Boiffin, E. Malezieux and D. Picard

17 Will Yield Barriers Limit Future Rice Production?   281
    J.E. Sheehy

18 New Crops for the 21st Century   307
    J. Janick

19 Plant Biotechnology: Methods, Goals and Achievements   329
    U. Sonnewald and K. Herbers

20 Transgenic Plants for Sustainable Crop Production   351
    B. Keller and E. Hütter Carabias

Part 5: Position Papers

21 Crop Science: Scientific and Ethical Challenges to Meet Human   369
    Needs
    L.O. Fresco

22 Declaration of Hamburg   381
    J.H.J. Spiertz

Index   385

Contributors

M. Bänziger, Centro Internacional de Mejoramiento de Maiz y Trigo (CIMMYT), PO Box MP163, Harare, Zimbabwe

J. Baudry, Institut National de la Recherche Agronomique, SAD-Armorique, 65 rue de Saint-Brieuc. F-35042 Rennes Cédex, France

H.J. Bohnert, Department of Plant Biology and of Crop Sciences, 1201 West Gregory Drive, Urbana, IL 61801, USA

J. Boiffin, Institut National de la Recherche Agronomique (INRA), 147 rue de l'Université, F-75338 Paris Cédex 07, France

R.A. Bressan, Department of Horticulture, Purdue University, Horticulture Building, West Lafayette, IN 47907-1165, USA

K.G. Cassman, Department of Agronomy, University of Nebraska, PO Box 830915, Lincoln, NE 68583-0915, USA

W.W. Collins, International Potato Center, PO Box 1558, Lima 12, Peru

D.J. Connor, Department of Crop Production, The University of Melbourne, Victoria 3010, Australia

M. Cooper, School of Land and Food Sciences, University of Queensland, Brisbane, Qld 4072, Australia

G.O. Edmeades, Pioneer Hi-Bred Int. Inc., 7250 NW 62nd Avenue, Johnston, IA 50131, USA

P.J. Edwards, Geobotanical Institute, ETH Zürich, Zürichbergstrasse 38, CH-8044 Zürich, Switzerland

L.O. Fresco, Agriculture Department, Food and Agriculture Organization of the United Nations, Viale delle Terme di Caracalla, 01100 Rome, Italy

H.H. Geiger, Institut für Pflanzenzüchtung, Saatgutforschung und Populationsgenetik, Universität Hohenheim, D-70593 Stuttgart, Germany

J.E. Habben, Pioneer Hi-Bred Int. Inc., 7250 NW 62nd Avenue, Johnston, IA 50131, USA

F. Heidhues, Institut für Agrar- und Sozialökonomie in den Tropen und Subtropen, Universität Hohenheim, D-70593 Stuttgart, Germany

K. Herbers, SunGene GmbH & CoKGaA, Corrensstrasse 3, D-06466 Gatersleben, Germany

A. Hilbeck, Geobotanical Institute, ETH Zürich, Zürichbergstrasse 38, CH-8044 Zürich, Switzerland

R.F. Hurrell, Laboratory for Human Nutrition, Institute of Food Science, ETH Zürich, PO Box 474, CH-8803 Rüschlikon, Switzerland

E. Hütter Carabias, Institut für Pflanzenbiologie, Universität Zürich, Zollikerstrasse 107, CH-8008 Zürich, Switzerland

J. Janick, Center for New Crops and Plant Products, Horticulture and Landscape Architecture, Purdue University, West Lafayette, IN 478907-1165, USA

B. Keller, Institut für Pflanzenbiologie, Universität Zürich, Zollikerstrasse 107, CH-8008 Zürich, Switzerland

R. Lafitte, IRRI, PO Box 3127 MCPO, 1271 Makati City, Philippines

C. Löffler, Pioneer Hi-Bred Int. Inc., 7250 NW 62nd Avenue, Johnston, IA 50131, USA

E. Malezieux, Centre de Coopération Internationale en Recherche Agronomique pour le Développement (CIRAD), BP 5035, 34032 Montpellier Cédex, France

R.J. Nelson, International Potato Center, PO Box 1558, Lima 12, Peru

J. Nösberger, Institute of Plant Sciences, ETH Zürich, CH-8092 Zürich, Switzerland

C.K. Ong, International Centre for Research in Agroforestry (ICRAF), United Nations Avenue, Gigiri PO Box 30677, Nairobi, Kenya

F. Papy, Institut National de la Recherche Agronomique, SAD APT, BP 01, F- 78850 Thiverval-Grignon, France

F.W.T. Penning de Vries, International Board for Soil Research and Management (IBSRAM), PO Box 9-109, Jatujak, Bangkok 10900, Thailand

D. Picard, Institut National de la Recherche Agronomique (INRA), Centre de Recherche de Versailles, Route de Saint Cyr, 78026 Versailles Cédex, France

M. R. Rao, 11, ICRISAT Colony-I, Akbar Road, Cantonment, Secunderabad-500 009, AP, India

J.-M. Ribaut, CIMMYT, Apdo Postal 6-641, Mexico 06600, Mexico

U.R. Sangakkara, Faculty of Agriculture, University of Peradeniya, Peradeniya 20400, Sri Lanka

J.E. Sheehy, International Rice Research Institute (IRRI), MCPO Box 3127, Makati City 1271, Philippines

U. Sonnewald, Institut für Pflanzengenetik und Kulturpflanzenforschung, Leibnitz-Institut, Corrensstrasse 3, D-06466 Gatersleben, Germany

J.H.J. Spiertz, Crop Ecology, Department of Plant Sciences, Wageningen University, PO Box 430, 6700 AK Wageningen, The Netherlands

P.C. Struik, Crop and Weed Ecology Group, Department of Plant Sciences,

Wageningen University and Research Centre, Haarweg 333, 6709 RZ Wageningen, The Netherlands

H. Stützel, Institut für Gemüse- und Obstbau, Universität Hannover, Herrenhäuserstrasse 2, D-30419 Hannover, Germany

N.C. Turner, CSIRO Plant Industry, Centre for Mediterranian Agricultural Research, Private Bag No. 5, Wembley (Perth), WA 6913, Australia

R.J. Wilkins, Institute of Grassland and Environmental Research, North Wyke, Okehampton, Devon EX20 2SB, UK, and University of Plymouth, Seale-Hayne Faculty of Agriculture, Food and Land Use, Newton Abbot TQ12 6NQ, UK

C. Zinselmeier, Pioneer Hi-Bred Int. Inc., 7250 NW 62nd Avenue, Johnston, IA 50131, USA

Preface

This book contains 20 invited chapters written by renowned scientists from throughout the world, that resulted from the Third International Crop Science Congress held in Hamburg, 17–22 August 2000. After the Congress, the papers were adapted to meet the requirements for a stimulating textbook for advanced students or young professionals. The topics around the theme of the Congress, 'Meeting Future Human Needs', are crucial to crop scientists worldwide. The challenges raised in each chapter clearly show the tasks ahead for crop scientists interested in meeting the demands for food of a growing population in a sustainable way.

The subject matter presented in this book is organized into five general parts that correspond to the four programme themes of the Congress. The first provides an overview of the growing needs of humankind and stresses the constraints imposed by scarce natural resources and the actual genetic potential of crop plants. Part 2 focuses on biotic and abiotic stress in crops and cropping systems. The analysis of the stress situation from the molecular to the system level offers new insights that are a prerequisite for innovative approaches in agronomy and plant breeding. However, agricultural land use is not only the core activity for the production of food, but also a driving force for the diversity and stability of agro-ecosystems. Part 3 explains why regional differences in gene populations as well as biological diversity in agricultural ecosystems are crucial traits for sustainable production systems, while the potential of new technologies is developed in Part 4. Cropping systems can be designed for specific requirements on a more rational basis with the use of decision support systems. Biotechnology offers great opportunities for changing crops for the future. Finally, the book contains the Declaration of Hamburg, expressing the concern of crop scientists about the role of science and society in meeting the demands of future human needs, while a contribution from FAO analyses world agricultural trends from an ethical perspective.

J. Nösberger
Swiss Federal Institute of Technology, Zurich, Switzerland

H.H. Geiger
University of Hohenheim, Stuttgart, Germany

P.C. Struik
Wageningen University, The Netherlands

Foreword

Through the achievements made in crop science and production technology over the last decades, agriculture is now able to feed the majority of the world's population better than in the past. However, there is an increasing concern that the present knowledge, resources and technologies will not be adequate to meet the demands, once there are 8 billion people on this planet by about 2020. Challenges are to feed and to fulfil the needs of a growing population in a sustainable way. This requires a better and more comprehensive insight into ecologically sound crop production processes, especially in fragile environments and resource-poor countries. Furthermore, there is a need to integrate the newly acquired knowledge in the field of gene and information technology in the development of future crops and cropping systems.

Strengthening agricultural research and education at national and international levels is a prerequisite to fulfilling future human needs. There is a need for crop scientists worldwide to rethink their responsibility towards the global needs for food, rural development, and human health and well-being at the one side and the conservation and efficient use of scarce resources at the other. Crop science deals with problems that are consumer related, such as food quality and safety, but at the same time with sustainable use of land, water and genetic resources. The scope is from the gene to the field and from the crop to food and health.

It was a great honour to organize the Third International Crop Science Congress in Europe. The European Society for Agronomy (ESA) in cooperation with the German Societies for Agronomy and for Plant Breeding took the formal responsibilities. Many individuals contributed to the success of this Congress with participants from over 100 countries. The core group of the Programme Committee made an utmost effort to invite outstanding

scientists in the various fields to enrich the scientific quality of the programme. The proceedings cover the plenary and keynote papers of four
themes: food security and safety, biotic and abiotic stresses, diversity in
agroecosystems and future crops and cropping systems. It is highly appreciated that these papers could be published in a way that the proceedings may
serve as a textbook for advanced students and young professionals in crop
science.

Hartmut Stützel
President ESA 1998–2000

Hubert Spiertz
President ICSC – 2000

Congress Sponsors

The Congress organizers express their sincere thanks to the following companies and institutions who made it possible to organize this important 3rd International Crop Science Congress in Hamburg, and who have supported a number of participants from many, mainly poor, countries. Financial assistance was also provided by the Rockefeller Foundation.

Aventis Crop Science, France and Germany
ASA, Germany
Bayer AG, Germany; http://www.agro.bayer.com
Centre de Coopération Internationale en Recherche Agronomique pour le Développement (CIRAD), France
City of Hamburg, Germany
CMA, Germany
CTA, EU, The Netherlands; http://www.cta.nl
European Commission (EU), Belgium; http://europa.eu.int/com/research and http://www.cordis.lu/improving
European Society for Agronomy (ESA), France
Eiselen Foundation Ulm, Germany; http://www.eiselen-stiftung.de/index e.html
German Association for the Promotion of Integrated Cropping, (FIP), Germany
German Ministry of Food, Agriculture and Forestry, (BELF), Germany
German Ministry of Technical Cooperation (BMZ and DSE), Germany; http://www.dse.de
German Research Foundation (DFG), Germany
World Phosphate Institute (IMPHOS), Morocco
Institut National de la Recherche Agronomique (INRA), France

International Foundation for Science (IFS), Sweden; http://www.ifs.se
ISTA Mielke GmbH, Germany
Kali und Salz GmbH, Germany; www.kalisalz.de
KWS Saat AG, Germany; www.kws.de
Leventis Foundation, UK
Norddeutsche Pflanzenzucht Hans-Georg-Lembke KG (NPZ), Germany
Saaten Union GmbH, Germany
Saka-Ragis Pflanzenzucht GbR, Germany
Union zur Förderung von Oel- und Proteinpflanzen e.V. (UFOP), Germany
Wageningen University, The Netherlands; http://www.wageningen-ur.nl/
uk/organisation
Wintersteiger GmbH, Austria
The World Bank, USA
Zeneca Agro, Germany

Food Security? We Are Losing Ground Fast!

F.W.T. Penning de Vries

International Board for Soil Research and Management (IBSRAM), PO Box 9-109, Jatujak, Bangkok 10900, Thailand

Introduction

'Food security' is a complex concept. It implies 'physical and economic access to balanced diets and safe drinking water to all people at all times' (Swaminathan, 1986). This means that ample food is grown, processed and transported, and that everyone has either money to buy food or grow it. This chapter cannot but deal with the 'land' subset of food security issues, and with how 'land' relates to food production and to income generation. In particular, this chapter discusses the impacts of land degradation on national food self-sufficiency and household food security. It will first discuss the trends in total cultivated land area and in the quality of the cultivated land, and then translate them as consequences for food production and income generation.

About 60% of the world's land surface is suitable for grazing, half of which can also be used for arable cropping in a sustainable manner (3.4×10^9 ha). Nations are endowed with good land to very different degrees. The area of land suitable for cropping but still unused is still very significant in southern Africa and the Americas, but suitable unused land is already scarce in Asia and East Africa. Yet, the growing population, particularly in Asia, and the changing diets will lead to a much higher food demand in 2020. Moreover, most population growth will be in urban areas (FAO, 2000). This means that: (i) food production globally should double in the next 20 years; (ii) trade and reduction of urban poverty should make food accessible to the entire urban population; and (iii) farmers on degraded land should acquire new means of generating income and hence achieving food security.

In earlier analyses based on data from the 1960s, some scientists underlined that the global carrying capacity was very large and could meet any demand for food (Penning de Vries and Rabbinge, 1997), while others stressed that many poor countries continue to have insufficient income to generate that demand (IFPRI, 1994), or are more conservative about possibilities (Alexandratos, 1995). But new information shows that widespread land degradation seriously threatens global food security, even if money is available, as it lowers the food production and carrying capacity, and moreover reduces the capacity of resource-poor farmers to generate income. Implications for crop science are discussed.

Trends in the Area of Land Cultivated

Irreversible degradation

To produce the food an average human being consumes, 0.05–0.5 ha of land is required. But agricultural land can become degraded completely and irreversibly by various processes, including soil erosion, nutrient mining, salinization and pollution. Among the immediate causes are agricultural practices, such as cultivation without manure or fertilizer, overgrazing and deforestation, and often abetted by an unfavourable climate. Unfavourable socio-economic conditions determine that farmers often cannot avoid such practices. On the basis of a review of more than 80 case studies with data from the 1980s, Scherr (1999) estimated that 16% of all agricultural land in developing countries (total 0.85×10^9 ha) is seriously degraded, meaning that crops can no longer be grown profitably and that restoration is economically impossible. She derived the global average annual loss for the past five decades to be 0.3–1.0% ($5–8 \times 10^6$ ha) of arable land, and calculated the global loss of agricultural productivity due to the cumulated degradation as high as 5–9%. The rate of loss of productive land is increasing (ADB, 1997). Scherr expects that degradation will force an additional 0.15–0.36×10^9 ha out of production by 2020. This is as much as 10–20% of all land currently cultivated. For an estimate of the future impact of degradation, the conservative and global average of 0.5% loss of agricultural land per year is applied.

Infrastructure

Building of roads and infrastructure often occurs at the expense of cultivated land. A significant portion of land suitable for agriculture is already covered by roads, houses, industries, golf courses and other recreational facilities. In OECD (Organization for Economic Cooperation and Development) countries, the fraction is 10% or more, and in developing countries generally 5% or less, but it generally covers the very best soils (Young, 1998). City encroachment on agricultural land increases the pressure on the remaining land.

Housing, roads and other infrastructures require in the order of 0.025 ha per capita (Young, 1998). This fraction is also rising: the annual loss of agricultural land in China is estimated at 0.5% (Xu, 1994). Also, in OECD countries suburbanization continues, and Evans (2000) cites the same relative rate (equal to 8×10^6 ha year^{-1}) for a global average, and stresses its importance. Compensation by urban agriculture is modest. For extrapolation into the future, the value of 0.1% annual loss is used, similar to the 40-year average.

Acquiring new land

Approximately 1.5×10^9 ha of land is currently cultivated, and another 1.7×10^9 ha is used for grazing and forestry (FAO, 2000). Even though we are losing land area in some regions, more is brought under cultivation elsewhere, and the total area of cultivated land continues to increase in Africa, but is almost stable in Asia (Fig. 1.1). Scherr (1999) and Young (1998) estimated that only an equivalent of 10–30% of currently cropped land is actually still available. Since land resources and populations are very unevenly spread, this implies that resource-poor farmers in some countries are already cultivating marginal lands, the exploitation of which cannot be sustainable: they can extract a living from the soil, but degrade it completely within 5–20 years in many cases. The overall 'degrade and pollute now, pay later' attitude towards environmental problems should therefore change quickly (ADB, 1997). Young (1998) calculated that some countries, including Bangladesh and Pakistan, with 50% or more of the poor population

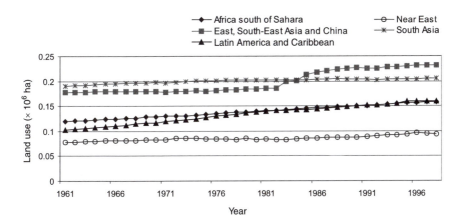

Fig. 1.1. Developments in land use globally, in sub-Saharan Africa (AFR), Middle East and North Africa (MENA), South Asia (SA), South-East and East Asia and the Pacific (EAP), and South plus Central America (LAC), in 10^6 ha (FAO, 2000).

involved in agriculture, have a negative land balance already. It is therefore expected that the cultivated area in Asia will start to contract.

Non-food crops

Some arable land is used for non-food crops. FAO data show that this fraction amounts to approximately 30% (sum of total arable area minus all food crops), and includes tree plantations for timber and paper (10%, Evans, 2000), crops for fibre, flowers and pharmaceuticals. The coarse data show that this percentage became smaller in past decades at the global level (FAO, 2000), and nearly halved in Asia and also, surprisingly, in Africa. While this trend may reverse in the future, expansion of non-food crops does not appear to threaten food production capacity. On the contrary, it provides more income to strengthen food security.

Energy crops

Crops used for energy generation but without any contribution to food production deserve separate attention. The global need for energy is huge, and since oil reserves are dwindling and the Kyoto protocol urges countries to reduce their net output of the greenhouse gas CO_2, energy plantations could theoretically provide a solution. Consumption of non-metabolic energy varies from 20×10^9 J per capita year^{-1} (traditional living in Africa) to 225×10^9 and more (OECD countries). Rapid growth in energy consumption is expected in many developing countries. Energy crops (such as willow coppice) in sustainable plantation systems in Europe yield $5-15 \times 10^3$ kg ha^{-1} year^{-1} of combustible dry matter, or $70-210 \times 10^9$ J ha^{-1} year^{-1}. In the humid tropics with fertilizer the figure may be double, but poor soils in a sub-humid climate yield less. These data imply that if all energy for human use (heating, transport, cooking, etc.) were generated by energy crops, every individual would need 0.2–2.0 ha of plantation. Minimum land area for green energy is clearly an order of magnitude larger than minimum area for food and infrastructure. This could make energy crops clear competitors with food crops for land area. Fortunately, energy crops are not attractive financially: cost per unit of electricity generated from biomass in Europe would be three times as much as electricity from coal (but nearly the same when externalities and an existing subsidy are included; Meuleman and Turkenburg, 1997). Moreover, economy of scale requires this technology to be applied at a large scale ($>1 \times 10^5$ ha), bringing the option to grow energy crops beyond the choice of individual farmers. Rising energy prices could invite large-scale energy crops in areas with a large demand for electricity, insufficient hydropower, and a low capacity to generate income through other crops. Such a development would resemble the use of crops to substitute oil through sugar and alcohol. Although technically this worked well in Brazil and elsewhere, the process has a high minimum throughput and brings a low return to investment. It is therefore expected that energy crops will not provide significant competition to food crops in the next 20–40 years.

The balance: a flight forward

Approximately 1.5×10^9 ha of land is cropped, mainly with food crops (FAO, 2000). Even though we are losing land in parts of many countries, more is brought under cultivation, and the total area of cultivated land is increasing slowly. However, more productive land may have been irreversibly lost in the past 10^4 years than is currently being cropped (Rozanov and colleagues, cited in Scherr, 1999). Ponting (1991) points at the decline of major civilizations, e.g. the Sumerians in southern Mesopotamia, among others, due to unsustainable land and water management. Humanity appears to be slowly consuming the natural resource of 'land'. Even though this process has been ongoing for millennia, it cannot continue much longer since most land and water reserves are nearly fully engaged.

The next 20 years

Extrapolation of the data by Scherr (1999, Table 5), where the category 'severe degradation' corresponds with loss of productive capacity and about two-thirds of all degraded land is 'severely degraded', we find that in Africa 28–35% of the land that was suitable for agriculture in the 1960s will be out of production by 2020 and in Asia 19–26% – the same as the average for the entire world. Judicious practices, including better crops and proper use of fertilizers and water, could stop degradation, and reverse it. Unfortunately, the socio-economic environment in many developing countries (unfavourable prices, markets, land tenure) often still provides insufficient inducement for adoption of these practices (Craswell, 2000). When these conditions do not improve, land degradation will remove 10–20% of land currently cultivated out of production by 2020. Building and infrastructures will remove another 5%. Provided that the area with non-food crops (including energy) does not expand, then 15–25% of the agricultural land will go out of food production between 1990 and 2020. In practice, severely degraded land can no longer be replaced fully by 'new' land on which agriculture could be sustainable. The unequal distribution of good soils and opportunities for non-farm income will force some countries to reach the limit soon, or they have reached it already and use unsuitable land (Young, 1998). Signs that this is happening include land use conflicts around conservation areas, low farm income on the recently acquired lands, rising national imports and accelerated degradation. Particularly when legislation around natural resources remains underdeveloped and land tenure is not arranged, much suffering by resource-poor land users is to be expected.

Figure 1.2 summarizes the developments in land area in five key regions of the world: sub-Saharan Africa (AFR), Latin and Central America (LAC), Middle East and North Africa (MENA), South Asia (SA), and South-East and East Asia (EAP). The figure shows land degraded, land in use and suitable land, all relative to the total area 'suitable for agriculture' before human impact started. We see that in all regions, land for agriculture is progressively consumed, and that the reserves of suitable land are getting smaller. In MENA and SA, land is already being used on which agriculture is unlikely to develop in a sustainable manner, and this fraction is growing. The data for these figures are calculated with the overall averages of rates of change mentioned before, while for the year 2020, a scenario of continued degradation was chosen. The implications for global carrying capacity are not easy to compute, as it relates also to climate and availability of irrigation water. However, very roughly, a loss in land area is proportional to the loss in potential food production.

Large countries can enhance production on their best lands and maintain self-sufficiency. Many (57) developing countries, however, are small and half of them already experience high (0.16–0.30 ha per capita) or very high land pressure (<0.15), so that conserving farmland quality must be a strategic food security concern (Scherr, 1999).

Trends in the Quality of Cultivated Land

Land loses value for agricultural production if it is not adequately managed. Degradation refers to removal of nutrients (soil mining) or soil and nutrients (erosion), salinization and compaction. Nutrient depletion occurs when crops are harvested and nutrients are not returned to the field, or replaced by manure or fertilizers. This can lead to acidification and is common on poor soils in areas with marginal agriculture. The most destructive process, erosion, occurs on sloping land when rain hits soil when it is vulnerable after tillage (Turkelboom, 1999), or from building infrastructures (Enters, 2000). It removes soil with nutrients from slopes and deposits it in valleys, where the sediments may enhance agriculture or silt up waterways and reservoirs. Salinization occurs when salt, brought by water to the surface, is not flushed down or out regularly. Land in several large irrigation schemes suffers seriously from salinization.

Light and moderate soil degradation affects agriculture in three ways (Fig. 1.3): (i) it reduces maximum crop yield (by affecting the water holding capacity of the soils for rainfed cultivation); (ii) it lowers actual yields if little fertilizer is used (weaker input use efficiency, less profit, increasing weeds); and (iii) it increases the risk of crop failure. These factors make production

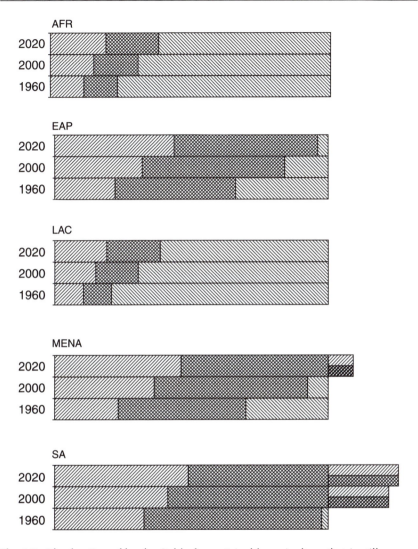

Fig. 1.2. The fraction of land suitable for sustainable agriculture that is still available (right), the land currently in use for arable and permanent crops (centre), and the area degraded to an extent that is not cultivable and recovery is uneconomical (left) is shown for all regions. The lower set of bars indicates the situation in 1960, the middle set the current situation (2000), and the upper set the situation for 2020 for a medium scenario of development described in the text. The bar is split when more land is 'used' than is 'available' for sustainable agriculture. See p. 6 for abbreviations.

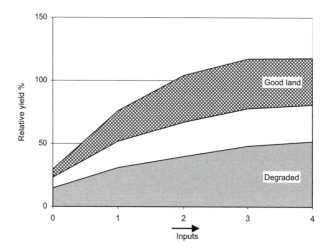

Fig. 1.3. Two hypothetical sets of curves of crop yield in rainfed conditions as a function of the level of inputs (labour, water, fertilizer, crop protection). The set on degraded land shows lower maximum yields, reduced input efficiency and higher risk.

costs higher (Scherr, 2001), generation of income more difficult, and investment in agriculture less attractive. Indirectly, land degradation tends to decrease rural household food security.

While it is impossible to be precise about the extent and rate of loss of soil quality due to the multitude of processes and causes, the shortage of specific data, and the sometimes localized appearance of the problems, a brief overview will show that the global community is suffering large physical and economic losses.

Global

The publication of the Global Assessment of Land Degradation (GLASOD) made it possible to provide an estimate of the extent of human-induced soil degradation (Oldeman *et al.*, 1991), and it is still the main study that underlies many analyses. One of these suggests an aggregate global loss of 11.9–13.4% in productivity due to human-induced degradation since the mid-1940s (Crosson, 1995); 22% of the area suitable for agriculture is damaged, and the average global cumulative loss of potential productivity is 5–9% (Oldeman, 1994).

Degradation brings huge losses in the production capacity of the land resource. For example, the total annual cost of erosion from agriculture in the USA only is about US$44 $\times$ 10^9, about US$250 per hectare of cropland

and pasture. On a global scale, annual displacement of 75×10^9 t soil causes a $\$4 \times 10^{11}$ reduction in value of the global natural resource (Eswaran *et al.*, 2001) – even considering important off-site effects. Continued nutrient depletion undermines the long-term sustainability of vast areas of land. Gruhn and Goletti (1998) estimated that by 2020, the annual global net nutrient removal will reach more than 350×10^9 kg. Global inorganic fertilizer production in 2001 is only 157×10^9 kg. The projected 2020 supply of fertilizer falls short of covering the gap, let alone restoring the already lowered level of soil nutrients.

Africa

Overall yield reduction caused by past soil erosion in Africa is estimated to be 9%, and if the current rates continue, it will amount to 16.5% by 2020 (Bridges and Oldeman, 2001). This is based, among others, on Lal (1995) who suggests a yield reduction of 2–5% for each mm of lost soil (apart from its value as sediment). Agricultural productivity in sub-Saharan Africa derives its value of 7% directly from nutrient depletion of its soils (Drechsel and Penning de Vries, 2001). Human-induced desertification is experienced on 33% of the global land surface and affects more than 10^9 persons, half of whom live in Africa (Beinroth *et al.*, 2001). Half of the rainfed lands in Africa experience at least a 10% loss in productive potential, and irreversible productivity losses of at least 20% due to erosion occur in large parts of 11 countries (Dregne, 1990). GLASOD estimated cropland productivity loss to be 25% (Oldeman *et al.*, 1991). The gross discounted future loss from degradation varied from under 1% to 18%, and the gross discounted cumulative loss (which assumed a continued process of degradation over time) for five countries ranged from under 1% to 36% to 44% of the annual gross domestic product (AGDP) (Scherr, 2001).

Asia

The expert survey, Assessment of Soil Degradation in South and South-East Asia (ASSOD), found 'moderate' (or worse) impacts of soil degradation on 10% of all lands, and that serious fertility decline or salinization affected at least 15% of arable land (van Lynden and Oldeman, 1997). The annual cost of soil degradation in South Asia was estimated to be US$$9.8–11.0 \times 10^9$, the equivalent of 7% of aggregate AGDP. In South Asia, annual loss in productivity is estimated at 36×10^9 kg cereal equivalent valued at US$5400 million by water erosion and US$1800 million due to wind erosion (Beinroth *et al.*, 2001).

Uncoupling production and consumption

Degradation is also related to the trend of uncoupling food producers and consumers. This started long ago and is accelerating because farmers have become specialists who sell all or most of their produce and buy other food items. Urban populations grow rapidly: nearly all population growth of 2 billion persons in the next decades will be in cities, particularly in Asia (FAO, 2000), and (inter)national food trade will continue to grow. Trade is positive in that it provides consumers with a larger diversity and minimal cost. But trade also gives rise to the ecological problem of incomplete recycling of crop nutrients (Miwa, 1990). Large quantities of crop nutrients are mined from rural soils and transported in processed and unprocessed food items to cities or feedlots. For Bangkok, which already gives much attention to a clean environment, Faerge et al. (2001) found that while the annual influx of N in food is 19.4×10^6 kg, and 1.8×10^6 kg of P, only 7% (N) and 10% (P) is recycled to agriculture. The remainder accumulates in land and water, pollutes canals and rivers and leads to health problems, or is burned or denitrified. Such a low degree of recycling is common among large cities. Cities in OECD countries avoid health problems by waste burning and dumping, but also do not comply with the ecological requirement of recycling. The magnitude of the problem in developing countries is not yet evident, but it ought to be addressed now that we are on the verge of a strong growth in size and numbers (ADB, 1997). Since nutrients are not returned to the region of origin, and in many degraded lands not fully replaced by fertilizers (as their use is not warranted due to low produce price), an imbalance is created between urban and rural areas, internationally. Due to export, production capacity declines and the value of the land as a provider of non-marketable ecosystem services, on gross average double that of marketed services (Constanza et al., 1997), is reduced.

Food security

Loss of soil quality is of particular importance for household food security in marginal areas, because acquiring a fair income becomes more and more difficult when degradation progresses. Whenever land is without clear tenure, people do not invest, even if other conditions are favourable. This and other issues should be addressed at political levels. But there is insufficient awareness and urgency because degradation is, by political standards, a slow process and its effects are postponed because of compensation in various ways (new land, extra fertilizer use, etc.). National food security is not threatened by degradation of land quality, as food can be bought from other countries, but food self-sufficiency is vulnerable.

Counteracting Losses of Natural Resources

Reports and documents about degradation are not clear in their definitions. This leads to uncertainty and variability. For example, a large multi-institutional study (WRI, 2000) reports that two-thirds of agricultural land is degraded; Scherr (1999) mentioned 38%. Yet it is clear that the problem is urgent, and that many of the actions recommended are beyond soil and crop science. Scherr (1999) concluded that:

> Degradation appears not to threaten aggregate global food supply by 2020, though world commodity prices and malnutrition may rise. . . . The area of soil degradation is extensive, and the effects of soil degradation on food consumption by the rural poor, agricultural markets and, in some cases, national wealth are significant. ... Active policy intervention will be needed to avert the consequences of soil degradation and harness land improvements to broader development efforts.

Pagiola (2001) adds that:

> Widespread land degradation not only affects agricultural production and watershed-scale natural resources, but may also have global environmental impacts. Terrestrial ecosystems are an important carbon sink, and declining above- and belowground biomass due to degradation can reduce this sink. Land degradation may also reduce biodiversity by forcing farmers to clear additional land or reducing native vegetation in agricultural areas.

In an optimistic view, Greenland *et al.* (1998) state that with the political will, and the necessary commitment to education and research, lowered food security could be prevented. Our analyses add that loss and degradation of agricultural land seriously reduce the options of many nations for food self-sufficiency, and cause many farmers to learn new skills and to try to earn an off-farm income. Degradation of agricultural lands makes farm products more expensive, or eliminates any profit and forces many farmers to leave their income-generating farming.

Implications for Crop Science

The challenge is to increase both farm productivity and sustainability. However, the related requirements are sometimes conflicting at the physiological, agronomic and economic levels, so that an exclusive focus on one aspect will not yield the optimum solution.

Land degradation would be less if more biomass were returned to the land or if more roots were left after the harvest. This would help to maintain, or even increase, the level of soil organic matter. In addition, this would increase C-sequestration. Perennial plants are more suitable to build up root mass than annual crops, and prevent and suppress erosion better. A larger root biomass may be achieved with new varieties that recycle more nutrients from leaves and stems to roots shortly before the harvest (IRRI, 1991). The requirement for more biomass and nutrients (returned) in root systems is contrary to the search for the maximum yield.

Land degradation invites crop scientists to address more production processes under severe stress conditions, such as drought, nutrient shortage and salinity. This includes eco-physiological research and identification of non-food crop species that do better under stress. Grasslands are still less exploited, and research is welcome on how to use and manage them better for food production.

National food self-sufficiency would be enhanced if the productivity of food crops on good land were greater. This warrants research on further yield increases. But household food security on rural and marginal land (hundreds of millions of hectares) will be unaffected if poor lands do not generate more income.

Applying chemical fertilizer to the land can prevent nutrient depletion. In many countries there is a significant resistance to chemical fertilizer. In the long run, this is a threat to food security. This may be overcome by better practices and by example. Introduction of N-fixation for new crops could be interesting, but it probably still remains in the distant future and will only be really helpful if poor farmers have economic access and do not resist genetically modified organisms (GMOs).

Lastly, much more attention is required to recycling between production and consumption sites, even though this will increase cost. One problem is that low value waste, which is very variable, is often contaminated, and has to be upgraded and transported over significant distances to poor farmers. (Peri)-urban farmers might provide an option, but processing closer to the rural farm could also help.

Acknowledgements

Suggestions by E. Craswell, R. Leslie, R. Lefroy and anonymous reviewers are kindly acknowledged.

References

ADB (1997) *Emerging Asia: Changes and Challenges*. Asian Development Bank, Manila.
Alexandratos, N. (ed.) (1995) *World Agriculture: Toward 2010, an FAO Study*. John Wiley & Sons, New York for the Food and Agriculture Organization.
Beinroth, F.H., Eswaran, H. and Reich, P.F. (2001) Land quality and food security in Asia. In: Bridges, E.M., Hannan, J.D., Oldeman, L.R., Penning de Vries, F.W.T., Scherr, S.J. and Sambanpanit, S. (eds) *Response to Land Degradation*. Oxford, New Delhi (in press).
Bridges, M. and Oldeman, R. (2001) Food production and environmental degradation. In: Bridges, E.M., Hannan, J.D., Oldeman, L.R., Penning de Vries, F.W.T.,

Scherr, S.J. and Sambanpanit, S. (eds) *Response to Land Degradation*. Oxford, New Delhi (in press).

Constanza, R. and d'Arge, R. (1997) The value of the world's ecosystem services and natural capital. *Nature* 387, 253–260.

Craswell, E.T. (2000) Save our soils – research to promote sustainable land management. *The Food and Environment Tightrope*. The Crawford Fund, ACIAR, Canberra, pp. 86–96.

Crosson, P. (1995) Future supplies of land and water for world agriculture. In: Islam, N. (ed.) *Population and Food in the Early Twenty First Century: Meeting Food Demands of an Increasing Population*. IFPRI, Washington, DC.

Drechsel, P. and Penning de Vries, F.W.T. (2001) Land pressure and soil nutrient depletion in sub-Saharan Africa. In: Bridges, E.M., Hannan, J.D., Oldeman, L.R., Penning de Vries, F.W.T., Scherr, S.J. and Sambanpanit, S. (eds) *Response to Land Degradation*. Oxford, New Delhi.

Dregne, H.E. (1990) Erosion and soil productivity in Africa. *Journal of Soil and Water Conservation* 45, 8–13.

Enters, T. (2000) *Methods for the Economic Assessment of the On- and Off-site Impacts of Soil Erosion. Issues in Sustainable Land Management*, 2nd edn. IBSRAM, Bangkok.

Eswaran, H., Lal, R. and Reich, P. (2001) Land degradation: an overview. In: Bridges, E.M., Hannan, J.D., Oldeman, L.R., Penning de Vries, F.W.T., Scherr, S.J. and Sambanpanit, S. (eds) *Response to Land Degradation*. Oxford, New Delhi.

Evans, L.T. (2000) The food and environment tightrope – will we fall off the tightrope? *The Food and Environment Tightrope*. The Crawford Fund, ACIAR, Canberra, pp. 149–154.

FAO (2000) FAOSTAT Database. Food and Agriculture Organization, Rome. http://apps.fao.org/default.htm

Færge, J., Magid, J. and Penning de Vries, F.W.T. (2001) Urban nutrient balancing approached for Bangkok. *Ecological Modeling* (in press).

Greenland, D.J., Gregory, P.J. and Nye P.H. (1998) Summary and conclusions. In: Greenland, D.J., Gregory, P.J. and Nye, P.H. (eds) *Land Resources: On the Edge of the Malthusian Precipice?* CAB International, Wallingford, pp. 1–7.

Gruhn, P. and Goletti, F. (1998) Fertilizer, plant nutrient management and sustainable agriculture, usage, problems and challenges. In: Gruhn, P., Goletti, F., (eds) *Proceedings of the IFPRI/FAO Workshop on Plant Nutrient Management, Food Security and Sustainable Agriculture: The Future Through 2020*. IFPRI and FAO, Washington and Rome, pp. 9–22.

IFPRI (1994) *World Food Trends and Future Food Security, 1994*. Food Policy Report, International Food Policy Research Institute, Washington, DC.

IRRI (1991) Systems simulation at IRRI. *IRRI Research Paper Series 151*, International Rice Research Institute, Los Baños.

Lal, R. (1995) Erosion–crop productivity relationships for the soils of Africa. *Soil Science Society of America Journal* 59, 661–667.

Lynden, G.W.J. van and Oldeman, L.R. (1997) *The Assessment of the Status of Human-induced Soil Degradation in South and Southeast Asia*. UNEP/FAO/ISRIC, Nairobi/Rome/Wageningen.

Miwa, E. (1990) Global nutrient flow and degradation of soils and environment. *Transactions of the 14th International Soils Science Society Congress, Kyoto, Japan*, Vol. 5, pp. 271–276.

Oldeman, L.R. (1994) An international methodology for an assessment of soil degradation and georeferenced soils and terrain database. In: *The Collection and Analysis of Land Degradation Data*. RAPA Publication 1994/3, FAO Regional Office for Asia and the Pacific, Bangkok, pp. 35–60.

Oldeman, L.R., Hakkeling, R.T.A. and Sombroek, W.G. (1991) *World Map of the Status of Human-induced Soil Degradation*, 2nd edn. ISRIC/UNEP, Wageningen.

Pagiola, S. (2001) The global environmental impacts of agricultural land degradation in developing countries. In: Bridges, E.M., Hannan, J.D., Oldeman, L.R., Penning de Vries, F.W.T., Scherr, S.J. and Sambanpanit, S. (eds) *Response to Land Degradation*, Oxford Press, New Delhi.

Penning de Vries, F.W.T. (1999) Land degradation reduces maximum food production in Asia. In: Horie, T. and Geng, S. (eds) *World Food Security and Crop Production Technologies for Tomorrow*. Graduate School Agriculture, Kyoto University, Kyoto, pp. 17–24.

Penning de Vries, F.W.T. and van Keulen, H. (1995a) Natural resources and limits of food production in 2040. In: Bouma, J. and Kuyvenhoven, A. (eds) *Eco-regional Approaches for Sustainable Land Use and Food Production*. Kluwer Academic Press, Dordrecht, pp. 65–87.

Penning de Vries, F.W.T., van Keulen H. (1995b) Biophysical limits to global food production 2020. International Food Policy Research Institute, Washington, DC, *Brief no. 18*; and in: *Economic Planning in Free Societies* 32, 3–4 (1996).

Penning de Vries, F.W.T., Rabbinge, R. (1997) Potential and attainable food production and food security in different regions. *Philosophical Transactions of the Royal Society, London B*. 352, 917–928.

Ponting, C. (1991) *Green History of the Earth*. Cambridge University Press, Cambridge.

Scherr, S.J. (1999) Soil degradation. A threat to developing country food security by 2020? *Food, Agriculture and Environment Discussion Paper 27*. IFPRI, Washington, DC.

Scherr, S.J. (2001) The future food security and economic consequences of soil degradation in the Developing World. In: Bridges, E.M., Hannan, J.D., Oldeman, L.R., Penning de Vries, F.W.T., Scherr, S.J. and Sambanpanit, S. (eds) *Response to Land Degradation*. Oxford Press, New Delhi.

Swaminathan, M.S. (1986) Building national and global nutrition security systems. In: Swaminathan M.S. and Sinha S.K. (eds) *Global Aspects of Food Production. Natural Resources and Environment Series*, Vol. 20. Tycooly Publishing Ltd, London, pp. 417–449.

Turkelboom, F. (1999) On-farm diagnosis of steepland erosion in Northern Thailand. PhD thesis 339, University Louvain, Belgium.

WRI (2000) *World Resources 2000–2001*. World Resources Institute, Washington, DC. www.wri.org

Xu, C. (1994) Needs and priorities for the management of natural resource: a large countries perspective. In: Goldsworthy, P.G. and Penning de Vries, F.W.T. (eds) *Opportunities, Use and Transfer of Systems Research Methods in Agriculture to Developing Countries*. Kluwer Academic Publishers, Dordrecht, pp. 199–211.

Young, A. (1998) *Land Resource Now and For the Future*. Cambridge University Press, London.

The Future of World, National and Household Food Security

2

F. Heidhues

Institut für Agrar- und Sozialökonomie in den Tropen und Subtropen, Universität Hohenheim, D-70593 Stuttgart, Germany

Introduction

The coming decades will pose daunting challenges for policy makers and the international agricultural science community mandated to solve the complex problem of providing adequate food for everyone. Between 70 and 80 million people will be added to the world's population every year between now and 2020. This will increase the world's current population of 6 billion by a third to reach almost 8 billion.

To produce and provide the food needed for those additional 2 billion people is possible but difficult. There is general consensus that it cannot merely come from expanding cultivated area; the lion's share of future food production growth will have to come from productivity increases. The International Food Policy Research Institute (IFPRI) and the Food and Agriculture Organization (FAO) estimate this share to be between 75% and 80% (Pinstrup-Andersen *et al.*, 1999). Without any doubt, this is the more complex, the more difficult way of increasing food production. It requires agricultural research to generate a steady flow of technological innovations and adapting them to local ecological conditions. Crop science research has an important role to play (Cassman, Chapter 3, this volume). Similarly, the institutional and socio-economic framework needs strengthening; it requires educating people, changing their behaviour, modifying institutions, formulating and implementing policies often against well entrenched interests and political power structures. This is the science intensive path to productivity growth under the constraint of preserving the natural resource base. Increased investments in agricultural research, both in the natural science area as well as in socio-economic, institutional and policy analysis, are needed to meet this challenge.

© 2001 CAB *International. Crop Science*
(eds J. Nösberger, H.H. Geiger and P.C. Struik)

World Food Prospects

There is general consensus that the overall production potential is adequate to provide the food demanded for the growing world population, provided that technological innovations will continue to be generated and applied in a way that preserves the natural resource base. In fact, actual food prices have declined during the last decades (with a temporary interruption in the mid-1990s) and are projected to remain steady or fall slightly in the next two decades, as IFPRI's global food model projects. However, those declining food prices are an inadequate indicator of food security as prices reflect only that part of food needed that commands the purchasing power necessary to enter the market. The food needs of the poor and hungry without the necessary income to enter the market are not reflected in market prices. Also, the decline in food prices seems to be slowing and perhaps has come to a halt due to the observed continuing slowdown in crop-yield growth.

Demand

Population growth, income increases and changing consumption patterns are the driving forces behind food demand increases. World population growth (Table 2.1), of which 98% is occurring in the developing countries, continued at unabated rates until the mid-1990s. Around 1995 the population curve seems to have reached the turning point and growth rates have started to decline.

Table 2.1. World population, 1995 and 2020. From United Nations, World Population Prospects: The 1998 Revision (New York: UN, 1999) in Pinstrup-Andersen *et al.*, 1999.

Region	Population level 1995 (millions)	2020[a] (millions)	Population increase 1995–2020 (millions)	(%)	Share of increase (%)
Latin America and the Caribbean	480	665	185	38.5	10.1
Africa	697	1187	490	70.3	26.7
Asia, excluding Japan	3311	4421	1110	33.5	60.5
China	1221	1454	233	19.1	12.7
India	934	1272	338	36.2	18.4
Developed countries	1172	1217	45	3.8	2.5
Developing countries	4495	6285	1790	39.8	97.5
World	5666	7502	1836	32.4	100.0

[a] Medium-variant population projections.

While Asia will have the largest absolute increase in population, growth rates will be highest in Africa, despite HIV/AIDS with its devastating effects on African economies and societies. Developing countries' growing populations will continue to be an important driving force behind increasing food demand, although less so than in the past. Food demand driven by population growth is projected to grow at 1.8% p.a. as compared with 2.3% p.a. in the 1980s.

As people in poor countries tend to spend a high share of their additional income on food, income growth is a second important determinant of food demand, and poverty is the key factor in explaining food insecurity. Prospects for overall economic growth in developing countries look favourable, although regionally very different. IFPRI's global food model (version July 1999) assumes total income in the developing world to increase at an average rate of 4.3% p.a. between 1995 and 2020. Far behind the other regions is sub-Saharan Africa, whose per capita income is projected to increase during this period from US$280 to US$359. This means that sub-Saharan Africa's per capita income on average will still be below the poverty line with one dollar a day (Table 2.2). Thus, food security in many parts of Africa will remain extremely fragile.

Urbanization is changing food demand patterns in an important way. While most of the poor and food insecure still live in rural areas, urbanization is progressing rapidly. Urban population growth is expected to be much higher resulting in cities overtaking rural areas in absolute numbers of people around 2015. Urban people tend to have more diversified diets, substituting rice and wheat for coarse grains, and eat more fruits, vegetables

Table 2.2. Income levels and growth, 1995–2020. From IFPRI IMPACT simulations, July 1999 in Pinstrup-Andersen *et al.*, 1999.

Region	Annual income growth rate 1995–2020 (per cent)	Per capita income level	
		1995	2020
		(1995 US$ per person)	
Sub-Saharan Africa[a]	3.40	280	359
Latin America and the Caribbean	3.59	3590	6266
West Asia and North Africa	3.83	1691	2783
South-East Asia	4.44	1225	2675
South Asia	5.01	350	830
East Asia	5.12	984	2873
Developed countries	2.18	17390	28256
Developing countries	4.32	1080	2217
World	2.64	4807	6969

[a] Excluding South Africa.

and meat products. Rapid urbanization with its effects on lifestyle, work organization, food preferences and hence food demand will require significant adjustments in food production and marketing, and food security policy.

Like urbanization, rising incomes will also encourage people to shift to more diverse diets. A particularly rapid growth of demand is foreseen for livestock products. On the basis of past trends IFPRI projects meat consumption to grow at 2.8% p.a. in the next two decades, compared with 1.8% p.a. for cereals. In response to the rapid expansion of developing countries' meat consumption, demand for feed grains is expected to grow more than twice as fast as demand for cereals for direct human consumption. These changes will also require adaptation of agricultural research priorities.

Supply of food

Food production

Future supply of food is likely to be based largely on productivity increase and this is innovation-driven. Decisive is how successful agricultural research will be in raising yields. Yield increases for cereals have slowed down, both in developed as well as in developing countries, as shown in Fig. 2.1 (also Alexandratos, 1995).

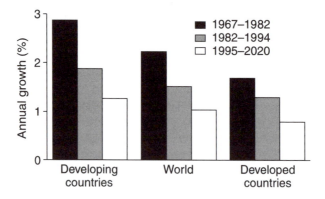

Fig. 2.1. Annual growth in cereal yields, in per cent per year, 1967–1982, 1982–1994, and 1995–2020. Source: 1967–1982 and 1982–1994, Food and Agriculture Organization of the United Nations, FAOSTAT database <http://faostat.fao.org>, accessed May 2000; 1995–2020, IFPRI IMPACT simulations, July 1999, in Pinstrup-Andersen *et al.* (1999).

The decline in yield growth rates reflects partly declining real prices for cereals. In developing countries institutional constraints on input supply, rural credit and infrastructure have become increasingly constraining on fertilizer use. Also, with agriculture being pushed into cultivating marginal lands yield growth becomes more difficult. In developed countries, the efforts of governments to control overproduction may have depressed yield growth. Investments in agricultural research have been insufficient to produce a continuous flow of yield-increasing technologies (Pinstrup-Andersen *et al.*, 1999). The controversial and often emotional debates about and actions against genetically modified foods have certainly hampered investments in innovation generation in these areas. To keep a steady flow of productivity-increasing innovations running, both public and private sector agricultural research needs to be directed towards increasing agricultural and particularly food production while preserving natural resources. Substantial and increased levels of investment in research of innovation generation and acceptance processes are necessary to avoid the continuing decline in yield levels and to turn the trend around. Also, new forms of cooperation between private and public research are needed, combining private sector interests with public sector needs.

Food trade

In food security international trade will play a key role. Most of the additional food will be needed in developing countries, and it is there where production is projected to increase much faster. IFPRI's impact model shows that between 1995 and 2020 cereal production in developing countries grows by 51%, as compared with 24% in developed countries (Pinstrup-Andersen *et al.*, 1999). Still, many developing countries will not achieve food self-sufficiency in the foreseeable future; they are shown in need of filling an increasing gap between food production and demand by rapidly rising imports. Particularly East Asia with fast rising incomes as well as South and West Asia with expected declining growth rates in yields are driving the rising import demand (Fig. 2.2).

The major sources for these cereal imports are likely to remain North America, the European Union (EU) and Australia. An unknown variable in this picture is future development in Eastern Europe and the countries of the former Soviet Union, particularly whether they will be entering the cereal export market to a significant extent. Also, the future of WTO negotiations will play a role in world cereal markets.

As a result of these projected developments, IFPRI's impact model projects world prices of major cereals to remain constant or slightly declining until 2020 (Fig. 2.3).

National and Household Food Security: a Complex Issue

Behind these overall developments are enormous differences between and within regions, calling for a more differentiated look at food security. In fact, world food security does not guarantee national food security, and national food security does not ensure that all households are food secure. Even

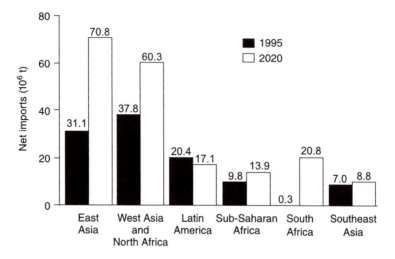

Fig. 2.2. Net cereal imports of major developing regions, 1995 and 2020. Source: IFPRI IMPACT simulations, July 1999, in Pinstrup-Andersen *et al.* (1999).

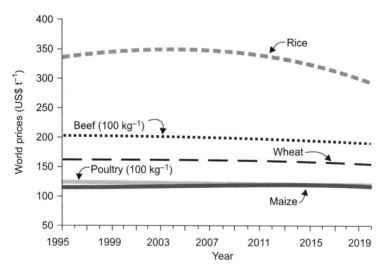

Fig. 2.3. World prices of major commodities, 1995–2020. Source: IFPRI IMPACT simulations, July 1999, in Pinstrup-Andersen *et al.* (1999).

within households distorted intra-household distribution may leave some members deficient in food although the household overall may have adequate access to food (Alderman *et al.*, 1995).

While the world as a whole is able to produce the additional food required to ensure food security for everyone, undernutrition* is likely to persist in large regions of the developing world. The highest number of people suffering under undernutrition today is in countries of South and East Asia, and during the coming two decades per capita food availability is expected to increase significantly in East Asia and moderately in South Asia. The highest incidence of undernutrition is found in sub-Saharan Africa, also the region where progress is expected to be slowest and where undernutrition is likely to persist. Food insecurity in the other developing regions is less severe, more localized and expected to be easier to handle.

A differentiated picture presents itself in looking at household food security which in large parts of the world remains critically inadequate. Household food security is conditioned by a complex set of factors that go far beyond the food production and trade domain. Household food security is not just a matter of producing sufficient food and ensuring its availability in adequate quantities and quality; it also requires creating the possibility for households to acquire the needed food and to access the knowledge necessary to use food properly. Thus, a household's food security is directly linked to poverty and a household's access to the food needed. Also, knowledge about adequate food diets, about hygiene and health and the interlinkages between them, is an important component of an expanded concept of food and nutrition security (von Braun *et al.*, 1998).

Poverty and household food security: causes and linkages

Household food security is closely interlinked with both the natural resource and infrastructure environment and the economic, political, socio-cultural and institutional setting. Poverty and household food security and their relationship to the household's environment interact in different ways, depending on location-specific factors. Therefore, they have to be analysed in a location- and time-specific setting (von Braun *et al.*, 1999).

The key factors determining household food security in an interacting way may be grouped into five categories:

- natural resources;
- innovation development and dissemination;
- population growth;
- infrastructural, socio-cultural and institutional setting, economic and policy framework;
- natural disasters, wars.

---

* Undernutrition and hunger are often used synonymously; both terms characterize a situation of inadequate access to food energy. They differ in the duration of a food deficit situation: hunger denotes a temporary shortage in food energy, while undernutrition defines a longer-term food deficit situation with impact on a person's health and anthropometric indicators.

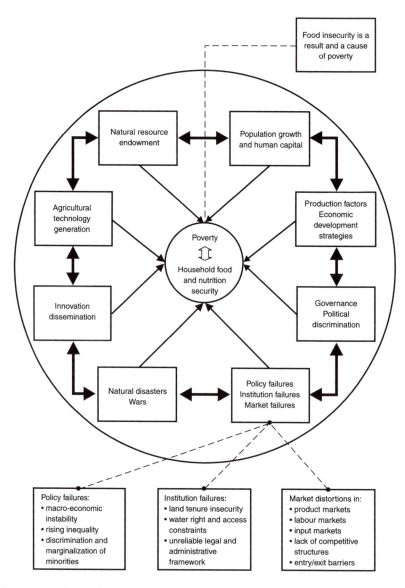

Fig. 2.4. Food security in context: causes and interrelationships.

Figure 2.4 highlights the major interrelationships. Actions to influence the factors and their impact on food security can be directed at any of the boxes in Fig. 2.4. Vitally important areas of action are natural resource conservation and innovation development and dissemination in the natural resource/agricultural productivity domain with the objective of raising the carrying capacity of natural resources. The poor and food insecure often live in agriculturally marginal areas with a fragile natural resource base. Mountainous regions with extensive hillside agriculture are particularly erosion prone and in danger of irreversible resource degradation processes (Penning de Vries, Chapter 1, this volume). They present a particular challenge to crops and natural resource research. The best of science available in these areas has to be mobilized for food security if the World Food Summit challenge of halving hunger worldwide by 2015 is to be met. Research to address food security issues deserves highest priority; it needs to include the marginal areas in its research agenda.

While enormous progress in these areas is necessary, it is not sufficient to ensure food security for the poor. A major focus in research and policy action needs to be placed on the socio-cultural, institutional, economic and policy factors and their relevance for poverty and food security. Identifying these factors and understanding their relevance and mutual interactions is a precondition for designing measures and policies that are effective in improving food security. Thus, research into socio-cultural, economic and institutional complexities is important and needs to complement natural science research work. It hardly needs to be elaborated that without water there cannot be irrigation. What is often less understood is that an institutional failure or an economic constraint, such as an insecure land tenure situation or lack of access to credit, can be just as limiting for innovation acceptance as a lack of water is for irrigation.

From the factors shown in Fig. 2.4 different levels of policy actions, apart from the intra-household actions, may be derived. The top group represents basic structural conditions: natural resources and population growth. The boxes in the mid-range denote important areas of action; on the left in agricultural research and innovation dissemination and on the right on economic development. They interact with each other as well as with the action areas shown in the bottom part of Fig. 2.4: policy, institutional, market and governance failures; they again are closely interdependent and decisively influence basic structural conditions. Both groups of factors, individually and in their interaction, together with the political framework determine a household's food security.

Natural disasters and wars, often perceived as exogenous factors or even *force majeure*, may well be seen as endogenous results of scarcities, failures in policies and flawed institutional set-ups (von Braun *et al.*, 1998). They often are only the outburst of a simmering fragile ecological and economic situation which may be caused by a small and, in itself, unimportant external shock.

Policy actions and programmes for household food security

The effectiveness and success of food security policy and programme interventions depend on the basic conditions, institutions and actors that determine food security. Food security policy is of a cross-cutting nature. Isolated programme interventions may bring some relief but are likely to fail in promoting long-term food security. Sustainable food security policy can only be successful if it integrates long-term population policies, economic development and productivity-enhancing strategies and natural resource conserving programmes with short-term actions to remove market distortions, to correct institutional failures, to design more effective policies and to move towards good governance. Thus, access to input, output, land and credit markets affects innovation acceptance and agricultural productivity. Better employment opportunities in less distorted labour markets can effectively enhance income and purchasing power of the poor and thus their food security. Access to financial markets can be instrumental in adopting innovations. Moreover, it has also been shown to be an effective way for poor households to bridge periods of temporary food stress; poor households with access to financial markets feel less need to sell their productive assets (such as animals or tools) and have been found to recover more quickly after periods of stress than households without access to the financial market; also their nutritional well-being suffered less (Zeller *et al.*, 1997). Programmes of physical and social infrastructure improvements (education, health, social and communication services) have an impact on labour productivity, employment and income, and also on mortality, fertility and migration which in turn affect population growth and a household's food situation (FAO-WFS, 1996). Thus, it is important to recognize that there is not one set of policies that is optimal for achieving food security. Depending on the location- and time-specific framework conditions and characteristics of the food security problem, there is a wide range of options from which the most effective ones are to be chosen. They are grouped and discussed here under the headings: 'Governance, development strategy and macroeconomic policy'; 'The agriculture and food security link'; and 'Food aid and specific food security/ Nutrition Programmes'.

Governance, development strategy and macroeconomic policy

GOVERNANCE As Fig. 2.4 tries to highlight, poverty and food insecurity are a multi-dimensional problem. One of the most important contributions of the development discussion during the 1990s is the recognition that poverty is not just defined by material well-being, but also by the choices an individual has and the freedom of developing individual capabilities. Amartya Sen in his ground-breaking work *Development as Freedom* (1999) has shown that freedom and democracy are constituent elements of development. There is increasing evidence indicating that poverty, ignorance and disease are highly correlated not only with low labour productivity but also with the poor's absence of freedom of choice and action (Dethier, 1999). The poor are often unable to control their own lives and have no way of influencing local or national decisions affecting them. Voicelessness and powerlessness are generally perceived by the poor as a major constraint on their life and, often, as a cause of their material poverty (Narayan *et al.*, 1999).

Governments have a central role to play in creating a framework that gives everyone the freedom to develop their own capabilities, i.e. to guarantee basic personal protection and security, to ensure fair and enforceable access to due legal process and to promote equity between different economic, social and ethnic groups in society. Governments and state institutions in many developing (and transition) countries often fail on all three tasks. Corruption, rent seeking and capture of the state by interest groups have been shown to perpetuate poverty and inequality. Transparency and participation of people, particularly of the poor, in decision making, and accountability of officials in the provision of public goods are key requirements of better governance that is essential for reducing poverty and food insecurity.

DEVELOPMENT STRATEGIES AND MACROECONOMIC POLICY Historical evidence and development experience have shown the striking relevance of the chosen development strategy for growth and poverty reduction. Big-push strategies emphasizing industrialization not only failed in generating growth and employment, they tended to polarize societies into a few rich and a majority of poor. Re-emphasizing the rural sector in the rural development paradigm and recognizing the complexity of the development problem and the multiplicity of actors in the process in the comprehensive development framework have been big steps in the direction of poverty reduction, but much remains to be done. Again, a general strategy cannot address the multifaceted issues of poverty in all countries; a country- and often region-specific approach geared to address the specific characteristics of the poverty situation is called for.

Structural problems and adjustment policies in the 1980s have amply demonstrated the important effects of macroeconomic policies on the poor and their food security (Pinstrup-Andersen, 1990). Macroeconomic stability, a liberal and competitive market and trade environment and efficiency promoting infrastructure and public sector policies have proved to be vitally important for integrating the poor in the development process. Still, there is a need to complement these policies with special support measures for the extreme poor, particularly for those who cannot participate in the labour market (discussed later).

Fluctuations in agricultural production, markets and trade are inherently linked to agriculture and can aggravate food insecurity. Rising food prices have been shown to bear particularly hard upon the poor (Teklu *et al.*, 1991). Storage and food-trade policies require renewed emphasis, given the volatility of food markets, the uncertainties of international trade and the amplified risks of international financial markets (Haug, 1999). It is against this background that governments often feel that a certain amount of storage under public control is essential for food security.

Storage needs are dependent upon country-specific parameters, such as cropping pattern, climate risks, sectoral diversification, infrastructure and ease of access to international markets. In determining food storage needs and in designing policies it is important to take account of the administrative capacities and to weigh the benefits of stabilized prices against the costs of the resources tied up in storage. In this context it is worth emphasizing that food insecurity and undernutrition, apart from their ethical and human dimension, in purely economic terms, are enormously costly and 'the largest worldwide waste of potential economic resources' (FAO-WFS, 1996). A benefit–cost comparison has to value properly the higher productivity of people now and in the future as a result of better nutrition and a more secure food supply against the momentary fiscal costs of stabilization expenses.

The agriculture and food security link

In formulating specific agricultural and food security policies different types of linkages can play a role.

DIETARY LINK Agriculture is obviously important as it produces virtually all the food. A healthy diet requires a diversified supply of food from fruits, vegetables and staples to high protein products. This has important implications for agricultural science. For the *urban* population the market is intermediate between producer and consumer. The variety of supply and market access determines the possibility of acquiring a diversified diet. Thus, infrastructure, market and trade policies need to take into account not only an adequate supply of staple foods but also the needs for diversified diets. In *subsistence agriculture* households' diets are more directly linked to farmers' cropping pattern; they can hardly be modified by varying market supply. Production and technology promotion policies need to be designed in support of diverse cropping patterns and multiple dietary needs. Also, plant-breeding research directed at raising nutritional value and micronutrient content can have favourable effects on diet quality (Bouis, 1995).

GROWTH–EMPLOYMENT LINK Placing highest priority on agricultural development and promoting agricultural production and productivity can benefit food security, if it directly or indirectly increases income and food consumption of the poor. These policies may affect the poor via two channels: higher incomes and lower food prices. Boosting agricultural production not only increases employment and income in agriculture, it also spreads to other sectors via income and consumption increases and multiplier–accelerator effects (Hazell and Ramasamy, 1991). The larger the share of agriculture in the economy, the more important the role of agriculture as an engine of overall growth and poverty-oriented economic development becomes.

Lower food prices play a key role in securing food for the poor who spend the overwhelming part of their income on food. Lower food prices are mainly driven by innovations and, therefore, agricultural research plays a most vital role. The development promoting effect of agricultural growth deserves emphasis as too often it is overlooked, particularly in analytical approaches based on partial equilibrium models. At the policy level, development institutions and governments need to re-recognize the key importance of agriculture for overall growth and development and for poverty reduction in particular. Fighting poverty and food insecurity without proper attention to agricultural growth is like trying to sweep water uphill.

THE AGRICULTURAL PRODUCTIVITY–ENVIRONMENT–HEALTH LINK Agricultural productivity increase is often linked to increased irrigation and extended use of fertilizers, pesticides and other chemicals. Irrigation may spread water-borne diseases if proper measures are not taken; increased health problems reduce work capacity and labour productivity. Overuse of chemicals is of concern in technologies that rely on irrigation and high intensity packages of pesticides and fertilizers. Reduced water quality downstream may limit its use for household consumption and pose serious health risks with subsequent effects on labour productivity and mortality.

THE TECHNOLOGY–SOIL DEGRADATION LINK Many mountainous regions in low-income countries are exposed to rapidly rising population pressure, pushing agriculture into hillside cultivation with often heavy soil degradation processes set in motion. While farmers recognize the erosion danger, they often underestimate the damage done to soil fertility as the simultaneous introduction of higher yielding varieties combined with increased fertilizer use tends, at least initially, to over-compensate the erosion-caused fertility decline; despite heavy erosion, yields are still increasing, making it difficult to raise awareness of the longer-term consequences of such agricultural practices. Extension and technology policies are challenged to turn farmers' attention to the long-term consequences.

Food aid and specific food security/nutrition programmes

Food security can be enhanced not only by policies oriented towards the development of agriculture, economic growth and employment but also by food aid and food-related income transfers. Food aid has played a critical role for some countries, both as a recurrent resource transfer and as a bridging measure in times of emergency. However, it is not a reliable source of food for poor countries. When world market prices for food rise, the supply of food aid from donors tends to decline. Thus, it has been observed in the mid-1990s, when international cereal prices increased by 30 to 40%, that food aid supplies dropped to about half of their early 1990s' level (FAO, 1998; FAO-WFP, 2000). Also, liberalization of agricultural markets in the main grain exporting regions set in motion under the Uruguay trade agreement have resulted in lower surpluses and food stocks. Food aid to individuals at critical times in their life cycle or at stress periods during the year can have a significant positive impact on developing their physical and mental potential and thus on their long-term food security. In situations of natural disasters and wars survival may hinge on access to food aid.

The importance of food aid has increased as the number and complexity of emergencies has risen (FAO-WFS, 1996), and it may well increase further in the future. The expected rising need and declining resources available point out the need for food aid programmes to become more effective and efficient in the future. Also criticisms that some programmes in the past have been wasteful, inefficient and counterproductive to long-term food security call for better management and coordination of food aid programmes and their integration into general development policy.

Specific food security/nutrition programmes including food-price subsidies and food stamps are widely used instruments. In order to control costs and limit access to those in need, food-price subsidy policies are often combined with rationing and targeting efforts. It has proven difficult to achieve household food security for all through rationed food subsidization programmes. Often corruption, rent seeking and the appropriation of programmes by powerful interest groups have undermined their effectiveness. Self-targeting by using commodities predominantly of interest to the poor and geographical delineation of programmes to poor regions or neighbourhoods are attempts to control costs and limit their scope to those in real need of food.

Specific nutrition interventions are often integrated into more comprehensive nutrition–education–health programmes. This approach seems logical given the multifaceted nature of nutrition problems. It has also proven to be an effective and efficient way of addressing complex nutrition problems.

Food security under increasing resource scarcity

While raising agricultural productivity in food insecure regions remains a top priority, preserving the natural resource base is a precondition for long-term agricultural development. With rising pressure on natural resources the dangers of overusing land and water, reducing biodiversity through inappropriate use of agrochemicals and polluting land and water have become apparent. Clearing of forest lands has contributed in many countries to severe water shortages and flooding. Poor farmers trying to secure survival of their family today tend to lose sight of the needs of tomorrow. Their high time preference for present consumption reduces their efforts needed for preserving natural resources in the long run. In economic terms the costs of exploiting natural resources and of using up the environment beyond sustainability levels are not internalized; the market has not been able to attach monetary cost values to these effects.

Industrialized countries have started to deal with markets' inability of internalizing environmental costs through imposing rules and regulations on the one hand and by levying taxes or paying incentives on the other. Also, 'moral persuasion' has been used to change behaviour. In poor countries where often the administrative structures are too weak to enforce rules and regulations effectively and where tax and incentive systems do not work, environmental degradation has proceeded unabated. Experiences in South-East Asia have demonstrated that resource conservation policies work best if they involve the local community; programmes directed against farmers' interests have largely failed.

As economic and environmental policy instruments are too limited to ensure sustainable agricultural development, agricultural research can make an important contribution to sustainability objectives. To support sustainable use of natural resources, agricultural research needs to take the above issues into account in its agenda.

Agricultural research has come a long way in reducing large-scale application of pesticides through integrated pest management. Biodiversity and environmental concerns call for strengthened efforts in this direction. Also, biotechnology research that increases resistance to pests and diseases will reduce the application of agrochemicals and, thus, can promote biodiversity and land and water quality. Similarly, research in soil–plant relationships and in soil conservation technologies could make valuable contributions to improved nutrient conservation and recycling strategies, erosion control and moisture conservation. Economizing water use and improving water management are increasingly important tasks, given the rising water scarcity in many regions. Both research to develop more water-efficient technologies and cropping systems as well as more effective water pricing mechanisms deserve a high priority in the interest of sustainable water management.

Farmers' attitudes and constraints must play a role in developing agricultural innovations. Involving farmers through participative methods in setting research priorities and designing research programmes is important.

Conclusion for International Agricultural Research

What then is needed to face future food security challenges and to develop solutions to these complex and multifaceted problems?

1. Building strengthened and adequately funded national and international research systems with the proper balance between specialized and interdisciplinary research and effectively interlinked internationally is a top priority. Public and private sector research need to be brought together complementing each other according to their specific roles and comparative advantages. Private sector research typically serves its enterprise's objectives of profitability, competitiveness and market share. Where eliminating hunger and poverty is at stake, public sector research has a leading role to play. Modern communication and transport systems allow the bringing together of specialists of different disciplines worldwide to address multi-faceted issues of food security and nutrition in an integrated way.

2. Food needs are continuing to increase. Given the limited possibilities of expanding cultivated area, raising natural resource productivity is of utmost importance. The trend of declining rates of yield increases is a matter of concern. The big potential that modern science offers needs to be mobilized for food research. In plant breeding, molecular biology promises new advances in yields and resistance characteristics. Also, research for improving production technologies and natural resource management offers considerable potential. Research needs to take the public's concern about the impact of biotechnology on health and the environment seriously. It needs to address these concerns to build up trust: without public confidence in these technologies it will be hard to mobilize public support for them.

3. Given that poverty is mostly found in marginal areas, research must include in its agenda crop improvement for marginal areas. There is ample evidence that redistribution of food from richer to poorer regions is conditioned by many constraints. Marginal areas must be directly the subject of research if it is to address poverty and reduce food insecurity. A particular emphasis is to be placed on mountainous regions. They are of key importance for the global ecosystem, and more than 10% of the world's population depend directly on these regions for their livelihood.

4. Research has to take into account farmers' problems and their priorities. Therefore, farmers need to be brought into the research process. The Global Forum of Agricultural Research has called for a research agenda where priorities are set with a focus on farmers' perspectives (GFAR, 2000). Research design and dissemination should involve the intended users and beneficiaries, particularly farmers. Besides this call for farmers' participation in agricultural research, participation itself is a research issue. At what stage in the research, innovation and dissemination process farmers' participation is appropriate and what are the constraints, benefits and costs, needs to be carefully assessed, if participation is to become a widely accepted and effective approach in agricultural research.

5. Research results have to be proven to be economical. That is a necessary precondition of acceptance, but not sufficient. The above discussion has highlighted the complexity of the food security issue. Many of these issues originate in cultural, social and political constraints. These are not simple issues to deal with. Causes and interlinkages need to be understood if effective measures and policies are to be formulated. This requires research. International agricultural research must include these issues in its mandate.

6. Removing food insecurity is also a matter of political will. How to generate political commitment and how to translate it into effective policy actions is a research issue. Too little attention has been given to these questions, where political science and law need to be brought into the research process for food security.

References

Alderman, H., Chiappori, P.-A., Haddad, L., Hoddinott, J. and Kanbur, R. (1995) Unitary versus collective models of the household: is it time to shift the burden of proof? *World Bank Research Observer* 10, 1–19.

Alexandratos, N. (ed.) (1995) *World Agriculture: Towards 2010. An FAO Study.* John Wiley & Sons, Chichester, UK.

Bouis, H.E. (1995) Breeding for nutrition. *Federation of American Scientists, Public Interest Report* 48(4), 1–16.

von Braun, J., Bellin-Sesay, F., Feldbrügge, T. and Heidhues, F. (1998) Verbesserung der Ernährung in Entwicklungsländern: Strategien und Politikempfehlungen. *Forschungsberichte des Bundesministeriums für Wirtschaftliche Zusammenarbeit und Entwicklung*, Band 123. Weltforum Verlag, Köln.

von Braun, J., Teklu, T. and Webb, P. (1999) *Famine in Africa. Causes, Responses, Prevention.* The Johns Hopkins University Press, Baltimore, Maryland.

Dethier, J.-J. (1999) Governance and poverty. *Quarterly Journal of International Agriculture* 38(4), 293–314.

FAO – World Food Summit (1996) Vol. 2, Technical Background Documents 6–11, and Vol. 3, Technical Background Documents 12–15. FAO, Rome.

FAO (1998) *The State of Food and Agriculture. Rural Non-Farm Income in Developing Countries.* FAO, Rome.

FAO – WFP (World Food Programme) (2000) 1999 Food Aid Flows. Online, URL: http://www.wfp.org/reports/faf/99/table_of_contents.htm (31/05/2000).

GFAR – Global Forum on Agricultural Research (2000) Dresden Declaration of May 25, 2000.

Haug, M. (1999) Globalisation of financial markets – is poverty a concern? *Quarterly Journal of International Agriculture* 38 (4), 277–292.

Hazell, P.B.R. and Ramasamy, C. (1991) *The Green Revolution Reconsidered. The Impact of High-Yielding Rice Varieties in South India.* Johns Hopkins University Press, for IFPRI, Baltimore, Maryland.

Narayan, D., Patel, R., Schafft, K., Rademacher, A. and Koch-Schulte, S. (1999) *Can Anyone Hear Us? Voices from 47 Countries.* Poverty Group, PREM, World Bank, Washington, DC.

Pinstrup-Andersen, P. (ed.) (1990) Macroeconomic Policy Reforms, Poverty, and Nutrition: Analytical Methodologies. *Cornell Food and Nutrition Policy Program* Monograph No. 3. Cornell Food and Nutrition Policy Program, Ithaca, New York.

Pinstrup-Andersen, P., Pandya-Lorch, R. and Rosegrant, M.W. (1999) World Food Prospects: Critical Issues for the Early Twenty-First Century. *2020 Vision Food Policy Report.* IFPRI, Washington, DC.

Sen, A. (1999) *Development as Freedom.* Alfred Knopf, New York.

Teklu, T., von Braun, J. and Zaki, E. (1991) Drought and Famine Relationships in Sudan: Policy Implications. *Research Report No. 22*, IFPRI, Washington, DC.

Zeller, M., Schrieder, G., von Braun, J. and Heidhues, F. (1997) Rural Finance for Food Security for the Poor. Implications for Research and Policy. *Food Policy Review 4*, IFPRI, Washington, DC.

Crop Science Research To Assure Food Security

3

K.G. Cassman

Department of Agronomy, University of Nebraska, PO Box 830915, Lincoln, NE 68583-0915, USA

Introduction

The research agenda for crop science in the 21st century will depend largely on whether policy makers believe that present conditions of global food surplus will continue, or food scarcity recurs. Accurate prediction of food supply and demand is therefore pivotal to prioritization of research. While recognizing the linkages between biophysical and socio-economic factors that influence food security, the purpose of this chapter is to identify the key biophysical constraints and scientific issues that must be addressed to ensure adequate global and regional food supplies in a sustainable fashion.

Whether there is food scarcity or surplus depends on the answers to several questions. How much food will be needed to meet demand? Are available land and water resources sufficient to sustain productivity increases? At what rate will scientific breakthroughs lead to development of ecologically sound crop and soil management technologies that advance yields of the major food and forage crops? Three factors will play a major role in determining the answers to these questions: (i) the rate of population growth; (ii) the rate of economic growth, especially in developing countries; and (iii) the amount of investment in research to improve productivity and environmental quality of agricultural systems. Because there is considerable uncertainty about each of these factors in the next 20–30 years, several scenarios will be evaluated.

Forecasting Global Food Demand

Both econometric and biophysical models have been used for quantitative assessment of food security scenarios. The IMPACT model, developed by economists at the International Food Policy Research Institute (IFPRI) (Rosegrant *et al.*, 1995), is a robust model that simulates future global food demand and supply based on: (i) population growth and food consumption patterns, the latter driven by predicted increases in income; (ii) yield and production area of the major crops; (iii) livestock production and feed conversion efficiency; (iv) commodity prices determined by the market-clearing point; and (v) the amount of investment in agricultural research and extension and the efficiency with which this investment results in greater on-farm productivity. Recent analyses with the IMPACT model to 2020 predict a 'baseline' scenario that requires a compound annual rate of increase in global maize (*Zea mays*, L.), rice (*Oryza sativa*, L.) and wheat (*Triticum aestivum*, L.) yields of 1.1%. These three crops account for a majority of all calories in human diets.

Compared with FAO estimates of area planted to these three crops from 1996 to 1999 (http://apps.fao.org), the IMPACT baseline scenario predicts little increase in area cropped to rice and wheat to 2020. In contrast, maize area is predicted to increase by 10 million ha, which is equivalent to about 35% of maize area in the USA. During the past 35 years, food demand has been met by increasing both cropped area (Table 3.1) and yield (Fig. 3.1). In the next 30 years, crop scientists must commit to the goal of no net increase in cropped area. Although such a commitment is justified by the need to preserve remaining natural ecosystems and to protect air and water quality from further deterioration, it places greater pressure on scientific and technological advances to sustain productivity gains on existing agricultural land. Ecological intensification of existing agricultural area is required to minimize human influences on global nutrient cycles and natural resources, including wildlife and biodiversity (Vitousek *et al.*, 1997; Cassman, 1999).

Table 3.1. Total area cropped to maize, rice and wheat from 1966 to 1999.

Crop	Period	Area[a] (10^6 ha)
Maize	1966–1969	112
	1996–1999	140 (+25%)
Rice	1966–1969	129
	1996–1999	152 (+18%)
Wheat	1966–1969	220
	1996–1999	225 (+ 2%)

[a] Mean area cropped during each of the indicated 4-year periods. Values in parentheses are the % increase between 1966–1969 and 1996–1999.

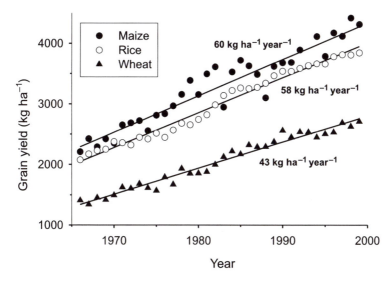

Fig. 3.1. Global yield trends of maize, rice and wheat from 1966 to 1999. Estimates of the rate of yield gain were obtained by linear regression of yield versus year. Yield data were obtained from the FAOSTAT database (http://apps.fao.org).

Specifying no increase in maize area requires an annual gain of 1.3% in maize yield to 2020 versus 1.1% in the baseline scenario of the IMPACT model.

Forecasting Global Food Supply

Yield gains and yield gaps

Projections made by the IMPACT model suggest little difficulty in achieving food security. In fact, decreasing cereal prices are predicted in the baseline scenario to 2020. Other studies have evaluated biophysical limits to global food production capacity by simulation of crop yield potential using regional databases for climate, soils, and water resources (Penning de Vries *et al.*, 1995). Like the econometric model, these simulations also predict adequate food production capacity at global and regional scales. Critical assumptions in the Penning de Vries *et al.* (1995) study are that crops utilize all available arable land and water resources as required to meet food demand and that plant growth is not limited by nutrients or pests.

Although econometric studies typically express yield trends as exponential rates of gain, it is notable that yield trends for the major cereals are decidedly linear (Fig. 3.1). Hence, the rate of gain on a percentage basis

Table 3.2. Linear rates of gain in grain yield relative to the trend-line yields in 1966 and 2000 for maize, rice and wheat in selected countries.

Crop	Region/ country	Trend-line yield[a] (kg ha^{-1})		Rate of gain (kg ha^{-1} year^{-1})	Proportional rate of gain[a] (%)	
		1966	2000		1966	2000
Maize	World	2290	4340	60	2.6	1.4
	Africa	1240	1600	11	0.9	0.7
	Brazil	1140	2570	42	3.7	1.6
	China	1560	5350	112	7.2	2.1
	USA	4740	8310	105	2.2	1.3
Rice	World	2050	4010	58	2.8	1.4
	China	2810	6640	113	4.0	1.7
	India	1355	3000	48	3.5	1.6
	Indonesia	1840	4330	109/0[a]	5.9	0
	Japan	5550	6350	24	0.5	0.4
	Thailand	1720	2340	18	1.0	0.8
Wheat	World	1340	2800	43	3.2	1.5
	Australia	1070	1820	22	2.1	1.2
	China	850	4127	97	11.6	2.4
	France	3260	7490	124	3.8	1.7
	India	900	2700	53	5.9	2.0
	USA	1890	2780	26	1.4	0.9

[a] Linear regression of yield on year from 1966 to 1999 was used to estimate trend-line yields in 1966 and 2000. Dividing the rate of gain, in kg ha^{-1} year^{-1}, by the trend-line yield provides the rate of gain as a percentage of the estimated yield in those years. In Indonesia, a spline linear regression indicated different slopes for initial years versus recent years (see Fig. 3.2). Primary data obtained from FAOSTAT (http://apps.fao.org).

decreases with time (Table 3.2). For example, the annual rate of gain in maize yield on a global basis was equivalent to 2.6% of the trend-line yield in 1966 but only 1.4% in 2000. Assuming there is no net expansion of cropped area, maize supply will not keep pace with demand, which is projected to increase 1.3% per year to 2020. A linear rate of yield gain of 63 kg ha^{-1} year^{-1} would be required to meet demand, which is less than the rate of gain in global maize yield in the past 34 years (Fig. 3.1). In contrast, linear rates of gain in rice and wheat yields since 1966 are sufficient to meet predicted global demand in 2020 if these rates are maintained.

Large variation in rates of yield gain is evident in different countries (Fig. 3.2). This unevenness raises three concerns. First, current rates of gain are not sufficient to satisfy demand in some developing countries. In Africa, for example, the increase in maize demand is much greater than the rate of yield increase since 1966, which is only 0.7% of present yield levels (Table

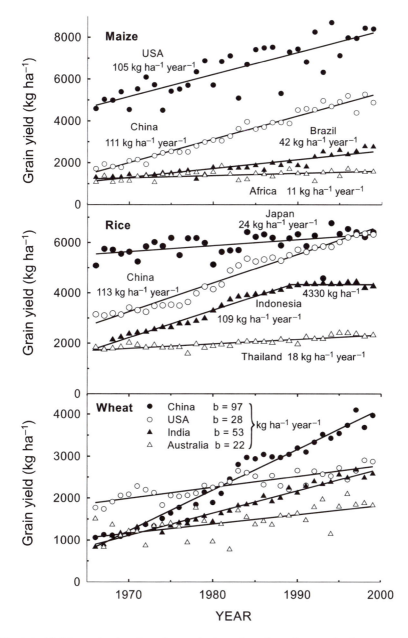

Fig. 3.2. Yield trends of maize, rice and wheat for selected countries from 1966 to 1999. Estimates of the rate of yield gain were obtained by linear regression of yield versus year. Yield data were obtained from the FAOSTAT database (http://apps.fao.org).

3.2). Without a substantial increase in the rate of yield gain, most African countries will require cereal imports to meet demand. A second concern is the relationship between natural resource endowments and rate of yield gain. Primary beneficiaries of the seed and fertilizer technology of modern farming practices were countries and regions with favourable climates, good soils and irrigation. Examples of favourable regions include maize, rice and wheat production areas in China and rainfed maize environments in the USA, where annual yield gains have exceeded $100 \text{ kg ha}^{-1} \text{ year}^{-1}$ (Fig. 3.2). In less favourable environments, the rate of gain is much slower. Rice production in Thailand, wheat production in Australia and the USA, and maize production in Brazil are examples of rainfed environments where drought, poor soil quality, or both, place severe constraints on yields even when access to modern seed and fertilizer technology is not a limitation. Slowest gains have occurred in Africa where a harsh biophysical environment, poor infrastructure, and lack of investment in research and extension have combined to limit improvements in grain yields.

A third concern is the sluggish yield advance and apparent yield plateau in several favourable production domains where farmers were early adopters of modern farming technology. Rice in Japan, which is produced in irrigated systems on generally fertile soils, is one example (Fig. 3.2). Another is the levelling-off of rice yields in Indonesia where 75% of rice is produced with irrigation. In Japan, average farm yields are presently about 80% of the climate-adjusted yield potential in rice production environments (Mathews *et al.*, 1995). Yield growth has also levelled off at about the same 80% yield potential threshold in Korea and Zhejiang province of China (Cassman and Dobermann, 2000). In both Korea and Japan, farmers are paid prices well above world market levels so that low commodity prices are not to blame. Instead, it appears that average yields begin to plateau as they approach the climate-adjusted yield potential.

A plateau in average farm yields well below the genetic yield potential ceiling occurs because it is not possible for farmers to implement the degree of precision in crop management required to achieve maximum possible yields on a commercial scale. Assuming that 80% of the climate-adjusted yield potential approximates a realistic upper threshold for the maximum *average* yield on a regional or countrywide basis, the difference between average farm yields and the 80% yield potential value represents the 'exploitable yield gap'. Although the rice yield trend in China does not appear to be levelling off, average yields above 6.0 t ha^{-1} have only recently been attained (Fig. 3.2). At issue is whether China can maintain the same linear rate of yield increase as average yields exceed 6.0 t ha^{-1}, or whether yields will level off. Accurate estimation of the climate-adjusted yield potential for rice in China is required to answer this question.

The yield plateau in Indonesia is more difficult to interpret because 25% of rice production occurs in rainfed systems that are not favourable environments for high yields. However, yield trends in West Java, where rice is almost entirely produced in irrigated systems, indicate that yields have lev-elled off at about 5.1 t ha^{-1} since the late 1980s (Cassman and Dobermann, 2000). A key issue for food security in Indonesia is the size of the exploitable yield gap for irrigated rice and whether present average yields are approach-ing 80% of the climate-adjusted yield potential.

Yield potential and stress tolerance

Exploiting the current yield gap is crucial to the goal of sustaining yield gains on existing cropland. Evans (1993) defines yield potential as the yield of a cultivar when grown in environments to which it is adapted, with nutri-ents and water non-limiting, and pests, diseases, lodging and other stresses effectively controlled. Solar radiation and temperature regime are the pri-mary determinants of yield potential. In rainfed systems, a water-limited yield potential can be defined along similar lines when other abiotic and biotic factors are not limiting. Crop simulation models can provide reason-able estimates of climate-adjusted yield potential for a given environment if historical weather data are available (Mathews *et al.*, 1995).

Although the definition of yield potential is straightforward, it is difficult to quantify yield potential trends in crop varieties and hybrids over time. Yield trends based on side-by-side comparisons of new and old varieties or hybrids confound improvement in yield potential *per se* with improvements in insect and disease resistance of more recently released varieties against rapidly evolving pest populations. Likewise, changes in soil properties and crop management can place older releases at a disadvantage because they were selected for performance under different environmental conditions. Yield levels of the best varieties or hybrids in such side-by-side comparisons must also attain yield potential levels to ensure that factors other than solar radiation and temperature do not limit expression of genetic yield potential. Because of these confounding factors, strong evidence is lacking to support the contention that yield potential of maize hybrids widely used in the north-central USA has increased since the early 1970s (Duvick and Cassman, 1999), or that the yield potential of modern rice varieties has increased since the release of 'IR8' in 1966 (Peng *et al.*, 1999). Although an increase in yield potential *per se* is difficult to document, it is quite clear that breeders have made significant improvements in stress resistance.

In contrast to rice and maize, there is strong evidence that breeders have made steady progress in raising the yield potential of wheat, although improvements in disease resistance appear to contribute at least as much to yield gains as the increase in yield potential *per se* (Sayre *et al.*, 1998). Removing the benefits of improved disease resistance in the Sayre *et al.* (1998) study gives an increase in yield potential of bread wheat that is considerably less than the 1.1% annual rate of increase in projected wheat demand.

Drought, deficiencies of macronutrients, and weeds are the primary limitations to improved crop yields in less favourable environments. Progress in crop genetic improvement to better withstand drought and nutrient deficiencies has been limited by conservative relationships between crop biomass production and transpiration or nutrient content (Tanner and Sinclair, 1983; Greenwood et al., 1990). One exception has been progress in selecting open-pollinated maize varieties for greater tolerance of drought and low N supply environments. Understanding the physiological importance of more synchronous anthesis and silking events and other traits associated with reduced sensitivity to stress were pivotal to achieving progress (Bolanos and Edmeades, 1996; Bänziger and Lafitte, 1997). But the magnitude of yield increase possible from improved germplasm alone is relatively small. Substantial improvements in yield depend on increasing the supply of water and nutrients in tandem with use of the best-adapted crop varieties (Waddinton and Heisey, 1997).

Food Surplus or Deficit?

While predictions from econometric or biophysical models uniformly paint an optimistic picture of global food security in the foreseeable future, linear rates of yield gain for the major cereals indicate that global and regional shortages of maize will occur unless the rate of gain can be increased substantially. Moreover, the yield plateau evident in some of the most productive rice-cropping systems of Asia suggests declining rates of gain in rice yields in areas where farmers were early adopters of modern seed and fertilizer technology. Uncertainty about trends in the yield potential of rice and maize varieties and hybrids also raises concerns about sustaining yield gains as average farm yields approach the climate-adjusted yield potential ceiling, thus reducing the size of the exploitable yield gap. These trends run counter to the more optimistic predictions of global food security derived from econometric and biophysical models.

Other factors that have a large influence on the food supply–demand balance are also difficult to predict. Competition for water resources (Postel, 1998) and the degree of industrialization of livestock production in developing countries will have large effects on predicted cereal demand and food prices (Delgado et al., 1999). Decreasing investment in public sector agricultural research and increasing investment by the private sector will influence both the scope and focus of the global research agenda. Private sector research, by nature, must emphasize short-term objectives with a high probability of profit. Basic research on complex traits, such as yield potential or N-use efficiency, or on ecological studies of crop and soil management to optimize both productivity and environmental quality, receives less emphasis. Also at issue is the rate at which research investment is transformed into greater yields on-farm. Increasing reliance on private sector research and the associated emphasis on intellectual property rights may affect food security outcomes if the free exchange of information and germplasm is hindered (Evenson, 1999). Finally, the rate of adoption of transgenic crops and associated productivity benefits are uncertain because of delays in acceptance due to public concerns about food safety and ecological impact from widespread use.

Because of these uncertainties, research priorities to assure food security will be considered under different scenarios: (i) continued global food surplus; (ii) global food scarcity; (iii) regional food security in resource-poor environments; and (iv) intensification of pasture and range systems.

Research Priorities with Continued Food Surplus

Humanity will enjoy the luxury of a food surplus scenario if farmers have access to adequate water and land resources, soil quality is not degraded, research investment is sufficient, and the yield potential of the major cereal crops increases at a rate that maintains an exploitable yield gap in high production systems. This scenario also requires appropriate national policies and efficient markets at regional and global levels. Assuming these conditions are met, crop science research is largely relieved of the need to focus resources on increasing cereal production *per se*. Instead, technologies are needed that provide farmers with an adequate profit in the face of low commodity prices. For example, the baseline scenario of the IMPACT model predicts that prices for maize, rice and wheat in 2020 will decrease by 2%, 8% and 10%, respectively, from average values in 1992–1994, because cereal production is forecast to stay ahead of demand (Delgado *et al.*, 1999). Under this scenario, farmers in developed countries must focus on achieving greater profit to stay in business. In developing countries, increased farm income is crucial for raising standards of living in rural areas and to reduce the rate of urban migration. In both developed and developing countries, pressure to increase farm size will continue in order to maintain income levels with low commodity prices.

High priority research targets include reduced production costs, greater crop end-use value, and diversification away from commodity grain production systems to minimize oversupply. The private sector will continue to have a comparative advantage for basic and applied research to develop crop cultivars with greater end-use value because of economic incentive and capital resources. Application of biotechnology to accelerate conventional breeding efforts provides a wide range of opportunities to develop transgenic cultivars with specific end-uses. Among the promising targets for increased value are nutraceuticals, improved taste and textures, greater nutritional value for humans and livestock, and unique oils, starch and proteins for specific food products and industrial uses. If energy prices increase substantially, crops grown solely to produce bio-fuels may become an economically feasible option under a food surplus scenario.

Greater profit is achieved by increasing the efficiency with which both external inputs (e.g. fertilizer, irrigation water, seed, pesticide and fuel) and indigenous resources (e.g. incident radiation, soil nutrient stocks, soil moisture and rainfall) are used to produce economic yield. Conservation tillage and no-till systems that increase the proportion of rainfall infiltrating into soil and reduce surface evaporation and erosion are examples of management practices that significantly improve the utilization efficiency of indigenous resources. Soil testing to determine fertilizer requirements is a practice that improves efficiency of an external input. Like most production practices that improve production efficiency, both conservation tillage and soil testing protect environmental quality through positive effects on soil quality and reduced hazard of degrading the quality of water resources.

Despite past successes, continued gains in crop production efficiency in high production systems must come from more precise coordination of all management practices because production factors strongly interact in their effects on yield. The goal of strategic research to improve production efficiency, as described by de Wit (1992), 'should not be so much directed towards the search for marginal returns from variable resources, as towards the search for the minimum of each production resource that is needed to allow maximum utilization of all other resources'. 'Precision agriculture' is a general term that describes optimal timing and placement of a field operation or input, either uniformly applied to the whole field or by site-specific management to account for within-field differences in soil properties, input requirements and pests.

While theory predicts increased input use efficiency with site-specific versus uniform management and concomitant environmental benefits, especially at high yield levels (Cassman and Plant, 1992), successful implementation of site-specific management requires accurate data about the spatial variability in soil properties, crop physiological status, pest and disease incidence, and exact knowledge of crop response to this variability. Unfortunately, present understanding of ecophysiological processes governing crop response to interacting environmental factors is not sufficiently robust to make accurate predictions of site-specific input requirements, or the outcome from their application. This knowledge gap is the primary limitation to adoption of site-specific management in large-scale agriculture.

In developing countries, precision agriculture will be needed to sustain yield increases as average yields approach the 80% yield potential threshold in high production systems, such as irrigated rice and rice–wheat systems in Asia. Because field size is typically less than 0.5 ha, precision agriculture must involve field-specific practices uniformly applied to small fields. For example, recent on-farm studies of irrigated rice systems in Asia document tremendous field-to-field variation in native soil N supply within small production domains where soil properties are relatively uniform (Adhikari *et al.*,

1999; Olk *et al.*, 1999). Field-specific N management can significantly improve N fertilizer efficiency, which is presently very poor. In contrast, most government and regional extension systems provide blanket fertilizer recommendations for an entire district or region. Given the magnitude of field-to-field variation, blanket fertilizer recommendations result in inefficient nutrient use because of nutrient deficiency, excess and imbalance.

The need to achieve field-specific nutrient management in hundreds of millions of small rice, wheat and maize fields in low-income countries represents a daunting challenge. Field-specific pest management will also be needed to protect crops from insect pests and diseases while minimizing pesticide use. Developing the scientific capacity, technology transfer mechanisms, and farmer education to allow diagnosis of limiting factors, prediction of expected yields and input requirements for field-specific management will be critical components of global food security.

In summary, progress towards implementing precision agriculture in both developed and developing countries will depend on research investment in production ecology, crop physiology, soil science and information technology to enable more accurate predictions of crop response to its environment. Deeper understanding of genotype–environment interactions is also fundamental to crop improvement efforts using conventional and molecular approaches. Increased competition for water resources and higher irrigation costs will provide incentives for substantial improvements in water use efficiency through greater precision in matching irrigation amount and uniformity to crop needs in time and space. Potential benefits from transgenic crops are most likely to contribute to higher end-use value, greater resistances to insects and diseases, and reduction in production costs if public concerns about food safety and environmental impact can be addressed.

Research Priorities Under a Food Scarcity Scenario

Research to increase cereal yield potential and the rate of increase in average farm yields in the major cereal production systems would be high priorities under a scenario of global food scarcity. Closing the exploitable yield gap will require development of precision agriculture approaches as described in the previous section although the emphasis would be somewhat different. Under a food scarcity scenario, the dual goals of sustaining yields near the yield potential threshold and reducing the negative environmental impact of management practices required to achieve high yields would be crucial (Cassman, 1999). High-yielding crops must acquire and utilize relatively large amounts of water and N because of tightly conserved relationships between crop yield and transpiration (Tanner and Sinclair, 1983) and N requirements (Greenwood *et al.*, 1990). Hence, farmers have little margin for error in providing adequate amounts of these resources to ensure that growth is not limited. If commodity prices rise as a result of food scarcity, economic incentives to optimize the efficiency of input use decrease. In addition, improved soil quality becomes important because the indigenous soil nutrient supply capacity and water holding capacity provide a crucial buffer against imperfections in management to achieve yields near yield potential levels.

Increases in genetic yield potential are also crucial for food security under the food scarcity scenario. Yield potential trends of the major cereals must be better quantified, and the physiological traits responsible for these trends identified (Duvick and Cassman, 1999; Peng *et al.*, 1999). Utilizing this knowledge to accelerate crop improvement by both conventional and molecular approaches is required. The International Rice Research Institute project on increasing rice yield potential illustrates the magnitude of the challenge. It was initiated in 1990 as a multi-disciplinary effort including geneticists, agronomists, physiologists, plant pathologists and entomologists. Despite a substantial effort, acceptable varieties with greater yield potential have not yet been developed, nor has increased yield potential of the new plant ideotypes been documented under field conditions. Improved understanding of source and sink limitations to rice yield potential and greater resistance to lodging are required to better target breeding efforts on key plant traits (Sheehy *et al.*, 2000a).

Ultimately, research on yield potential must consider manipulation of the fundamental processes governing the efficiency with which intercepted radiation is converted to biomass. Although research to date has generally indicated little improvement in conversion efficiency, Loomis and Amthor (1999) suggest that the maximum observed values of radiation use efficiency (RUE) fall below theoretical values. They identify a number of potential research targets to increase conversion efficiency including photoprotection mechanisms, enzymatic properties of rubisco, and respiration. Other scientists have proposed incorporating the C_4 photosynthetic capacity into C_3 cereals to improve RUE, which would require genetic engineering to introduce genes for enzymes of the C_4 pathway and for leaf anatomy (Sheehy *et al.*, 2000b). Although the path to greater crop yield potential remains uncertain, continued reliance on brute-force breeding to achieve yield advances will not suffice. Success will depend on improved physiological insight and use of biotechnology tools to complement conventional breeding efforts. Success also will depend on greater research investment towards the explicit goal of increasing crop yield potential than is presently given.

Research Priorities for Regional Food Security in Resource-poor Agroecosystems

Major rainfed cropping systems in less favourable environments include: (i) maize-based systems on sloping land in South and Central America; (ii) maize, sorghum and millet systems on poor soils in subhumid and semiarid regions of sub-Saharan Africa; (iii) lowland rice systems in drought- and flood-prone areas of South and South-East Asia; (iv) upland rice systems in Asia, Africa, and South America; (v) slash and burn systems in the humid tropics of Africa, Asia and Central and South America; and (vi) wheat-based systems in semiarid regions of the North American Great Plains, Australia and Central Asia. Water relations, soil quality, or both water and soil constraints place severe limits on plant growth in each of these systems.

In developed countries, such as the USA, Canada and Australia, productivity increases in rainfed wheat systems have been achieved by expanding farm size, large-scale mechanization, improved varieties, judicious use of inputs, conservation tillage and improved rotations. Despite these improvements, the linear rate of gain in wheat yield has been much smaller than for irrigated wheat in China and India (Fig. 3.2), or in the more favourable rainfed wheat systems of northern Europe (data not shown). Research to sustain yield increases in less favourable environments must focus on optimizing yield, yield stability, input-use efficiency, and profit in the face of highly variable climatic conditions. Water and soil conservation from improved tillage and residue management practices, maintenance of soil fertility, improved germplasm with increased end-use value and greater tolerance of abiotic and biotic stresses, and the best match of genotype to environment are pivotal components. Robust decision-support tools are needed to help farmers implement responsive crop rotations and management practices that reduce risk and maximize profit by taking advantage of variations in climate, stored soil moisture, and prices of commodities and inputs.

In less favourable environments of developing countries, large-scale farming and mechanization are generally not feasible or desirable because of higher population density and lack of infrastructure and capital. Farming systems that include improved varieties, judicious use of inputs, conservation tillage, small-scale mechanization, and improved rotations offer promising options for resource-poor farmers. Integration of cropping systems with livestock production also provides opportunities for increased efficiencies and profits. In many cases, technologies that increase productivity have already been identified. Improved nutrient management for slash and burn systems in the humid tropics are one example (Nye and Greenland, 1960), but farmers lack capital, credit and access to required inputs and information to implement these practices. Markets are often lacking in which surplus production can be sold. Hence, crop and soil scientists alone cannot provide the solutions to raising productivity and creating the conditions that allow farmers to benefit from it. Concerted efforts to improve markets, infrastructure and policies supportive of resource-poor farmers are required.

Over the long term, sustaining subsistence agriculture that only meets minimum food needs but does not alleviate poverty will accelerate rural to urban migration and risk further environmental degradation. Local and regional food security are preconditions for economic development. Crop science research must therefore target substantial increases in productivity and profit while preserving the quality of the natural resource base.

Because numerous factors typically limit crop production in less favourable environments, research prioritization must rely on accurate identification of the extent and severity of the factors most limiting to crop production. The difficulty of making such an assessment should not be underestimated. An example of the challenge is illustrated by two studies conducted on rainfed lowland rice in mostly unfavourable environments of South Asia. One study estimated the severity of biophysical constraints to rainfed lowland rice in eastern India for the purpose of prioritizing the allocation of funding for biotechnology research to improve productivity in these systems (Widawsky and O'Toole, 1990). A detailed survey was sent to rice scientists in four lowland rice-producing states at 12 research stations. More than 100 rice scientists estimated the severity and extent of yield loss from all major constraints including adverse soil conditions, drought, flooding, high or low temperatures, insects, weeds and disease. Survey results identified five constraints that were estimated to account for 59% of the difference between actual on-farm yields and the attainable yield using the best available technology (Table 3.3). Deficiencies of N, P and K were not identified as being among the top five constraints in this study.

In contrast, Wade et al. (1999) conducted a series of field studies to estimate the yield response of rainfed lowland rice to N, P and K at 78 representative locations across India, Bangladesh, Thailand, Indonesia and the Philippines. Experiments were conducted both on-farm and at research stations. The mean yield response to optimal N, P and K fertilizer application was 80% above the control without applied nutrients (Table 3.3). Based on the difference between average yields presently obtained by farmers in these regions and the yields obtained with recommended fertilizer levels, Wade et al. (1999) concluded: 'Substantial increases in yield are possible in rainfed systems with application of appropriate (macro) nutrients, especially if used in conjunction with cultivars adapted to the targeted environment'. While the two studies are not directly comparable, the contrasting results concerning the importance of macronutrients as a yield-limiting factor suggest that scientist surveys alone, without validation under field conditions, are not adequate for accurate estimation of on-farm constraints. Moreover, while crop scientists must be active in leading these efforts, collaboration with soil scientists, economists and rural sociologists also is needed to effectively identify constraints and prioritize research.

Research Priorities for Forage and Grassland Systems

Livestock are an integral component of sustainable agriculture. They complement crop production enterprises by utilizing plant materials not suitable for human consumption, such as forages and by-products, and marginal

Table 3.3. Biophysical constraints to rainfed lowland rice production as estimated in two studies.

Constraint	Survey of rice scientists in eastern India[a] (% of total yield loss)	Field experiments on response to nutrients[b] (% yield increase)
Drought	26	–
Weeds	12	–
Diseases	11	–
Insects	5	–
Zn deficiency	5	–
Total	59	
Optimal, N, P and K	–	80

Eastern India study[a]		Nutrient response experiments[b]	
Estimated average yield	2151 kg ha^{-1}	Control plots (–NPK)	2250 kg ha^{-1}
Estimated attainable yield	4300 kg ha^{-1}	Optimal NPK	4050 kg ha^{-1}

[a] Estimated yield reduction from the attainable yield level that can be achieved with the best available technology under on-farm conditions. Estimates were obtained from a survey of rice scientists at 12 research stations in four states (Widawsky and O'Toole, 1990). The macronutrients N, P and K were not identified as a limiting factor by the rice scientists included in this survey study.
[b] Average yield response measured in replicated plots in field experiments conducted at 78 representative rainfed lowland rice sites in eastern India, Indonesia, the Philippines, Thailand and Bangladesh (Wade *et al.*, 1999).

land for grazing. Meat, eggs and dairy products provide nutritional benefits in human diets, especially for children. Recycled manure helps to maintain soil fertility. As standards of living increase in developing countries, demand for livestock products is projected to grow more rapidly than demand for cereals (CAST, 1999; Delgado *et al.*, 1999).

Animals presently consume one-third of global cereal grain production. Under a scenario of continuing food surplus, feed grain prices will remain low. Expanded use of feed grains in livestock rations would be promoted by low grain prices, which in turn provides economic advantages to industrial livestock production. Likewise, the price of forages and hay would remain relatively low, which provides little incentive for increases in the primary productivity of range and pasture systems. Under this scenario, research must emphasize the role of range and grassland systems in soil conservation, as habitat for both livestock and wildlife, and as refugia for biodiversity. This type of research presently receives greatest emphasis in funding allocated for the study of range and forage systems in developed countries. With good husbandry, low-input grassland systems can meet the multi-purpose goals of natural resource conservation, preservation of wildlife and plant biodiversity, and livestock production at low, but sustainable levels of productivity (CAST, 1999).

Under the food scarcity scenario, feed grain prices would rise and there would be greater pressure on range and grassland resources to meet the increased demand for livestock products. Because most range and grassland systems respond to more intensive management, substantial increases in forage yields can often be obtained by application of nutrients, such as N and P, even in semiarid environments (Breman and de Wit, 1983). Fencing to permit more intensive control of grazing periods and animal numbers, and seeding with improved forage species are other practices that improve productivity (CAST, 1999). Basic research on the biochemical and biophysical determinants of forage digestibility hold promise for guiding conventional breeding and genetic engineering for improved forage quality. Perhaps the greatest research challenge under a scenario of increasing grain prices is the preservation of environmental benefits provided by range and grassland systems in the face of intensified management. Technologies to increase productivity must be investigated in concert with longer-term effects on wildlife habitat, biodiversity and soil conservation.

The Challenges Ahead

Accurate prediction of the food supply–demand balance is fundamental to prioritization of research in crop science to ensure regional and global food security. The latter depends in large part on productivity trends of a few major cropping systems in favourable environments, while local and regional food security depend on substantial improvements in productivity of cropping systems in less favourable environments, especially in developing countries.

Three variables must be estimated to predict food production capacity and its variability in a cropping system. The genetically determined crop yield potential and the size of the exploitable gap between average farm yields and the yield potential ceiling are two of these parameters, while year-to-year variation in the size of the exploitable yield gap caused by climatic variation is the third. Together these parameters determine global and regional food production capacity, the land area required to produce it and the degree of uncertainty in meeting food demand as a result of climatic variation. Despite the importance of these variables as determinants of food security, few studies have attempted to estimate them in the major cropping systems that represent the foundation of our human food supply. Instead, efforts to predict global and regional food security have been largely guided by economists whose predictions rely on projection of past trends.

Crop and soil scientists can help to improve these predictions by providing more accurate estimates of the biophysical constraints to crop productivity and of the investment in research and extension that is required to alleviate them. Rapid progress in the fundamental sciences and information technology provide exciting opportunities to ensure regional and global food security for generations to come. With continuing trends of reduced public investment in research and extension worldwide, however, research prioritization is crucial. Crop and soil scientists must become 'activists' in developing an appropriate framework to improve prediction of sustainable food production capacity and to improve the prioritization of research to ensure food security.

Acknowledgements

The author is grateful to Chuck Francis, Paul Singleton, John Sheehy, Jim Specht and Len Wade for their comments and suggestions on an earlier draft of this manuscript.

References

Adhikari, C., Bronson, K.F., Panaullah, G.M., Regmi, A.P., Saha, P.K., Dobermann, A., Olk, D.C., Hobbs, P. and Pasuquin, E. (1999) On-farm soil N supply and N nutrition in the rice–wheat system of Nepal and Bangladesh. *Field Crops Research* 64, 273–286.

Bänziger, M. and Lafitte, H.R. (1997) Efficiency of secondary traits for improving maize for low-nitrogen target environments. *Crop Science* 37, 1110–1117.

Bolanos, J. and Edmeades, G.O. (1996) The importance of the anthesis–silking interval in breeding for drought tolerance in tropical maize. *Field Crops Research* 48, 65–80.

Breman, H. and de Wit, C.T. (1983) Rangeland productivity and exploitation in the Sahel. *Science* 221, 1341–1347.

Cassman, K.G. (1999) Ecological intensification of cereal production systems: yield potential, soil quality, and precision agriculture. *Proceedings of the National Academy of Sciences USA* 96, 5952–5959.

Cassman, K.G. and Dobermann, A. (2001) Evolving rice production systems to meet global demand. In: Rockwood, W. (ed.) *Proceedings of a Symposium on Rice Research and Production in the 21st Century.* International Rice Research Institute, Los Baños.

Cassman, K.G. and Plant, R.E. (1992) A model to predict crop response to applied fertilizer nutrients in heterogeneous fields. *Fertilizer Research* 31, 151–163.

CAST (Council for Agriculture, Science and Technology) (1999) *Animal Agriculture and Global Food Supply. Task Force Report No. 135.* Council for Agriculture, Science and Technology. Ames, Iowa.

Delgado, C., Rosegrant, M., Steinfeld, H., Ehui, S. and Courbois, C. (1999) *Livestock to 2020: The Next Food Revolution. Food, Agriculture, and the Environment Discussion Paper 28.* International Food Policy Research Institute, Washington, DC.

Duvick, D.N. and Cassman, K.G. (1999) Post-green revolution trends in yield potential of temperate maize in the North-Central United States. *Crop Science* 39, 1622–1630.

Evans, L.T. (1993) *Crop Evolution, Adaptation, and Yield.* Cambridge University Press, Cambridge.

Evenson, R.E. (1999) Intellectual property rights, access to plant germplasm, and crop production scenarios in 2020. *Crop Science* 39, 1630–1635.

Greenwood, D.J., Lemaire, G., Goose, G., Cruz, P., Draycott, A. and Neeteson, J.J. (1990) Decline in percentage of N of C3 and C4 crops with increasing plant mass. *Annals of Botany* 66, 425–436.

Loomis, R.S. and Amthor, J.S. (1999) Yield potential, plant assimilatory capacity, and metabolic efficiencies. *Crop Science* 39, 1584–1596.

Mathews, R.B., Horie, T., Kropff, M.J., Bachelet, D., Centeno, H.G., Shin, J.C., Mohandass, S., Singh, S., Zhu Defeng and Moon Hee Lee (1995) A regional evaluation of the effect of future climate change on rice production in Asia. In: Mathews, R.B., Kropff, M.J., Bachelete, D. and van Laar, H.H. (eds) *Modelling the Impact of Climate Change on Rice Production in Asia.* CAB International, Wallingford, pp. 95–139.

Nye, P.H. and Greenland, D.J. (1960) *The Soil Under Shifting Cultivation. Technical Communication 51.* Commonwealth Bureau of Soils, Slough.

Olk, D.C., Cassman, K.G., Simbahan, G., Sta Cruz, P.C., Abdulrachman, S., Nagarajan, R., Pham Sy Tan and Satawathananont, S. (1999) Interpreting fertilizer-use efficiency in relation to soil nutrient-supplying capacity, factor productivity, and agronomic efficiency. *Nutrient Cycling in Agroecosystems* 53, 35–41.

Peng, S., Cassman, K.G., Virmani, S.S., Sheehy, J. and Khush, G.S. (1999) Yield potential trends of tropical rice since the release of IR8 and the challenge of increasing rice yield potential. *Crop Science* 39, 1552–1559.

Penning de Vries, F.W.T., van Keulen, H. and Rabbinge, R. (1995) Natural resource limits of food production in 2040. In: Bouma, J., Kuyvenhoven, A., Bouman, B.A.M., Luyten, J.C. and Zandstra, H.G. (eds) *Eco-Regional Approaches for Sustainable Land Use and Food Production.* Kluwer Academic Publishers, Dordrecht, pp. 65–87.

Postel, S.L. (1998) Water for food production: will there be enough in 2025? *BioScience* 48, 629–637.

Rosegrant, M.W., Agcaoili-Sombilla, M. and Perez, N.D. (1995) *Global Food Projections to 2020: Implications for Investment. Food, Agriculture and the Environment Discussion Paper 5.* International Food Policy Research Institute, Washington, DC.

Sayre, K.D., Singh, R.P., Huerta-Espino, J. and Rajaram, S. (1998) Genetic progress in reducing yield losses to leaf rust in CIMMYT-derived Mexican spring wheat cultivars. *Crop Science* 38, 654–659.

Sheehy, J.E., Mitchell, P.L., Dionora, M.J.A., Peng, S. and Khush, G.S. (2000a) Nitrogen increases radiation conversion factor and unlocks the yield barrier in rice. *Plant Production Science* (in press).

Sheehy, J.E., Mitchell, P.L. and Hardy, B. (eds) (2000b) Redesigning rice photosynthesis to increase yield. *Proceedings of the Workshop on the Quest to Reduce Hunger: Redesigning Rice Photosynthesis.* International Rice Research Institute, Los Baños and Elsevier Science (in press).

Tanner, C.B. and Sinclair, T.R. (1983) Efficient water use in crop production: research or re-search? In: Taylor, H.M., Jordan, W.R. and Sinclair, T.R. (eds) *Limitations to Efficient Water Use in Crop Production.* American Society of Agronomy, Madison, Wisconsin, pp. 1–27.

Vitousek, P.M., Mooney, H.A., Lubchenco, J. and Melillo, J.M. (1997) Human domination of earth's ecosystems. *Science* 277, 494–499.

Waddinton, S.R. and Heisey, P.W. (1997) Meeting the nitrogen requirements of maize grown by resource-poor farmers in Southern Africa by integrating varieties, fertilizer, crop management and policies. In: Edmeades, G.O., Banziger, M., Mickelson, H.R. and Pena-Valdivia, C.B. (eds) *Proceedings of a Symposium on Developing Drought- and Low N-Tolerant Maize*. CIMMYT, Mexico, DF, pp. 44–57.

Wade, L.J., Amarante, S.T., Olea, A., Harnpichitvitaya, D., Naklang, A., Wihardjaka, A., Sengar, S.S., Mazid, M.A., Singh, G. and McLaren, C.G. (1999) Nutrient requirements in rainfed lowland rice. *Field Crops Research* 64, 91–107.

Widawsky, D.A. and O'Toole, J.C. (1990) Prioritizing the rice biotechnology research agenda for Eastern India. The Rockefeller Foundation, New York.

de Wit, C.T. (1992) Resource use efficiency in agriculture. *Agricultural Systems* 40, 125–151.

Modifying the Composition of Plant Foods for Better Human Health

4

R.F. Hurrell

Laboratory of Human Nutrition, Institute of Food Science, ETH Zürich, PO Box 474, CH-8803 Rüschlikon, Switzerland

Introduction

When considering plant foods in relation to human health, it should be remembered that plant foods may also have a health value in addition to their nutritional value. All the nutrients essential to man, except vitamin B_{12} and vitamin D, can be obtained from plant foods. They contain the carbohydrates, protein and fat necessary for energy production. They contain the amino acids and fats necessary to make body structures such as muscles and cell membranes, or to make the enzymes, hormones or prostaglandins that help control the main body reactions. Plant foods contain the 12 minerals and trace elements that are necessary as enzyme co-factors or as essential components of body structures such as bone and blood, and they contain 11 of the 13 vitamins which regulate and facilitate chemical reactions such as oxidation, reduction, hydroxylation, methylation, etc., that control our metabolism. Unfortunately, diets based mainly on plant foods often lack high quality protein or are deficient in specific minerals and vitamins. Regular consumption of staple plant foods such as cereals, roots or pulses, in the absence of animal tissue, can lead to serious deficiency diseases. On the other hand, plant foods contain many different phytochemicals which can have a major beneficial effect on health. These include the phytooestrogens, phytosterols, carotenoids, glucosinolates and many different phenolic compounds.

In the developing countries, poverty leads to food shortage and undernutrition and many populations survive largely on plant-based diets with little or no meat or dairy products. In the industrialized countries, relative affluence leads to overconsumption of food and especially to the overconsumption of animal foods at the expense of plant foods. These eating patterns play a major role in the aetiology of deficiency diseases in developing countries, and diet is one of the several risk factors important in the aetiology of chronic diseases in industrialized countries. The other risk factors include genetic predisposition, exercise, smoking and stress. In developing countries, chronic hunger due to food shortages leads to much stunting and wasting in children and there are major deficiencies of several micronutrients (FAO, 1997). Vitamin A deficiency leads to blindness and a weakened immune system in young children. Iodine deficiency leads to goitre and cretinism, Fe deficiency leads to anaemia, and Zn deficiency leads to reduced growth and lower immune function. In industrialized countries micronutrient deficiencies are rare; however, we find much heart disease, cancer, diabetes, osteoporosis, obesity, cataract, dementia and other chronic diseases which are associated with excess calorie consumption or diets high in saturated fat and sodium. Diets rich in fruit, vegetables or cereals help prevent such diseases and this beneficial effect is thought to be related to certain minor plant components (Subar *et al.*, 1992; Steinmetz and Potter, 1996).

This chapter first discusses current knowledge of the relationship of minor plant components to chronic diseases and examines the possibility and the usefulness of modifying the level of certain phytochemicals in plant tissues. Micronutrient deficiencies in developing countries are then discussed and modifying the nutrient composition of staple crops is proposed as a potential strategy to improve public health.

Minor Plant Components

Phytochemicals are non-nutrients with a health benefit. The major interest has been focused on the many flavonoids and other phenolic compounds in fruits and vegetables (Parr and Bolwell, 2000), carotenoids such as lycopene and lutein (Clinton, 1998; van den Berg *et al.*, 2000), glucosinolates from *Brassica* spp. (Mithen *et al.*, 2000) and the phytooestrogens (Potter *et al.*, 1998) and phytosterols (Westrate and Meijer, 1998) from legumes such as soybean. Flavonoids and carotenoids are powerful antioxidants which quench free radicals. They are suggested to protect against the damage to the DNA bases which could initiate cancers, and to prevent the oxidation of the LDL-cholesterol which is thought to be a major risk factor in cardiovascular disease. Glucosinolates activate enzyme systems in the liver which detoxify the carcinogens and thus prevent cancer. Isoflavones from soybean have a weak anti-oestrogenic effect and have been shown to be beneficial against hormone-related cancers such as breast and prostate cancer and against bone loss. Finally, the phytosterols reduce cholesterol absorption and thus lower serum cholesterol which is a risk factor in cardiovascular disease. It should be possible to increase the phytochemical levels in edible plant tissues by conventional breeding techniques or by genetic engineering; however, current knowledge is still too limited to recommend which of the many plant components is protective and at which concentration it should be present in the diet (Grusak and DellaPenna, 1999).

Many steps are needed to demonstrate the health-promoting effect of specific phytochemicals before manipulating their levels in plants. Firstly, epidemiological studies comparing dietary intake with disease prevalence are needed to identify the beneficial phytochemicals. A mechanism for disease prevention must be proposed which is first tested in *in vitro* studies. The *in vitro* studies must demonstrate a physiologically relevant effect such as an antioxidant effect, preventing DNA damage or the activation of a specific enzyme. Animal studies and then human studies must be made to measure the absorption, metabolism and physiological effect of the phytochemical compound *in vivo*. Finally, it is necessary to make long-term intervention studies in man showing a positive effect on a disease biomarker (e.g. serum cholesterol) or better still a reduction of disease prevalence. This is the most difficult step since disease prevalence can be influenced by many different factors.

Table 4.1 summarizes current evidence for the health promoting effect of different phytochemicals (Lyndsay and Clifford, 2000), and indicates that much more research should be made before attempting to modify their levels in plant foods. The best evidence of a positive health effect has been obtained when feeding phytosterols. The proposed cholesterol-lowering effect of phytosterols has been demonstrated *in vitro* as well as in animal and human studies, which have also investigated both absorption and metabolism (Piironen *et al.*, 2000). As yet, there are no data demonstrating less cardiovascular disease with high phytosterol consumption. For carotenoids, flavonoids, isoflavones and glucosinolates, only *in vitro* and animal studies have been reported. The human studies on absorption and metabolism have not been completed, there is often no data, little data or incomplete data on the proposed physiological effect, and no consistent data on disease prevention. Several intervention studies have been made with high dose supplementation of β-carotene and results have been contradictory. In one large supplementation trial, contrary to expectation, high dose β-carotene supplementation was found to increase the incidence of lung cancer in smokers (ATBC Study Group, 1994).

Phytosterols are perhaps the only phytochemicals to have a clearly defined health benefit via their inhibition of cholesterol absorption. These compounds occur mainly as minor components of vegetable oils. The main phytosterol in our diet is β-sitosterol (50–95% total sterols) with other minor sterols including campesterol and stigmasterol. β-Sitosterol has a very similar chemical structure to cholesterol, differing only in the side chain. The typical content in vegetable oils such as soy, maize, oat or wheatgerm varies from 150 to 2600 mg 100 g^{-1} for a typical daily intake of some 200 mg. Phytosterols reduce cholesterol absorption by reducing its solubilization in the lipid micelles. Less than 5% of the phytosterol is absorbed, most of which is rapidly excreted in the bile, although some is incorporated in the LDL-particle.

Table 4.1. Current evidence for health promoting effects of phytochemicals.

Phytochemicals	Proposed mechanism	In vitro	Animal studies	Absorption/ metabolism	Human studies	
					Physiological effect	Disease prevention
Phytosterols	Cholesterol-lowering	✓	✓	✓	✓	No data
Carotenoids	Antioxidant	✓	✓	Not complete	Not complete	Inconsistent results
Flavonoids	Antioxidant	✓	✓	Not complete	No data	No data
Isoflavones	Oestrogenic	✓	✓	Not complete	Not complete	No data
Glucosinolates	Enzyme activation	✓	✓	Not complete	Not complete	No data

✓ Information published and physiological effect established.

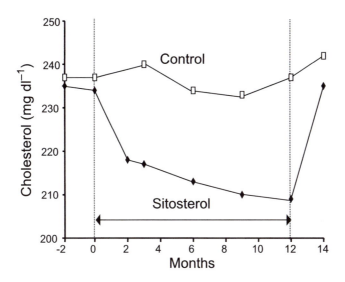

Fig. 4.1. Changes in serum cholesterol levels before, during and after consumption of a cholesterol-lowering margarine containing added sitosterol (Miettenen *et al.*, 1995).

Benecol is a cholesterol-lowering margarine containing added phyto-sterol. It is a rapeseed oil margarine plus 12.5% of hydrogenated sitosterol. In the study of Miettenen *et al.* (1995) (Fig. 4.1), 102 mildly hypercholes-terolaemic subjects consumed 8 g of margarine (containing 1g sitosterol) three times per day with meals. At the beginning of the study, the mean serum cholesterol level was close to 240 mg dl^{-1} and it remained at this level in the control group for the duration of the study. In the sitosterol group, total serum cholesterol fell to 210 mg dl^{-1} after some months and remained at this level until the study finished at 12 months. After 1 year, LDL-cholesterol was reduced by 14%, which would be expected to reduce the risk of cardiovascular disease by 15–40% (Law *et al.*, 1994). Oil-seed plants could be usefully selected to increase the phytosterol content of their oils either for extraction of the phytosterol or for direct consumption.

Micronutrient Deficiencies

In relation to the micronutrient deficiencies, Fe is the micronutrient with the greatest magnitude of deficiency and 2 billion people, or one third of the world's population, are reported to be at risk of Fe deficiency. This compares with 1.5 billion people who are reported to be at risk of iodine deficiency, and some 250 million children under 5 years who are at risk of vitamin A deficiency of which 250,000–500,000 each year suffer from irreversible blindness (FAO, 1997). The magnitude of Zn deficiency is uncertain, since there are no adequate status measurements, but it is predicted to be widespread.

The three major intervention strategies that are commonly used to correct for micronutrient deficiencies are food fortification, supplementation with pharmaceutical doses, and nutrition education followed by diet modification (Lachance and Bauernfeind, 1991). Diet modification has high personal costs and is considered to have only moderate impact since people do not like to change their eating habits or cannot afford to change them. Food fortification is considered to have moderate capital expenditure, low running costs and high population coverage and, if widely consumed foods such as salt or wheat flour are fortified, it is considered the best long-term approach to prevention. Supplementation involves high doses of nutrients: high costs and poor compliance often limit its usefulness. In contrast to food fortification, supplementation is a therapeutic strategy most useful for short-term interventions in highly deficient groups such as pregnant women although high dose vitamin A supplements are often provided every 4–6 months to young children at risk of vitamin A deficiency. A possible fourth strategy to correct micronutrient deficiencies could be an agricultural strategy involving plant breeding or genetic engineering techniques which could be used to increase the Fe, Zn and β-carotene content of plant foods (Bouis, 1996; Graham *et al.*, 1999; Frossard *et al.*, 2000; Ye *et al.*, 2000).

Vitamin A

Vitamin A deficiency affects many pre-school children and leads to xerophthalmia – dryness of the conjunctiva of the eye – which coupled with infection, can lead to total blindness. Vitamin A deficiency also leads to impaired immunity and increased infection, and many children die from relatively minor diseases such as measles and diarrhoea. The deficiency is due to the consumption of low vitamin A diets often based on rice and cassava. Rice, like all plant foods, contains no vitamin A (retinol) *per se* but also contains no pro-vitamin A (β-carotene). Typical rice diets contain little or no meat, dairy products or eggs, so children consume little vitamin A *per se*. These diets also lack dark-green vegetables, such as spinach, or orange/yellow fruits and vegetables, such as mango or pumpkin, which contain pro-vitamin A (β-carotene). Most cases of vitamin A deficiency are found in the rice-eating countries of South Asia and Africa.

Recently Ye *et al.* (2000) used genetic engineering techniques to produce rice grains containing β-carotene. The β-carotene biosynthetic pathway was inserted into the rice endosperm via the *Agrobacterium*-mediated introduction of the genes for three enzymes: phytoene desaturase and lycopene β-cyclase from daffodils, and carotene desaturase of bacterial origin. The β-carotene was expressed in the grain at $1.6–2.0 \mu g\ g^{-1}$ to give 'golden' rice. Assuming that 6 μg β-carotene is necessary to give 1 μg retinol equivalents (RE) (National Research Council, 1989), it can be calculated that 1.2 kg of 'golden' rice per day would be necessary to provide all the recommended vitamin A intake of a 4–6-year-old child (400 μg RE). Even a relatively high rice intake of 200 g day^{-1} would provide only 70 μg RE or some 20% of recommended vitamin A intake for this age group. Although the creation of 'golden' rice without doubt opens a new chapter in world nutrition, the β-carotene level of this first transformation is relatively low and must still be demonstrated to be bioavailable to man. Attempts should be made to increase the β-carotene.

Iron and zinc

A major factor in Fe and Zn deficiency is the low bioavailability of Fe and Zn from diets based on cereals and legumes. These plant-based diets are often high in phytic acid, low in ascorbic acid and low in meat products. Phytic acid inhibits Fe and Zn absorption, ascorbic acid enhances Fe absorption, and meat products are high in bioavailable Fe and Zn (Fairweather-Tait and Hurrell, 1996). Low Fe absorption can lead to reduced levels of haemoglobin and reduced levels of Fe enzymes which causes fatigue, impeded brain development (reduced IQ) in infants and children, more premature births, increased perinatal mortality, increased maternal deaths at childbirth (100,000 per year), and a reduced immune function (Hurrell, 1999). Zn deficiency reduces Zn enzyme levels and causes reduced growth in children, reduced immune function which increases diarrhoea and death in the young (Davidsson, 1999). There are agricultural solutions to increasing the level of Fe and Zn in edible plant tissues and these include the application of fertilizers, conventional plant breeding, and genetic engineering techniques (Frossard *et al.*, 2000).

Although Fe forms about 6% of the earth's crust, it is mainly present as oxides which are poorly absorbable by plants. Many plants therefore suffer from Fe deficiency when grown on the calcareous and alkaline soils which cover 25–30% of the earth's surface. Soil fertilization, however, is relatively unsuccessful since added Fe compounds become insolubilized in the same way as the native soil Fe. Foliar fertilization is useful to correct for Fe deficiency in plants, but cannot be considered as a method to increase Fe content (Frossard *et al.*, 2000). Plants also become Zn deficient when grown on calcareous or alkaline soils, or on the highly weathered tropical soils which cover vast areas of India (50%), China (30%), Turkey and Australia. Application of Zn fertilizer to Zn-deficient soil, however, can increase the yield of major food crops such as wheat, maize, sorghum and beans and can also greatly increase the Zn content of the grain (Rengel *et al.*, 1999; Table 4.2). Increasing Zn fertilization from 0–3.2 mg Zn kg^{-1} soil progressively increased the Zn content of the wheat grains from 9.1 to 145 mg kg^{-1}. Yield

Table 4.2. Influence of Zn fertilization on yield and zinc content of wheat (Rengel et al., 1999).

Zinc fertilization (mg Zn kg⁻¹ soil)	Grain yield (g dry wt per plant)	Grain Zn content (mg Zn kg⁻¹ dry wt)
0	1.0 ± 0.17	9.1 ± 0.4
0.05	2.20 ± 0.13	9.9 ± 0.6
0.2	2.24 ± 0.16	14 ± 0.0
0.8	2.51 ± 0.30	83 ± 4.0
3.2	1.70 ± 0.30	145 ± 5.0

also increased gradually up to 0.8 mg kg⁻¹ but fell slightly at the highest Zn addition.

There is also much potential for increasing both the Fe and the Zn content of food staples by plant breeding. In a recent screening of germplasm, a high variation in the Fe and Zn content was reported in the edible parts of staple foods within cultivars of a given species. The Fe content of wheat cultivars varied from 25 to 56 mg kg⁻¹ and the Zn content from 25 to 64 mg kg⁻¹. The variation in Fe and Zn contents of rice cultivars was of the same magnitude as in wheat although the Fe content was lower in rice (Frossard et al., 2000; Table 4.3). Wild varieties of wheat and rice have higher levels of these trace elements than the modern varieties. The variation in maize was particularly high with some grains containing very high levels of Fe and Zn. The Fe and Zn concentration also varied widely in the common bean, cassava and yam, with some yam varieties being particularly high in Fe. It should be noted that the bioavailability of Fe and Zn in these staple plant foods is still unknown as is the form in which the Fe and Zn are present within the plant. Some research groups are already searching for the genes

Table 4.3. Potential to increase Fe and Zn content in food staples by plant breeding (Frossard et al., 2000).

	Variability of Fe and Zn in plant foods (mg kg⁻¹ dry wt)		
	n	Fe	Zn
Wheat	170	25–56	25–64
Rice	1138	7–23	17–52
Maize	126	13–160	11–95
Common bean	>1000	34–89	21–54
Cassava	162	3–48	4–18
Yam	73	7–187	7–30

determining high Fe and Zn levels. Such genes have been identified on chromosomes 6A and 6B in wheat (Cakmak *et al.*, 1999).

There are several other genetic engineering approaches which could be used to increase the Fe content and Fe bioavailability in plant foods. The Fe content could be increased by increasing Fe uptake from the soil or by increasing Fe storage within the edible plant parts (Grusak and DellaPenna, 1999). Increasing the levels of siderophores, chelating agents, reducing agents, enzymes or transporter proteins in roots could increase Fe uptake. Samuelsen *et al.* (1998) recently reported on a transgenic tobacco plant, expressing a yeast ferric reductase gene, which increased leaf Fe by 50%. Similarly increasing the concentration of the storage proteins phytoferritin and metallothionein could increase the content of Fe and Zn, respectively. Goto *et al.* (1999) reported a two- to threefold increase in Fe in transgenic rice expressing the soybean ferritin gene. A more successful approach might be to combine both strategies and to increase both absorption and storage. Phytate reduction could improve the bioavailability of both Fe and Zn. A 65% phytate reduction in a maize mutant with virtually unchanged agronomic traits has been reported recently (Ertl *et al.*, 1998). It should also be possible to introduce heat- and acid-resistant phytases into cereal grains. These phytases should withstand cooking and degrade phytic acid in the human gut during digestion. Finally, another option would be to introduce Fe absorption enhancers, such as ascorbic acid, cysteine-containing peptides or haemoglobin, into the edible plant tissues (Fairweather-Tait and Hurrell, 1996).

Lucca *et al.* (2000) recently reported on a combination of three genetic engineering approaches for improving the bioavailability and level of Fe in rice grains. Firstly, a thermostable phytase from *Aspergillus fumigatus*, active over the pH range found in the gastro-intestinal tract, was introduced using an Agrobacterium-mediated transformation. The phytase had been developed to withstand food processing (Pasamontes *et al.*, 1997). In addition, a ferritin gene from *Phaseolus vulgaris* and a gene expressing a cysteine-rich metallothionein-like protein from rice were also introduced. All three genes were expressed in the endosperm. The phytase activity increased sevenfold to a concentration similar to that of rye, which is highest of the cereal grains. Phytic acid in uncooked rice was completely degraded during a simulated intestinal digestion; however, the phytase was not as thermostable after expression in the rice endosperm and was destroyed when the rice was cooked. The Fe content, however, was increased from 10 to 22 μg g^{-1} and cyst(e)ine content from 45 to 320 mg g^{-1} protein. The influence of these changes on Fe bioavailability still needs to be investigated; however, studies of this type, although still preliminary, open up a new approach to combating Fe deficiency.

Conclusions

It would seem possible to modify the composition of plant foods so as to improve human health. In industrial countries, phytochemicals such as flavonoids, isoflavones, glucosinolates and phytosterols are reported to benefit human health and their concentration in edible plant tissues could be increased by genetic engineering or plant breeding techniques. However, many more human metabolic and intervention studies are still needed in order to define specific compounds, levels, mechanisms of action, benefits and potential harmful effects. Levels of phytochemicals in plants could only be increased if they did not harm the plant.

The main nutrient deficiencies in developing countries are Fe, I, Zn and vitamin A. It is possible to increase the Fe, Zn and β-carotene (pro-vitamin A) content in staple foods by agricultural techniques such as fertilizer application, plant breeding or genetic engineering. Improving the nutrient composition of plant foods could in the future become a sustainable strategy to combat Fe, Zn and vitamin A deficiencies in human populations replacing or complementing other strategies such as food fortification or nutrient supplementation.

Challenges

Plant foods can be used as a vehicle to improve human nutrition and human health. The important challenges concerning the minor plant components are to identify the specific phytochemicals which are physiologically active in man and to demonstrate in human studies a beneficial effect on biomarkers of chronic disease or better still a reduction in the disease itself. After the recent demonstration that Fe, Zn and β-carotene levels can be increased in plant foods by plant breeding or genetic engineering techniques, the challenge is now to show that they can be increased to nutritionally useful levels and that the additional micronutrient is bioavailable. This has not yet been demonstrated.

References

ATBC Study Group (1994) The effects of vitamin D and β-carotene on the incidence of lung cancer and other cancers in male smokers. *New England Journal of Medicine* 330, 1029–1035.

Bouis, H. (1996) Enrichment of food staples through plant breeding: a new strategy for fighting micronutrient malnutrition. *Nutrition Reviews* 54, 131–137.

Cakmak, I., Ozakan, H., Braun, J.J., Welch, R.M. and Römheld, V. (1999) Zinc and iron concentrations in seeds of wild, primitive and modern wheats. In: *Improving Human Nutrition Through Agriculture: the Role of International Agriculture Research*. A workshop hosted by the International Rice Research Institute, Los

Baños, Philippines and organized by the International Food Policy Research Institute, 5–7 October.

Clinton, S.K. (1998) Lycopene: chemistry, biology, and implications for human health and disease. *Nutrition Reviews* 56, 35–51.

Davidsson, L. (1999) Zinc. In: Hurrell, R.F. (ed.) *The Mineral Fortification of Foods*, 1st edn. Leatherhead Publishing, Surrey, pp. 187–197.

Ertl, D., Young, K. and Raboy, V. (1998) Plant genetic approaches to phosphorus management in agricultural production. *Japan Environment Quarterly* 27, 299–304.

Fairweather-Tait, S.J. and Hurrell, R.F. (1996) Bioavailability of minerals and trace elements. *Nutrition Research Reviews* 9, 295–324.

FAO (Food and Agriculture Organization of the United Nations) (1997) *Preventing Micronutrient Malnutrition. A Guide to Food-Based Approaches*. International Life Science Institute, Washington, DC.

Frossard, E., Bucher, M., Mächler, F., Mozafar, A. and Hurrell, R. (2000) Potential for increasing the content and bioavailability of Fe, Zn and Ca in plants for human nutrition. *Journal of the Science of Food and Agriculture* 80, 861–879.

Goto, F., Yoshihara, T., Shigemoto, N., Toki, S. and Takaiwa, F. (1999) Iron fortification of rice seeds by the soybean ferritin gene. *Nature Biotechnology* 17, 282–286.

Graham, R.D., Senadhira, D., Beebe, S., Iglesias, C. and Montasterio, I. (1999) Breeding for micronutrient density in edible portions of staple food crops: conventional approaches. *Field Crops Research* 60, 57–80.

Grusak, M.A. and DellaPenna, D. (1999) Improving the nutrient composition of plants to enhance human nutrition and health. *Annual Review of Plant Physiology and Plant Molecular Biology* 50, 133–161.

Hurrell, R.F. (1999) Iron. In: Hurrell, R.F. (ed.) *The Mineral Fortification of Foods*, 1st edn. Leatherhead Publishing, Surrey, pp. 54–93.

Lachance, P.A. and Bauernfeind, C.J. (1991) Concepts and practices of nutrifying foods. In: Lachance, P.A. and Bauernfeind, C.J. (eds) *Nutrient Additions to Food: Nutritional, Technological and Regulatory Aspects*. Food and Nutrition Press, Trumbull, Connecticut, pp. 19–86.

Law, M.R., Wald, M.J. and Thompson S.G. (1994) By how much and how quickly does a reduction in serum cholesterol concentration lower risk of ischaemic heart disease? *British Medical Journal* 308, 367–373.

Lindsay, D.G. and Clifford, M.N. (eds) (2000) Critical reviews produced within the EU concerted action: 'Nutritional Enhancement of Plant-based Food in European Trade' (NEODIET). *Journal of the Science of Food and Agriculture* 80, 795–1137.

Lucca, P., Hurrell, R. and Potrykus, I. (2001) Genetic engineering approaches to improve the bioavailability and the level of iron in rice grains. *Theoretical and Applied Genetics* 102, 392–397.

Miettenen, T.A., Pekka, P., Gylling, H., Vanhannen, H. and Vartiainen, E. (1995) Reduction of serum cholesterol with a sitostanol ester margarine in a mildly hypercholesterolemic population. *New English Journal of Medicine* 333, 1308–1312.

Mithen, R.F., Dekker, M., Verkerk, R., Rabot, S. and Johnson, I.T. (2000) The nutritional significance, biosynthesis and bioavailability of glucosinolates in human foods. *Journal of the Science of Food and Agriculture* 80, 967–984.

National Research Council (1989) *Recommended Dietary Allowances*, 10th edn. National Academy Press, Washington, DC.

Parr, A.J. and Bolwell, G.P. (2000) Phenols in the plant and in man. The potential for possible nutritional enhancement of the diet by modifying the phenols content or profile. *Journal of the Science of Food and Agriculture* 80, 985–1012.

Pasamontes, L., Haiker, M., Wyss, M., Tessier, M. and van Loon, A.P. (1997) Gene cloning, purification, and characterization of a heat-stable phytase from the fungus *Aspergillus fumigatus*. *Applied and Environmental Microbiology* 63, 1696–1700.

Piironen, V., Landsay, D.G., Miettenen, T.A., Toivo, J. and Lampi, A.-M. (2000) Plant sterols: biosynthesis, biological function and their importance to human nutrition. *Journal of the Science of Food and Agriculture* 80, 939–966.

Potter, S.M., Baum, J.A., Teng, H., Stillman, R.J., Shay, N.F. and Erdman, J.W. Jr (1998) Soy protein and isoflavones: their effects on blood lipids and bone density in postmenopausal women. *The American Journal of Clinical Nutrition* 68, 1375S–1379S.

Rengel, Z., Batten, G.D. and Crowley, D.E. (1999) Agronomic approaches for improving the micronutrient density in edible portions of field crops. *Field Crops Research* 60, 27–40.

Samuelsen, A.I., Martin, R.C., Mok, D.W. and Mok, M.C. (1998) Expression of the yeast FRE genes in transgenic tobacco. *Journal of Plant Physiology* 118, 51–58.

Steinmetz, K.A. and Potter, J.D. (1996) Vegetables, fruit, and cancer prevention: a review. *Journal of the American Dietetic Association* 96, 1027–1039.

Subar, S., Heimendinger, J., Krebs-Smith, S., Patterson, B., Kessler, R. and Pivonka, E. (1992) *A Day for Better Health: a Base-line Study of Americans' Fruit and Vegetable Consumption*. National Cancer Institute, Washington, DC.

van den Berg, H., Faulks, R., Fernando Granado, H., Hirschberg, J., Olmedilla, B., Sandmann, G., Southon, S. and Stahl, W. (2000) The potential for the improvement of carotenoid levels in foods and the likely systemic effects. *Journal of the Science of Food and Agriculture* 80, 880–912.

Weststrate, J.A. and Meijer, G.W. (1998) Plant sterol-enriched margarines and reduction of plasma total- and LDL-cholesterol concentrations in normocholesterolaemic and mildly hypercholesterolaemic subjects. *European Journal of Clinical Nutrition* 52, 334–343.

Ye, X., Al-Babili, S., Klöti, A., Zhang, J., Lucca, P., Beyer, P. and Potroykus, I. (2000) Engineering the provitamin A (β-carotene) biosynthetic pathway into (carotenoid-free) rice endosperm. *Science* 287, 303–305.

Facing the Growing Needs of Mankind – Grasslands and Rangelands

5

R.J. Wilkins

Institute of Grassland and Environmental Research, North Wyke, Okehampton, Devon EX20 2SB, UK, and University of Plymouth, Seale-Hayne Faculty of Agriculture, Food and Land Use, Newton Abbot TQ12 6NQ, UK

Background

Grasslands are a major form of land use throughout the world with FAO (1996) statistics indicating that 'permanent grassland' covers some 70% of agricultural land and 26% of total land area – some 3.4×10^9 ha (see Table 5.1). There are in addition grasslands sown in arable rotations and specific forage crops, giving a total area approaching 4×10^9 ha. The traditional role of grasslands has been to provide feed for the production of milk, meat

Table 5.1. Permanent grassland in different regions, 1994, FAO (1996) statistics.

	Permanent grassland		
	10^6 ha	As % agricultural area	As % total land area
Africa	884	84	30
North and Central America	362	57	17
South America	495	82	28
Oceania	429	89	51
Asia[a]	1036	67	34
Europe[a]	92	33	16
Russian Federation	87	40	5
World total	3385	70	26

[a] Excluding Russian Federation.

© 2001 CAB International. *Crop Science*
(eds J. Nösberger, H.H. Geiger and P.C. Struik)

and fibre by ruminant animals, with at least 80% of the energy consumed by ruminants being derived from these sources.

Several factors distinguish grasslands from other crops and have major effects on their prevalence and management and the nature of required research. Distinguishing factors include: (i) grasslands generally comprise mixtures of species, often including both grasses and legumes and in many cases also shrubs and trees; (ii) they may be natural or semi-natural vegetation and even when grasslands are resown, this may only be at intervals of many years; (iii) grasslands may be harvested on several occasions during the year at variable stages of growth, with harvesting by hand or machinery (for green feeding or for conservation as hay or silage) or by grazing animals, which may be of contrasting species and sizes and with different physiological requirements. Also, levels of nutrient inputs from fertilizers vary widely with N applications being from 0 to over 400 kg ha^{-1} annually with the input of N from biological fixation covering a similar range.

There is thus tremendous variation in grasslands, their management and productivity. The quantity of dry matter (DM) harvested by domesticated herbivores probably varies from 0.1 to approaching 20 t ha^{-1}, but net primary production is much higher, with Coupland (1993) estimating that only some 7% of the primary production was consumed by domesticated herbivores; this proportion would, though, be higher in intensive systems.

Grasslands have predominated in areas not suited to cropping and those where human population pressure is low (e.g. parts of South America) or levels of affluence permit the devotion of land to animal production, whilst importing crop foods (e.g. parts of north-west Europe).

Current Pressures

Focus on grasslands to support animal production has already been noted. Over the last 20 years, different pressures have led to other outputs becoming of much increased importance, with increased emphasis on multifunctional grasslands providing a range of environmental goods and recreation, as well as animal products.

In developed countries there has been little population growth. With concern for adverse effects on health of consumption of dietary fat and, particularly, products with a high ratio of saturated to unsaturated fats, there has been a fall in the demand for ruminant products. In the European Union (EU) policies have been introduced to limit ruminant production by quotas, parallel to compulsory setaside to limit the output of arable crops. At the same time concern by society for the environment has much increased, particularly in relation to the prevention of pollution of water and the atmosphere and the protection and enhancement of wildlife, biodiversity and the landscape. Over this period there has been increased evidence that intensive grassland systems have considerable pollution potential and generally reduce biodiversity.

There has been marked reduction in the political influence of the farming sector, whilst the influence of groups concerned about environmental protection has much increased. Generally high levels of affluence have led to increased preparedness to pay for environmental goods. Payments by government to farmers for adopting specific practices, and payments are made by individual members of society to environmental groups and to landholders for access and tourism.

In contrast, in developing countries, high rates of population growth, coupled with aspirations to increase daily consumption of energy and protein, have continued to increase the overall requirement for food. In many areas better land has been converted from grassland to cropping. This has produced tensions in the use of communal grazings. These are particularly important in developing countries, but there has been progressive breakdown of traditional restrictions to stocking in many common property areas, leading to overgrazing. The demand on rangeland areas for firewood collection, medicinal plants and for tourism has also increased. With farmers seeking to at least maintain livestock numbers, the pressure on the grassland resources has increased, with resulting increases in land degradation, particularly in Africa and Asia. A second important development has been increased demand for milk and milk products, particularly in South and East Asia. This provides increased opportunities for sown grasses and forages to contribute to intensive land use systems to satisfy this market requirement.

These pressures are already having major impacts and are likely to be major drivers for change over at least the next two decades, providing a sound basis for research and policy and for responses by farmers and land managers.

Concern for climate change effects will be of increased importance over this period. Grassland makes a major contribution to global C fluxes. Grassland soils are particularly important stores of C in soil organic matter, whilst ruminants on grassland make major contributions through methane emission.

The Responses

This section reviews likely responses and identifies key research areas. Developed and developing regions are considered, with finally a discussion of climate change covering the two regions.

Developed countries

Animal production potential *per se* will continue to decrease in importance, unless there is a substantial reduction in the grassland area. This could arise from increased land use for non-food crops – chemicals, pharmaceuticals, fibres, timber, biomass energy. In the medium term, the most likely major development is in biomass energy, which, given the appropriate political and economic signals, could occupy millions of hectares. It is more likely, however, that policy makers will give major attention to increasing the net environmental benefits from land and its use. Increased freedom of trade will, at least in Europe, lower real prices of basic animal products. Farmers will need to reduce unit costs of production, in order to maintain their competitive position, or increase income by the production of premium foods or the sale of alternative products or services. Consideration will be given to the possibilities of reducing production costs, reducing pollution risks, increasing market demand for animal products and management for biodiversity and landscape. A final section discusses rangelands in developed countries.

Reducing production costs

The two alternative approaches to reduce unit costs of production are: (i) to increase output per hectare and spread land costs over a larger total output (e.g. by using more fertilizers and higher-yielding varieties); or (ii) by maintaining (or reducing) yield, but achieving this with reduced inputs. For countries producing for the world market with a low cost base and without acute environmental problems (e.g. New Zealand), there may be scope for some increase in profitability by increasing inputs, but in areas with high densities of both livestock and humans (e.g. much of north-west Europe), there are already substantial environmental problems and it is more likely that the focus will be on input reduction.

Wilkins and Vidrih (2000) argued the need in that situation to exploit, as far as possible, long-term swards and forage legumes and to maximize utilization by grazing, rather than by stall feeding or the use of conserved forage. The technology is already available for the effective use of legumes in long-term swards, but management problems and limitations to nutrient intakes (particularly by animals with high output potential) restrict reliance on grazing. The understanding of the interactions between the grazed sward and the grazing animal, and factors limiting nutrient supply from grazed swards remain important research areas. Recent research indicates the possibility for identifying and producing cultivars with increased feed intakes at grazing (Orr *et al.*, 2000).

Reducing pollution risks

It is imperative that there is a reduction in pollution risk from intensive systems in the developed world, particularly in relation to the loss of N compounds and of phosphorus. Reduction in stocking intensity, implicit in some of the approaches discussed above, will give some reduction in pollution risk, but specific actions will be required to comply with current and probable

Table 5.2. Efficiencies (%) of component parts of specialized intensive dairy farming systems in The Netherlands on sandy soil: (i) technically attainable, and (ii) realized in practice by skilled farmers (from Jarvis and Aarts, 2000).

Component	Technically attainable		Average realized in practice	
	N	P	N	P
Soil: transfer from soil to harvestable crop	77	100	53	60
Crop: transfer from harvested crop to feed intake	86	92	71	75
Animal: conversion from feed to milk and meat	25	32	18	22
Slurry/dung and urine: transfer from excreta to soil	93	100	80	100
Whole farm: from inputs to outputs	36	100	16	27

future legislation in relation to loss of nitrate and phosphorus to water, and loss of ammonia (and possibly also the greenhouse gases, nitrous oxide and methane) to the atmosphere.

Jarvis and Aarts (2000) reviewed the efficiency of N transfer processes involved in intensive dairy farming systems (Table 5.2). Limits to what is considered to be technically achievable are indicated, but it is also clear that performance in practice is highly variable and generally much lower than that which is technically possible. There are research requirements to seek new ways to increase the efficiency of the key conversion processes and for development and extension to better achieve 'best practice' on farms. The crop scientist needs to develop grasses with more efficient capture of N and with a composition which leads to more efficient N capture by the animal. In most grazing situations this could be achieved by increasing the ratio of readily available carbohydrate to crude protein in the harvested herbage. An alternative approach is to reduce the rate of protein breakdown in the rumen, through altering protein structure or by reaction with tannins, as discussed by Wilkins and Jones (2000).

Research also has a major contribution to make by providing tools for more precise nutrient management. This can be progressed through models and through improved diagnostics. Models of N transformation processes through the soil, plant, animal and atmosphere can be developed to indicate precise requirements for fertilizer N to sustain the desired rate of herbage growth and to indicate feed supplements that may be required to optimize rumen function. Confidence and precision would be further enhanced by diagnostic kits to be used on farms to check factors such as nutrient availability in organic manures, soil and plant nutrient status and the efficiency of microbial protein synthesis in the rumen. These are all areas within which there has been considerable recent research progress.

New market opportunities

There are opportunities for higher-value animal products to be produced from grassland. The present major development has been organic production. Although accounting for only 1% or so of total production, this sector is growing rapidly in Europe with substantial, but variable, premiums being obtained for milk and meat. Legume-based swards are particularly important to sustain production, as discussed by Younie and Hermansen (2000), and restrictions on the use of other feeds makes high forage feeding value critically important in organic systems. It is unlikely, however, that substantial premiums will be received should there be substantial increase in organic production.

Organic production provides renewed focus on the complementary relationship between grassland and arable cropping. This was the key feature of ley farming, much advocated by Stapledon and Davies (1948). With alternation between grassland and arable cropping, there is an accumulation of soil organic matter during the grass phase, which breaks down during the arable phase supplying nutrients to help sustain crop yields. Such an approach is vital for maintaining efficient crop production in organic systems and has continued to be an important feature of low-input systems in eastern Europe and in Australia. There are research requirements to minimize the loss of nutrients during the transition from grassland to arable cropping. Severe restriction in organic systems in the use of chemicals to control intestinal helminth parasites makes it important to identify and develop alternative approaches for control, probably by exploiting plants with natural antihelminthic properties.

There are opportunities for an improved image to be delivered to consumers for ruminant products from grassland. With society increasingly requiring 'natural' production systems with minimum inputs of technical chemicals and industrial feeds, and with increased concern for animal welfare, grazing-based systems have many positive features. Recent research indicates possibilities for positive contributions to human health. Grass feeding substantially reduces the ratio of omega-6 to omega-3 polyunsaturated fatty acids in meat and milk, and conjugated linoleic acid (CLA) content in milk is much increased with grazing. The change in ratio of fatty acids is associated with reduced risk of coronary heart disease, whilst CLA is a potent anti-carcinogen. These discoveries open up important promotional and marketing opportunities for grass-fed ruminant products, which could lead to increase in the market or to premium prices (Wilkins and Vidrih, 2000).

There are clearly new research challenges on how best to deliver CLA and appropriate polyunsaturated fatty acids to animal products. There is already evidence of variation between species and varieties and effects of post-harvest wilting on concentrations of omega-3 fatty acids. Information is required on ways to reduce biohydrogenation of fatty acids in the rumen.

Biodiversity and landscape

Grassland worldwide provides important habitats for a wide range of biota. It is a key element in many food webs and provides refugia for insects, birds and mammals. Intensification and fragmentation have led, in many areas, to reductions in biodiversity in grasslands. Fuller (1987) noted that semi-natural grasslands in lowland Britain occupied less than 3% of the area present 50 years previously, that 95% of neutral grasslands lacked significant wildlife interest and only 3% had not been damaged by intensification. The changes may be attributed principally to reseeding, high fertilizer applications, intensive grazing and cutting immature grass for silage; these changes leading to uniform swards containing few plant species and thus a restricted range of resources for other biota. These developments adversely affecting biodiversity have, though, been the essence of good management from the viewpoint of agricultural production.

Concern for the loss of biodiversity and wildlife has led to government actions in many developed countries. In 1985 the EU introduced regulations to permit subsidies to be paid for farming practices favourable for the environment. This led to the introduction of Agri-Environmental Measures, with farmers being paid for following management prescriptions intended to maintain or enhance environmental features. In the UK over 15% of agricultural land is in Environmentally Sensitive Areas (ESA), within which farmers may receive subsidies for following decision rules. Much of the land is grassland, with the management prescriptions restricting the quantities of fertilizers and organic manures that can be applied, severely limiting use of herbicides and pesticides and limiting the times at which swards can be cut or grazed. Participation is voluntary, but there has been high uptake, providing an important additional income source for grassland farmers.

Whilst approaches of this type are likely to be an important feature in Europe over the next few years, several key questions are posed. How much land should be devoted to biodiversity management? What level of payment should be made? Do the decision rules imposed provide the desired results? How best is land managed for biodiversity integrated with that managed principally for agriculture?

Despite progress in environmental economics, it is still extremely difficult to put a monetary value on environmental benefits. In the EU, levels of payment have been fixed in relation to: (i) estimates of income likely to be foregone; and (ii) that needed to achieve a target participation rate. A scheme recently introduced in Wales involves a tender process, with farmers indicating what they would do for what price, with tenders being accepted on the basis of perceived best value for money up to the maximum funds available. This approach has some attractions, but may have high administrative costs.

Of major concern is that evaluation of the ESA scheme in the UK queries whether real benefits to the environment are being delivered (see Sheldrick,

1997), with marginal reductions in intensity of production having little impact on sward composition and biodiversity more generally. There is increased evidence that high levels of plant species diversity will not develop in situations with high soil P, resulting from previous intensive inputs, and in situations in which there is poor availability of new species in the seed bank and from neighbouring vegetation (Peeters and Janssens, 1998). Experiments in the UK have shown very slow increases in species numbers when grasslands are extensified with withdrawal of fertilizer inputs, in contrast to rapid changes when grassland is changed from extensive to intensive management. Research is required to improve understanding of factors determining vegetation change, so that rules for agri-environmental schemes can be improved to give more consistent benefits.

It appears difficult to deliver high levels of biodiversity and efficient agricultural production from the same area of land with areas managed for biodiversity likely to produce low yields of herbage of low feeding value (Peeters and Janssens, 1998; Tallowin and Jefferson, 1999). I believe that areas should be targeted and managed principally for one purpose or the other, depending on physical, chemical and biological characteristics and the relative demand for biodiversity and for agricultural products. There are, however, major questions of scale which require research to better determine required size and level of connectivity between different forms of land use (or grassland management) in order to deliver particular biodiversity benefits. The requirement for grassland managed for biodiversity to be defoliated by grazing at some times during the year contributes to the case for such areas to be managed in a farm context, using the livestock management skills of farmers and facilitating integration with other land areas and feed resources. This poses important research questions on the appropriate species and breeds of grazing animals and the best approaches in integrated systems of production involving areas managed principally for biodiversity, together with areas managed principally for agricultural efficiency.

Landscape issues will be of increasing importance, with research required on the functional ecology of landscapes and to increase understanding of what types of landscape are required by society, bearing in mind differences in perception of visitors and residents and by people differing in age and background. With tourism and recreational use of rural areas likely to increase as a source of revenue – and to produce pressures on land use and management – it is of high priority that research methodologies are further developed to provide an improved basis for the analysis and appraisal of landscapes, in order to help influence land use and planning decisions.

Rangelands in developed regions

The foregoing review has focused mainly on temperate areas with intensive management, but many of the arguments also apply to rangelands in developed countries which have traditionally been used for commercial rather than subsistence farming. Walker (1996) concluded that 'in most cases sustainable occupation (of rangelands) will depend on a mix of land areas, both spatially segregated and within the same piece of rangeland'. He illustrated the case for targeting management to different objectives according to range condition in relation to either livestock production efficiency or conservation value, assuming that economic rewards were available for both of these targets. He stressed the need for research to develop appropriate indicators of range condition and also the formidable problem of restoring degraded lands. He noted that degradation was widespread, with loss of productivity and valuable species and often encroachment by woody species, and that even with management intervention to protect from overgrazing, the time-frame for their recovery is extremely prolonged.

Developing countries

Increasing populations, both regionally and within grassland areas, together with substantial present land degradation, produce greater challenges than those facing grassland in developed countries. The goal will be to increase production and quality from a diminishing grassland area using sustainable systems which restore rather than degrade land.

The possibilities will vary with climatic circumstances, the extent of human population pressures and the demand for animal products. Consideration will be given to some of the possibilities in three distinct areas: (i) reasonable rainfall and moderate population pressure; (ii) land which is marginal for cropping; and (iii) dry rangelands.

Reasonable rainfall and moderate population pressure

In some areas, including parts of South America, there are good opportunities to increase production through the introduction of improved grass and legume species, improved plant nutrition and the complementary use of grasses and forages with crop residues. In drier areas major contributions have been made by *Stylosanthes* spp. and the shrub legume, *Leucaena leucocephala*, whilst in humid areas particularly promising results have been obtained with *Arachis pintoi* (Cameron *et al.*, 1993), a species which may become the 'white clover' of some of the humid tropical and sub-tropical regions.

Maraschin and Jacques (1993) drew attention to the large increases in production that may result from introduction of improved species in the sub-tropics of Brazil. Natural grasslands commonly give beef live weight gains of around 50 kg ha^{-1} annually. With improved management, that can be increased to 150 kg ha^{-1}, but introduction of improved grasses and legumes can increase output to over 1000 kg ha^{-1}.

In some areas, particularly in South and East Asia, there has been substantial market development for milk and milk products. This is stimulating interest in the introduction of high quality forages. In order to be competitive with cropping for direct consumption, sown grasses and legumes need to be grown intensively. With favourable water regimes, yield potentials are high and in sub-tropical areas C_3 and C_4 species may both be used in the system to give production through the year. Grasses and legumes will be used to complement straw and other crop residues in production systems and forage crops may be grown as catch crops. With crop residues being of low nutritive value, it is particularly important that other feeds are of high nutritive value in order to achieve reasonable rates of animal production. Large animal responses are likely to result from the inclusion in the diet of small quantities of high-quality forages, such as berseem. In view of the complex feeding systems that are likely to be used, modelling, with both biological and economic components, has the potential to make a major contribution to identifying appropriate land and resource use, and feeding strategies. It will be important for such systems to be developed without the adverse environmental effects which occurred with the intensification of systems in developed countries.

Land marginal for cropping

Increasing population pressures are leading to extension of cropping into marginal and sub-marginal areas, with land often being fallowed between crops. There appear to be good opportunities for introduction of forage legumes following cropping to produce ley farming systems. Saleem and Fisher (1993), discussing developments in sub-Saharan Africa, concluded that rapid-growing legumes gave benefits to soils and increased crop and livestock profitability. They drew attention, however, to problems in achieving adoption of this technology. There is high priority for further research on ley farming in areas that are marginal for cropping, with participatory involvement of producers.

Dry rangelands

The opportunities for improving production in dry rangeland are severely limited by climate and by structural and sociological factors. Narjisse (1996) stated that rangeland condition throughout the developing world is declining as a result of historical, demographic, economic, institutional and policy-related factors. In recent years pressure on rangelands has increased through competition with cropping, firewood harvesting and recreation, particularly ecotourism. Narjisse (1996) considered that these new developments should not be resisted, but embraced in more multi-functional land use. He noted the need to involve local people in planning ecotourism developments and to ensure appropriate returns for landholders. The record of many national and international initiatives to improve rangeland has been poor, a conclusion also reached by Niamir-Fuller (1999). Narjisse (1996) stressed that technology is only one component and the achievement of greater productivity with long-term sustainability depends on the ability to meet requirements in the legislative, policy, institutional, financial and technological areas. Inputs are required from land user groups in framing actions within all these areas. Effective technology transfer, rather than new research concepts, was considered to be crucial.

A key issue is land tenure. Narjisse (1996) noted the effectiveness of strong formal and informal control of common property management that had existed traditionally and given systems which were sustainable over centuries. This control has now largely collapsed to give open access (and over-exploitation). Where land had been privatized, with establishment of separate holdings and sedentarization, more limited movement of livestock had restricted efficiency of rangeland use.

Behnke and Scoones (1993) pointed out that the variable harsh climatic situation in rangelands gave non-equilibrium ecosystems which were highly variable. Traditional pasture land use patterns, incorporating communal tenure, large-scale exploitation and the tracking of volatile resources, are well suited to the sustainable use of rangelands. This analysis has encouraged further consideration of ways in which effective control can be re-established into communal tenure arrangements.

The importance of people-participation and exploitation of indigenous knowledge in rangeland development projects has been endorsed by the FAO. Emphasis is now placed on multi-disciplinary approaches including people-participation; the importance of local institutions and community regulatory mechanisms; sustainable range development; land tenure regulations; alternative income-generating activities; and risk management (Reynolds *et al.*, 1999).

There are though clear research requirements for dry rangelands. Foran and Howden (1999) stressed the need for development of systems and theory which will allow regional institutions and their stakeholders to integrate biophysical, social and economic dynamics and to compare development strategies over time frames spanning human generations. There is continued high priority for research to further understand soil, plant and animal processes and their interactions as they determine the sustainability of rangeland systems.

Technology development

In reviewing several models of technology development, Jiggins (1993) advocated participatory technology development as having widespread application. This approach seeks to join the power of organized science to farmers' own knowledge and experience in ways which strengthen a community's own capacity to experiment and progress. This principle is now widely accepted and is relevant to both developing and developed regions. It is particularly important in grasslands and rangelands, because of the intimate connections between soils, plants and animals, and the multiplicity of potential systems and interactions that occur spatially and seasonally.

It is important that participatory exercises do not become 'anti-science'. It is interesting to contrast the view of Behnke quoted by Niamir-Fuller (1999) that 'we have yet to demonstrate appropriate technologies for arid rangelands that are superior to the technologies that traditional pastoralists are already accustomed to', with that of Walker and Hodgkinson (1999) that 'there are many underused technologies available that would greatly enhance the management of grazing in rangelands'. Scientists involved in participatory exercises need to have good knowledge of basic principles and of technical possibilities which have arisen from research worldwide. There is need to ensure that new possibilities are not too rapidly rejected because of caution about their applicability by producers with rather narrow experience, and that research derived from a participatory process is conducted with appropriate rigour to enable confident conclusions to be drawn.

Climate change

Climate change will affect overall and seasonal patterns of grassland production and dictate changes in management and utilization, but I do not anticipate dramatic effects.

Most grasslands have a wide genetic base and are adapted for long-term survival and to withstand biotic and abiotic stresses. They have the potential to adapt progressively to changes in temperature, rainfall and CO_2 concentration. With sown grasses and legumes, the continuous nature of plant breeding in relation to the slow progress of climate change should ensure continued availability of species and varieties with reasonable adaptation to current conditions.

The impact on grasslands of actions introduced to restrict emissions of greenhouse gases is likely to be of much greater importance. The effects of grassland systems are being increasingly recognized both in relation to emissions of greenhouse gases, particularly CH_4 and N_2O, and to carbon balance and sequestration. Grassland contributes about 20% of terrestrial CO_2 fluxes, and C in soil organic matter under grassland is a similar order of magnitude to that in tree biomass (Minami et al., 1993).

The large area of grassland throughout the world, the large changes in C in soil organic matter as swards age or are converted into arable cropping and effects of grassland management have key effects on global C balances. Foran and Howden (1999) noted the opportunity for storing carbon in rangeland systems of 2.5–40 t C ha^{-1}, and 8 t C ha^{-1} with rehabilitation of degraded soils. The introduction of carbon trading may well provide substantial financial incentives to farmers to convert arable land to grassland, or to retain grassland, when it may otherwise have been converted into arable land. The introduction of shrubs and trees into grassland may be encouraged.

The emissions of methane in relation to animal output are extremely high in extensive rangeland systems. Greenhouse gas policies may provide financial incentive for reducing stock numbers in such systems, a development which would provide other benefits by reducing pressures on the rangeland resource.

Considerable information has been generated over the last 10 years to quantify emissions of greenhouse gases from grassland systems. This is facilitating technical, practical and financial evaluation of mitigation options and will help inform policy makers. There is need for more research to further quantify accumulation and depletion of carbon in grassland systems and effects of management in order to provide a sound basis for policy development.

Conclusions

Grasslands will continue to be a major land use, but their role will change substantially, particularly in developed countries. Grasslands and grassland systems are extremely adaptable, and management and intensity of use can respond to signals from the marketplace and from government.

In developed countries there will be some opportunity to produce and market high-value animal products from grassland, but overall the importance of maximizing animal production will decline. Systems will need to be made more nutrient tight in order to reduce risks of atmospheric and aquatic pollution. There will be increased emphasis on biodiversity, landscape and tourism, and requirements to better target areas for biodiversity enhancement, to increase understanding of the functional ecology of landscape and more generally to ascertain the real requirements of society from the grassland environment.

Increased population pressures provide particular problems in developing countries. There will, though, be some areas within which more intensive grassland-based systems will develop, particularly for milk production. There will be a challenge for these systems to be developed without the adverse environmental effects that occurred with intensification of grassland use in developed countries. In more marginal areas it is particularly important to involve all stakeholders in research and development projects. In areas that are marginal for cropping, grassland (particularly forage legumes), has the potential to play a positive role by substituting for fallow and increasing the production of feed for animals and supplying nutrients for crops in ley farming systems. In drier areas there may be little scope for increasing production, but a need to continue to focus on identifying and developing acceptable methods for more sustainable management. Actions to reduce greenhouse gas emissions may provide economic incentives to reduce stock numbers and manage grassland for C storage.

Developments will be strongly influenced by government actions, particularly in relation to environmental control and for the provision of environmental goods. Grassland situations are essentially multi-disciplinary and it will be particularly important to have effective linkages between research scientists, socio-economists, policy makers and land managers in the evolution of research programmes and in the exploitation of research results.

Several research requirements have already been highlighted but three particular areas of widespread relevance and importance are listed below:

1. Forage quality – for grassland used for animal feeding it will be particularly important to focus on quality. In developing countries the focus should continue to be on energy, protein and absence of anti-quality factors. In developed countries increased emphasis needs to be given to factors relating to human and animal health, with new breeding objectives likely to emerge from this work.

2. Vegetation dynamics and herbivory – although the objectives of grassland management may vary from maximizing harvested output to maximizing biodiversity, understanding of competition and the processes controlling sward composition remain key requirements for indicating opportunities for exploitation in more applied research and practice.

3. Carbon and nitrogen dynamics – improved understanding of C and N flows through grassland systems is of key importance in sustaining productivity, in restricting losses to the environment and enhancing storage in soil organic matter.

References

Behnke, R.H and Scoones, I. (1993) Rethinking range ecology: implications for rangeland management in Africa. In: Behnke, R.H., Scoones, I. and Kerven, C. (eds) *Range Ecology in Disequilibrium: New Models of Natural Variability and Pasture Adaptation in African Savannas.* Overseas Development Institute, London, pp. 1–30.

Cameron, D.F., Miller, C.P., Edye, L.A. and Miles, J.W. (1993) Advances in research and development with *Stylosanthes* and other tropical pasture legumes. *Proceedings of the Seventeenth International Grassland Congress,* Palmerston North, New Zealand and Rockhampton, Australia, pp. 2109–2114.

Coupland, R.T. (1993) Review. In: Coupland, R.T. (ed.) *Natural Grasslands. Eastern Hemisphere and Resume. Ecosystems of the World 8B,* Elsevier, Amsterdam, pp. 471–482.

FAO (1996) *FAO Production Yearbook, 1995.* Food and Agriculture Organization of the United Nations, Rome.

Foran, B. and Howden, M. (1999) Nine global drivers of rangeland change. *Proceed-*

ings of the Sixth International Rangeland Congress, Townsville, Australia, pp. 7–13.

Fuller, R.M. (1987) The changing extent and conservation interest of lowland grasslands in England and Wales: a review of grassland surveys 1930–84. *Biological Conservation* 40, 281–300.

Jarvis, S.C. and Aarts, H.F.M. (2000) Nutrient management from a farming systems perspective. *Grassland Science in Europe* 5, 363–373.

Jiggins, J. (1993) From technology transfer to resource management. *Proceedings of the Seventeenth International Grassland Congress*, Palmerston North, New Zealand and Rockhampton, Australia, pp. 615–622.

Maraschin, G.E. and Jacques, A.V.A. (1993) Grassland opportunities in the subtropical region of South America. *Proceedings of the Seventeenth International Grassland Congress*, Palmerston North, New Zealand and Rockhampton, Australia, pp. 1977–1981.

Minami, K., Goudriaan, J., Lantinga, E.A. and Kimura, T. (1993) Significance of grasslands in the emission and absorption of greenhouse gases. *Proceedings of the Seventeenth International Grassland Congress*, Palmerston North, New Zealand and Rockhampton, Australia, pp. 1231–1238.

Narjisse, H. (1996) The range livestock industry in developing countries: current assessment and prospects. *Proceedings of the Fifth International Rangeland Congress*, Vol. 2, Salt Lake City, pp. 14–21.

Niamir-Fuller, M. (1999) International aid for rangeland development: trends and challenges. *Proceedings of the Sixth International Rangeland Congress*, Townsville, pp. 147–153.

Orr, R.J., Cook, J.E., Atkinson, L.A., Clements, R.O. and Martyn, T.M. (2000) Evaluation of herbage varieties under continuous stocking. *Occasional Symposium No. 34*, British Grassland Society, pp. 39–44.

Peeters, A. and Janssens, F. (1998) Species-rich grasslands: diagnostic, restoration and use in intensive production systems. *Grassland Science in Europe* 3, 375–393.

Reynolds, S., Bartello, C. and Baas, S. (1999) Perspectives on rangeland development – the Food and Agricultural Organization of the United Nations (FAO). *Proceedings of the Sixth International Rangeland Congress*, Townsville, pp. 160–165.

Saleem, M.A.M. and Fisher, M.J. (1993) Role of ley farming in crop rotations in the tropics. *Proceedings of the Seventeenth International Grassland Congress*, Palmerston North, New Zealand and Rockhampton, Australia, pp. 2179–2187.

Sheldrick, R.D. (ed.) (1997) *Grassland Management in Environmentally Sensitive Areas. Occasional Symposium No. 32*, British Grassland Society.

Stapledon, R.G. and Davies, W. (1948) *Ley Farming*. Faber and Faber, London.

Tallowin, J.R.B. and Jefferson, R.G. (1999) Hay production from lowland semi-natural grasslands: a review of implications for livestock systems. *Grass and Forage Science* 54, 99–115.

Walker, B.H. (1996) Having or eating the rangeland cake: a developed world perspective on future options. *Proceedings of the Fifth International Rangeland Congress*, Vol. 2, Salt Lake City, pp. 22–28.

Walker, J.W. and Hodgkinson, K.C. (1999) Grazing management: new technologies for old problems. *Proceedings of the Sixth International Rangeland Congress*, Townsville, pp. 424–430.

Wilkins, R.J. and Jones, R. (2000) Alternative home-grown protein sources for ruminants in the United Kingdom. *Animal Feed Science and Technology* 85, 23–32.

Wilkins, R.J. and Vidrih, T. (2000) Grassland for 2000 and beyond. *Grassland Science in Europe* 5, 9–17.

Younie, D. and Hermansen, J. (2000) The role of pasture in intensive organic livestock farming. *Grassland Science in Europe* 5, 493–509.

Abiotic Stresses, Plant Reactions and New Approaches Towards Understanding Stress Tolerance

6

H.J. Bohnert[1] and R.A. Bressan[2]

[1]Departments of Biochemistry, Molecular and Cellular Biology, and Plant Sciences, The University of Arizona, Biosciences West, Tucson, AZ 85721–0088, USA; [2]Department of Horticulture, Purdue University, Horticulture Building, West Lafayette, IN 47907-1165, USA

Introduction

Among abiotic factors a number of conditions are meaningful stresses because they can be life threatening and in combination determine the distribution of plants and the productivity of crops (Boyer, 1982). Prolonged drought, high or fluctuating salinity, and low and freezing temperatures account for most production losses, but flooding, high light, ozone, ion deficiency or imbalance, heavy metals and soil structure are other factors that threaten plant life. For example, various USDA statistical analyses identified drought as the single most yield-reducing factor. Yield reductions due to drought, high salinity and other abiotic factors amount to two-thirds of all losses, much more than the reduction due to pathogens. Despite considerable breeding efforts, breeding for tolerance to water deficit has not had much impact in the face of affordable irrigation, low commodity prices, and the globalization of trade (Flowers and Yeo, 1995). Various strategies have been employed for environmental stress tolerance breeding (Table 6.1), which can now be assisted by the inclusion of 'genomic' approaches.

© 2001 CAB International. Crop Science
(eds J. Nösberger, H.H. Geiger and P.C. Struik)

Table 6.1. Classical and molecular breeding, and transgenic and reverse genetics strategies. The table juxtaposes procedural suggestions for breeding programmes relating to salinity tolerance with molecular, transgenic and reverse genetics tools that have recently become available. Transgenic approaches must include the choice of appropriate promoters for transgene expression. The strategy may include multiple gene transfers, or the generation of artificial chromosomes. Strategies rely on knowledge about the expression of all genes and pathways essential, relevant and coincident to stress responses in model organisms.

Classical breeding strategies	Molecular breeding/ transgenic strategies	Reverse genetics
Disregard tolerance, focus only on yield (Richards, 1992)	Enhance production capacity (resident biochemical pathways); QTL analysis (Quarrie et al., 1997)	Screen for changes in organ structure and developmental timing; utilize models
Design new phenotypes and generate mutants (Ramage, 1980)	Metabolic engineering (Roxas et al., 1997; Nuccio et al., 1999); T-DNA or transposon tagged Arabidopsis lines (Zhu et al., 1997; Shi et al., 2000) for tolerance determinants; screening for developmental (tagged) mutants and generation of transgenic lines	Screen for metabolite, growth regulator, or developmental changes in T-DNA tagged mutants
Screen/select within a phenotype (Norlyn, 1980)	QTL analysis (Zheng et al., 2000); marker-assisted breeding; microarray analysis (Richmond and Somerville, 2000)	Use Arabidopsis and rice as models based on sequence information (Hasegawa et al., 2000b)
Use 'wide' crosses (Rush and Epstein, 1981)	QTL analysis; utilize molecular markers; microarray analysis	Test concepts with Arabidopsis and its relatives
Develop halophytes (O'Leary, 1994)	Gene transfer to glycophytes; establish a halophytic model; find stress-relevant genes and halophytic protective pathways	Utilize control genes that regulate resource allocation, growth and development

For a discussion, see Flowers and Yeo (1995) and Bohnert and Jensen (1996).

Parallel to breeding, physiological studies have focused on plant responses to abiotic stresses, yet the two existed largely in separate arenas. In the search for stress tolerance mechanisms, plant biologists produced data in a correlative way in many different species. Biochemical and biophysical principles have been outlined long ago (Levitt, 1980). These principles have survived the test of time in such a way that they have been adopted as guiding principles by following generations of plant biologists, who moved from protein or enzyme analysis, to genetics, genes, gene expression studies and finally to transgenic analyses. The 1990s have seen rapid progress that resulted from quantitative genetics, molecular genetics and especially reverse genetics studies. The last decade has brought drastic changes in views about stress sensitivity, tolerance or resistance based on results from molecular genetics. Transcript abundance and stress-induced change in gene expression have been monitored in many species under different conditions. As of 2000, the sequence of the *Arabidopsis* genome is completed, and soon we will have a catalogue of plant gene expression exceeding half a million transcripts (http://www.ncbi.nlm.nih.gov/dbEST/index.html). While numbers are important, more important is that these transcripts, collected as ESTs (expressed sequence tags), describe frequency and type of transcribed genes in specific tissues, cells and organs, during development and under various stress conditions (http://stress-genomics.org). Accompanying these activities has been yet another breakthrough, whose impact may have even greater impact on stress biology and crop breeding. In *Arabidopsis thaliana* and corn, saturation mutagenesis has been reached (Krysan *et al.*, 1999). With these mutants experiments for the proof of concepts become possible.

With the instrumentation in place that allows large-scale DNA sequencing and soon also large-scale analysis of all proteins in a cell, utilizing bio-informatics tools will be essential to sort and understand the large amount of information. The extent to which biology is presently driven by new technology and bio-informatics requires a new mindset. Biology and crop sciences have arrived at a threshold that permits the analysis of plant stress responses on the level of entire genomes. ESTs, microarrays and the clustering of expression patterns, the availability of molecular markers and mutants, and the mapping of robust quantitative trait loci (QTLs) provide tools for assessing the contribution of genes to stress tolerance (Eisen *et al.*, 1998; Frova *et al.*, 1999; Kehoe *et al.*, 1999). Conclusions can now be drawn that supersede correlative evidence. A major effort will be necessary to integrate the diverse elements of information into comprehensive databases, such that QTL utilize genome sequences and ESTs as markers, or that data from cluster analysis of thousands of transcripts are reconciled with physiology. Now is the time to critically examine and categorize the principles established in the past, and to integrate results from physiological analyses with molecular genetics and genomics studies to aid breeding efforts aimed at improving environmental stress tolerance.

Abiotic Stress in a Societal Context

Plants are permanently exposed to their physical environment, with changes in seasonal and diurnal cycles, and extreme and stressful climatic variations. The foremost abiotic stress is water deficit, yet water is needed for both agriculture and a still increasing human population. The competition for resources poses problems in areas where water is an increasingly precious commodity, for example in Australia, countries of the Middle East, North Africa, the west and Midwest of the USA, parts of the Indian subcontinent and central Asia (http://www.undp.org/popln/fao/water.html). Even in areas with typically ample precipitation transient drought can lead to economic hardship. A statistical analysis by the USDA indicated that harvested yields of corn and soybean were lower by 15–25% in 3 years characterized by severe drought (1980, 1983, 1988) compared with the long-term average. At the same time, predictions of possible global climatic changes indicate that the distribution of rain might become more erratic than in the past. Lack of water prolongs the agricultural growing cycle, increases the vulnerability to pathogens, and ultimately results in decreased yield. In addition, agricultural practice jeopardizes productivity in irrigated land, because long-term irrigation leads to the build-up of sodium chloride and other salts in the soil, with up to half of the area under irrigation affected (Flowers and Yeo, 1995; Postel, 1999). What is happening in growing areas with elaborate irrigation schemes is reminiscent of events that led to the decline of ancient civilizations, the former 'fertile crescent' being the prime example. How will we provide a stable supply of food, feed and fibre for a human population that may exceed 9 billion within 50 years (Khush, 1999)?

Defining the Problem and Searching for Solutions

For continued vegetative growth and the development of reproductive organs under stress, plants must, above all, obtain water. To satisfy this essential need, each of the many adaptive mechanisms that have been detected must be subordinate to this goal. When stomata are closed to limit water loss, a series of events adjusts photosynthesis, carbon fixation and carbohydrate transport, initiating processes to maintain the integrity of the photosynthetic and carbon fixation apparatus (Long et al., 1994; Horton et al., 1996). Three stressors are relevant for water relations. The most important objective during drought episodes is the extraction of water from the soil. In high-salinity soils sodium must be distinguished from the essential macronutrient, potassium. During both drought and high-salinity conditions, the internal osmotic potential must be lowered to allow for water uptake from the medium, and sodium and chloride must be stored out of harm's way. At low temperature it must be assured that water arrives in

leaves and apical meristems. The search for solutions by physiological procedures has generated many data and some correlations which have in part
been tested by transgenic alterations and further physiological studies (Hare
et al., 1997; Bohnert and Sheveleva, 1998; Bohnert *et al.*, 1999; Hasegawa
et al., 2000b). As miniscule as the stress-ameliorating effects have been
resulting from genetic engineering – it is unlikely that the engineered plants
perform better than their progenitors in a farmer's field – they have
uncovered mechanisms. More educated strategies emerge from these pioneering inquiries.

Environmental Stress Adaptation Mechanisms

Many reports on stress adaptations provided a glimpse of how plants survive,
or even maintain some growth, during episodes of stress. Probable protective
mechanisms can be identified in a general sense. For a discussion, three
broad categories of responses can be distinguished although they are in
reality interconnected: (i) the immediate (or 'downstream') reactions; (ii)
stress sensing and signalling leading to adjustments for a new equilibrium;
and (iii) long-term developmental changes.

Protection of downstream reactions

Several mechanisms for which a protective role is clear can be placed in the
immediate or emergency response category. Important are scavenging of
radical oxygen species, controlled ion and water uptake, management of
accumulating reducing power through adjustments in carbon/nitrogen
allocation, and the accumulation of compatible solutes, which serve at least
in part in osmotic adjustment. These mechanisms have been discussed
(Ansell *et al.*, 1997; Amtmann and Sanders, 1999; Asada, 1999; Bohnert
et al., 1999; Hasegawa *et al.*, 2000b). All reactions are an enhancement of
metabolic processes that are taken from the repertoire of daily plant life (Fig.
6.1). They encompass, for example, the down-regulation of the light-
harvesting and water splitting complexes, enhancement of the water–water
cycle, the increased synthesis and action of radical scavenging enzymes and
non-enzymatic scavenging molecules, the storage of reducing power in a
reusable form (e.g. as proline, polyols or fructans) and the engagement of
proton pumping and ion and water transport systems. Also, the stabilization
of proteins and membranes through enhanced chaperone synthesis belongs
in this category of reactions (Schroda *et al.*, 1999). Although these mechanisms accomplish ordinary adjustments, reaction bandwidth, speed and magnitude seem to delineate species-specific boundaries. Several such emergency
measures have been enhanced in transgenic models with limited, albeit
recognizable, success (Hare *et al.*, 1997; Apse *et al.*, 1999; Bohnert *et al.*,

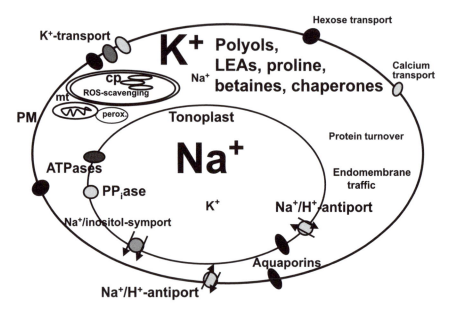

Fig. 6.1. Physiological and biochemical determinants of salinity stress tolerance. The schematic rendition of elements important for cellular aspects of stress tolerance includes ATPases (and inorganic phosphatase) generating proton gradients, various potassium transport systems, hexose and calcium transporters as well as sodium/proton antiporters and sodium/inositol symporters (Hasegawa *et al.*, 2000b). Also established is an important role for radical oxygen scavenging (in chloroplasts, mitochondria, peroxisomes and the cytosol) (e.g. Roxas *et al.*, 1997). Intracellular processes that support tolerance include protein turnover, altered endomembrane traffic, and the accumulation of solutes, such as polyols, proline, betaines, and induced synthesis of chaperones and LEA (late embryogenesis abundant) proteins. The size of symbols for potassium and sodium indicates their distribution in vacuole (sodium high) and cytosol (potassium high) indicating how sodium may be used as an osmoticum. (Modified after Hasegawa *et al.*, 2000b.)

1999), but prolonged stress requires additional mechanisms. This view assumes a common set of genes for stress responses, similar or identical for unicellular organisms, such as cyanobacteria or yeasts, and for glycophytes, halophytes or xerophytes. The differences between glycophytic and halophytic/xerophytic species seem to lie in how, how fast, or how emphatically emergency responses are initiated, elaborated or sustained.

Stress sensing and signalling

During the last decade, sensing and signalling components of abiotic stress signalling pathways have emerged (Shinozaki and Yamaguchi-Shinozaki,

1997). These pathways initiate both short-term emergency and long-term developmental adaptations. The signalling circuits that act short term seem to be evolutionarily conserved. The pathway that leads to osmotic stress tolerance in yeast (*Saccharomyces cerevisiae*) provides an example. In yeast, sensing and signalling through a MAPK (mitogen activated protein kinase) pathway leads to the synthesis of transcription factor(s) that recognize specific *cis*-elements in promoters of genes to be activated (Gustin *et al.*, 1998). This signal transduction circuit is known as the *Hog1*-pathway (high osmolarity glycerol). It leads to the synthesis of enzymes, which adjust carbon utilization such that glycerol accumulates (Ansell *et al.*, 1997). Glycerol reduces the osmotic potential of the cell interior, which allows the influx of water. Equally important, glycerol production adjusts the ratio $(NADH + H^+):(NAD^+)$. This contributes to redox control (Ansell *et al.*, 1997; Shen *et al.*, 1999), which tends to reduce the production of radical oxygen species in mitochondria. Evolutionary conservation is well documented by the ability of a plant MAPK homologue which effectively replaced the yeast HOG1 kinase under high osmotic stress conditions (Popping *et al.*, 1996). Other signalling components in plants, such as calcium-dependent protein kinases (Harmon *et al.*, 2000), have recently become known. In yeast, other targets of specifically induced transcription factors are genes and proteins that initiate the synthesis of chaperones, cell wall components, and proteins that alter growth characteristics. Yet another conserved feature is the redundancy of signalling pathways. Apart from the *Hog1* pathway, at least two other signalling pathways exist in yeast, and similar pathways have been revealed in plant cells (see Nelson *et al.*, 1998; Hasegawa *et al.*, 2000b, for reviews). The multiple pathways function at different levels, are connected, and respond to a variety of environmental stimuli through signalling molecules including calcium, abscisic acid (ABA), gibberellins, ethylene, phosphatidylinositols, cytokinins, brassinosteroids or jasmonates (Giraudat, 1995; Pical *et al.*, 1999; Knight, 2000; Mizoguchi *et al.*, 2000).

Differences in stress tolerance could be due to the diversification of biochemical hardware or to altered stress perception or signalling. Because higher plant speciation proceeded during the last approximately 150 million years, gene complement should be rather similar in all species (Somerville and Somerville, 1999). Thus, the *Arabidopsis* genome contains the same set of genes present in the genome of, for example, the halophytic *Mesembryanthemum*. Differences in stress perception and how the information is converted into a biochemical response most likely account for the degree of tolerance distinguishing species. Halophytes, for example, may have evolved distinct stress recognition or signalling pathways, and regulatory controls that confer stress protection. The expectation is for duplications or alterations in genes encoding signal transduction components (e.g. receptors, protein kinases, protein phosphatases, transcription factors) as the basis of differences.

Specifically for salt stress sensing, a series of 'salt-overly-sensitive' (*sos*)

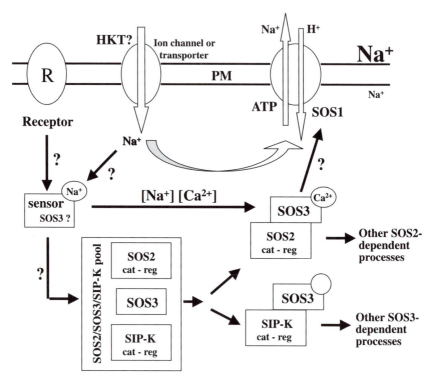

Fig. 6.2. Signal transduction elements. Signal transduction components involved in salinity stress signalling include two genes isolated from *Arabidopsis thaliana* mutants, which showed a salt-overly-sensitive (*sos*) phenotype (Zhu *et al.*, 1998). The SOS3 gene encodes a putative calcium sensor with homology to the regulatory subunit of calcineurin B (protein phosphatase 2B) (Liu and Zhu, 1998). SOS2 is a serine–threonine protein kinase with homology to, for example, the SNF1 kinase involved in stress signalling in yeast (Liu *et al.*, 2000). SOS2 and SOS3 interact, although a yeast two-hybrid screen did not result in such an interaction but rather produced a family of other kinases, termed SIP (for SOS3-interacting-proteins) (SIP-K in the figure) (Halfter *et al.*, 2000). The multiple SIPs may be responsible for salt stress signal amplification or diversification. A third gene, SOS1, has recently been identified as encoding a sodium/proton antiporter protein (*Nhe* family) tentatively located in the plasma membrane (Shi *et al.*, 2000). In this function SOS1 would transport sodium into the apoplast. A causal interaction between the SOS2/SOS3 aggregate and SOS1 has not yet been demonstrated.

Arabidopsis mutants have provided fundamental insights (Fig. 6.2). The approach to mutant generation has also provided information about the interaction and crosstalk between ionic, osmotic, temperature and ABA-dependent pathways (Zhu *et al.*, 1998; Xiong *et al.*, 1999). Some of the genes underlying the *sos*-specific phenotype have now become known. *SOS3* encodes a putative sodium sensor with homology to a regulatory subunit of yeast protein phosphatase 2B (calcineurin) (Liu and Zhu, 1998). *SOS3* interacts with *SOS2*, a serine/threonine protein kinase with homology to the yeast SNF1 kinase which is involved in nutrition-deficit and stress signalling (Halfter *et al.*, 2000). Possibly regulated through *SOS2* or the *SOS3/SOS2* complex is another component: a sodium/proton antiporter termed *SOS1* (Shi *et al.*, 2000). In the course of these studies another set of protein kinases, SIP (*SOS3*-*i*nteracting-*p*rotein), has been revealed, indicating the possibility for further diversification of signalling at the level of these proteins which bind to the putative calcium sensor SOS3 (Hasegawa *et al.*, 2000a).

Equally important is that the signalling pathways set in motion other responses that lead to long-term adaptations for abiotic stress tolerance involving changes in plant development, growth dynamics, vegetative–floral meristem transitions and flowering time, and seed production and maturation programmes. Such long-term adaptations in development provide perhaps the greatest prospect for engineering improved agricultural productivity under abiotic stress conditions.

Establishing a new developmental equilibrium

Long-term adjustments in development possibly constitute the most important category in stress preparedness. These are either constitutive or are set in motion by signalling pathways that respond to the immediate emergency and lead to signalling molecules that alter the pattern of development and growth so that the plant is protected from environmental extremes. The classical explanation of a stress-mediated decline in photosynthesis followed by a retardation of plant growth has been based on metabolic correlations, i.e. assuming altered development as a consequence of impaired metabolism (Munns, 1993). Dehydration and salinity stress initiate events, for example, altered pH, changes in ABA amount and distribution, altered ion composition, products of membrane lipid oxidation, or turgor changes. In response, plants move towards a state of minimal metabolic activity that persists until the stress is relieved (Netting, 2000). In this view, Ca^{2+} and ABA signalling lead to adjustments in proton gradients across membranes, and the associated K^t and water movements generate signals. Yet it is probable that other, as yet unknown, signalling pathways exist, which actively slow or stop meristem activity, cell expansion and growth. This seems likely based on the detection of many sequences, emerging from EST projects (see below), with homology to protein kinases or protein phosphatases in plants. Defining functions for these putative signal mediators will be a major challenge, which can only be successful by gene tagging and mutant analysis in plants. Current experiments mostly involve functional tests in yeast cells which do not express developmental functions that are of interest in plants. Suppression of growth by a stress appears to be an active process, rather than a passive reaction (Hasegawa *et al.*, 2000b). If this should be the case, the genes that are responsible can be found and characterized. In fact, genes that control cell expansion and entry into the cell division cycle have been described in yeast (e.g. Fassler *et al.*, 1997) and some plant counterparts of these have been identified (e.g. Cockcroft *et al.*, 2000). The phenotype of appropriate knock-out mutants of such genes with respect to growth under stress will be very revealing. Recently another pathway in yeast was found to regulate cell growth. Interestingly, the SLN (two component) receptor from yeast that transduces a signal generated by osmotic stress to activate genes controlling solute accumulation and intracellular osmotic balance was found to affect a separate signal process that controls cell growth through the SKN7 mediated branch pathway (Fassler *et al.*, 1997). In fact, the dual signalling nature of the SLN receptor appears to be designed to balance growth and osmotic adjustment such that when osmotic adjustment is activated, growth is repressed and vice versa. High concentrations of osmolytes outside the cell cause SLN to shift to the dephosphorylated state and induce the HOG (high osmolarity glycerol – a mitogen-activated protein kinase pathway) side of the pathway and simultaneously repress the SKN7 side of the pathway. Low osmolyte concentration outside results in the opposite, induction of HOG and repression of SKN7 (Fassler *et al.*, 1997). This duality of the SLN receptor may explain the long observed tight linkage between osmotic adjustment and reduced cell growth (Bressan *et al.*, 1990). Shinozaki and co-workers have reported the existence of an SLN-like gene in *Arabidopsis*, which is capable of functionally substituting for the yeast SLN gene (e.g. Sakakibara *et al.*, 2000). As yet, no counterparts in plants of the SKN7 or other components in the cell-growth part of the SLN-mediated osmotic signalling system of yeast have been reported. Their characterization – we are convinced that these genes exist – may prove interesting in the ability to control plant cell division/expansion during desiccation (osmotic) stress. Another interesting gene that may be part of stress-mediated control of cell growth is the transcription factor HB7. This desiccation-induced gene was recently shown to affect growth rate when overexpressed in *Arabidopsis* (Soderman *et al.*, 1999). It would not be surprising to find that an essential character such as cell growth is responsive to the environment through a number of different signal pathways.

Apart from the cessation or delay of growth, plants native to arid environments have evolved a host of phenological and morphological adaptations that are pertinent to growth and survival under dry, hot conditions. The adaptive features of plants native to these environments have been classified into three broad categories: escape, avoidance and tolerance (Levitt, 1980; Turner, 1986; Evans, 1993). Generally speaking, escape involves phenological characteristics that allow completion of the plant's life cycle before stress becomes too severe. Avoidance is accomplished by mechanisms that prevent the cells of the plant from actually experiencing a severe stress such as those that conserve water during desiccation (closed stomata) or insulate and protect sensitive tissue during cold (thick cell walls of bud scabs). True tolerance occurs only if the plant tissues actually experience the stress environment but display a physiological or biochemical process (e.g. osmotic adjustment) that prevents or reduces injury or deleterious effects.

It has been historically accepted that escape and avoidance characteristics are genetically complex (e.g. Flowers and Yeo, 1995), although breeding crop plants for such traits has been successful (O'Toole, 1989). The existence of genes for such traits is documented by many genetic analyses (Sharp *et al.*, 1994; Sarma *et al.*, 1998). However, even with the use of modern molecular marker tools, these characteristics must be selected in each individual crop. The recent application of genomics-based approaches toward genetic analysis of model organisms, especially *Arabidopsis thaliana*, had begun to cause many investigators to reconsider our abilities to understand and manipulate more complex traits such as avoidance and escape from drought. An extremely rapid increase in the number and type of mutants and corresponding genes being identified in *Arabidopsis* is now taking place. Many of these mutants and genes may be directly applicable to escape or avoidance of abiotic stresses. We provide a few examples of such mutants and genes that may be used to impact the ability of plants to be productive in environments with limited moisture available.

Many mutants affecting the duration of life cycle have been described in *Arabidopsis* and several genes responsible for the mutant phenotypes have been cloned (Ratcliffe *et al.*, 1998; Ogas *et al.*, 1999). Timing of the plant's life cycle with availability of soil moisture is an extremely important phenological trait that allows drought escape (Turner, 1986). In fact, timing of flowering with respect to the onset of desiccation has been observed often to be the most critical plant characteristic affecting productivity when moisture is severely limiting (Turner, 1986; Evans, 1993). Some mutations such as the leafy mutant in *Arabidopsis* may be directly applicable to shortening the life cycle of plants to escape drought stress since this gene has already been demonstrated to be functioning in an unrelated species (Weigel and Nilsson, 1995).

Plants may also avoid the deleterious effects of osmotic stresses by controlling water loss and absorption. The two most critical features impacting water use by plants are stomatal functioning and root development. Again, many mutants and corresponding genes have been described in *Arabidopsis* that impact these traits. Stomatal mutants have been described in *Arabidopsis* (e.g. Berger and Altmann, 2000). Coupled with populations of tagged insertion mutants, this technique could allow the rapid identification of many genes that impact stomatal behaviour including some which may be important to avoid excessive water loss during osmotic stress. Many *Arabidopsis* mutants with altered root morphology and development have been described (Malamy and Benfey, 1997). Many of these may be useful to evaluate the effect of specific changes in root characteristics on water use in stressful environments. Some mutants may be particularly helpful to elucidate the role of root structures such as the endodermis in hydraulic conductivity and its potential impact on drought avoidance (Wysocka-Diller *et al.*, 2000).

Several other important features of drought-avoiding plants may be studied using *Arabidopsis* genetics. Of particular interest may be traits such as leaf morphology for which several mutants and corresponding genes have been described (Berná *et al.*, 1999). Apart from mutants that exhibit the familiar narrow shapes of xerophyte leaves that reduce the area contributing to water loss (Kim *et al.*, 1999), some mutants of *Arabidopsis* demonstrate a constitutive leaf-rolling phenotype similar to the behaviour response of plants to desiccation (Dingkuhn *et al.*, 1989). The effect of this leaf characteristic on water use in these *Arabidopsis* mutants has never been determined.

An important consideration for all traits that may impact escape or avoidance from drought is the environmental plasticity of such characteristics. Plasticity of flowering time or leaf morphology or any other trait important to water use is common among plants native to arid environments. This plasticity allows the 'fine tuning' of the phenology of the plant with the variations in the environment (Turner, 1986). The fact that native plants are able to adjust phenological traits to coincide more exactly with moisture availability implies that expression of genes controlling avoidance traits is under a genetic control mechanism that responds to the environment. Plasticity of avoidance traits could be genetically engineered in a number of ways; the simplest would involve modification of genes that control avoidance (such as flowering time) with a desiccation stress controlled promoter.

Considering that abiotic stresses are generally accompanied by changes in developmental progression and morphology, mutational or transgenic approaches to alter development in crop species can be imagined, but the *Arabidopsis* mutants have not rigorously been investigated under the aspect of stress tolerance. This concept is now possible based on the completion of the *Arabidopsis* genome sequence and due to the fact that saturation mutagenesis in *Arabidopsis* has been achieved (Krysan *et al.*, 1999; Hasegawa *et al.*, 2000b). The ongoing high-throughput EST sequencing and microarray hybridization projects will provide additional information about genes that alter plant stress responses and development.

Genomics

The analysis of gene expression changes in response to abiotic stress is now possible on the whole-genome level, made possible by large scale EST sequencing, microarray analysis, and the mutants in *Arabidopsis*, which report altered gene expression under abiotic stresses (Zhu *et al.*, 1997; Krysan *et al.*, 1999; Hasegawa *et al.*, 2000b; Shi *et al.*, 2000). This will replace the gene-by-gene approach of the past, but the extent to which such analyses can be exploited for crop improvement strategies requires intense collaboration between breeders and molecular biologists.

Genome-wide analyses have been made possible because of the ease of DNA manipulation and the feasibility of analysing large sets of genes facilitated by advanced bio-informatics tools. The first step towards categorizing genetic complexity is gene discovery by large-scale, partial sequencing of randomly selected cDNA clones, ESTs. EST collections already exist for a number of organisms (http://www.ncbi.nlm.nih.gov/dbEST/dbESTsummary.html), and the speed is accelerating with which additions are deposited into the databases. By now, we know more than half of the basic set of genes in plants, if we assume 28,000 coding regions in total based on the available *Arabidopsis* genome sequence. At present, plant EST collections are biased towards high to moderate abundance classes of transcripts. Also, the analyses have not yet included all tissues, organs, cells, developmental states, various external stimuli or plant growth regulator treatments. Significantly, only a few studies, including small numbers of ESTs, have focused on ESTs from plants exposed to abiotic stresses (Umeda *et al.*, 1994; Pih *et al.*, 1997).

The set of transcripts that characterizes the stressed state is fundamentally different from the set of transcripts found in unstressed plants. This statement describes the first results of a comprehensive EST sequencing initiative with a stress focus. The project uses more than 40 cDNA libraries for specific cells, tissues and developmental stages. The libraries were generated with RNA from salt-stressed glycophytes, *Arabidopsis thaliana* and *Oryza sativa*, and also *Mesembryanthemum* and *Dunaliella salina*, which serve as halophytic model species (listed at: http://www.biochem.arizona.edu/BOHNERT/functgenomics/front2.html). Preliminary data sets are available at http://stress-genomics.org. As part of the gene discovery effort in maize, EST collections are also being established from cDNA libraries prepared from salt-stressed roots and shoots (http://www.zmdb.iastate.edu/zmdb/ESTproject.html). Similar projects are underway with a focus on drought tolerance-specific ESTs, and several initiatives which target crop species and drought are being implemented at various laboratories.

A sampling of approximately equal numbers of ESTs from well watered and salinity stressed *Mesembryanthemum* leaves suggests that stressed plants express approximately 15% more functionally unknown genes than unstressed plants. Also, the overlap of transcripts, comparing stressed versus unstressed states, is surprisingly small, indicating drastic changes in transcript populations. The preliminary analysis based on transcript abundance profiles and changes under stress also documented that ESTs related to salinity stress are vastly under-represented in the non-redundant plant databases. In leaves, differences between unstressed and stressed plants revealed pronounced down-regulation in transcript abundance for components of the photosynthetic apparatus, water channels, or the translation machinery, and strong up-regulation of transcripts whose products lead to a restructuring of the cell (e.g. proteases, components of the ubiquitin system, chaperones and chaperonins). Also increased are transcripts for functions that are known to lead to osmotic and dehydration stress adaptation, and to detoxification of oxygen radical species (Hasegawa *et al.*, 2000b). Another result is the detection of many transcripts that have no known function. In all EST analyses, the precise function of approximately 50% of the predicted proteins remains unknown (Somerville and Somerville, 1999). However, among the stress-dependently up-regulated *Mesembryanthemum* transcripts the percentage is higher, and many transcripts are 'novel' in a sense that they have not previously been detected in any other organism.

The amount of EST sequence data requires sorting before functional assignments can be attempted. One tool is provided by microarray analysis of large numbers of ESTs printed on slides and probed with fluorescence-labelled probes in hybridizations that compare the stressed and unstressed states. Microarrays are at present the tool of choice to gain insights into the complexity of gene function and regulatory control and to approach functional analysis of unknown sequences (Eisen *et al.*, 1998; Kehoe *et al.*, 1999). The immediate result is the identification of stress-regulated transcripts. The next step is establishing patterns of expression by which functionally unidentified sequences are clustered with the expression profiles of known sequences. The kinetics and changes of expression ('cluster analysis') provide clues about functions. Large-scale cDNA microarray analyses of salinity-stress responsive gene expression profiles are underway for *Mesembryanthemum*, rice and *Arabidopsis* (http://stress-genomics.org). Although the present analyses include only a small fraction of the total gene complement, they provide a starting point for providing information about the role of functionally unknown ORFs in stress adaptation. Closer analysis of the present data sets indicated that a number of strongly salt stress-upregulated ORFs in yeast (J. Yale and H.J. Bohnert, unpublished) have equally up-regulated counterparts in higher plants. Such comparisons among evolutionarily divergent organisms begin to reveal the gene complement outlining cell-based tolerance mechanisms – the 'emergency' or 'downstream' reactions defined before. It seems that this set of genes, which determines reactions of individual cells, is deeply rooted and largely conserved in function, if not in gene complexity, between cyanobacteria, yeasts, and glycophytic and halophytic plant species.

Concluding Remarks

The views expressed here are those of bench scientists, concerned mostly with trying to understand mechanisms, and not concerned with application. Breeding of tolerant or more tolerant crops requires different views and experimental strategies (Table 6.1). The last 10 years have immensely advanced the understanding of biochemical and physiological mechanisms.

At the same time, the introduction of molecular markers and new approaches towards measuring quantitative traits have enhanced plant breeding. The similarities and homologies between different organisms (for example, yeasts and plants) have both broadened our views and sharpened our focus to recognize essentials of abiotic stress tolerance. In molecular genetics, there is growing recognition of the homology that exists between genomes at the level of chromosomes, developmental programmes, and about the identity of many reactions to stress, which have in the past been hidden by morphological and ecological divergence. It is apparent that at least the most important genes for proteins that provide abiotic stress tolerance are present in all plants. In fact, the evolutionary appearance of entirely novel genes and mechanisms seems quite unlikely. Tolerance may be determined by how effectively, or how fast, the biochemical response machinery can be brought into play and how vigorously such signalling then leads to altered growth and development. Stress tolerance seems largely determined by stress sensing, signal transduction and the networking of sensory stimuli, rather than by the biochemical 'hardware' (Hasegawa *et al.*, 2000b; Mizoguchi *et al.*, 2000), still allowing for variability at the level of orders, families or even species in how the signalling and response pathways underlying tolerance are connected. A combination of molecular tools, engineering approaches and knowledge-based breeding, in collaborations between disciplines, seems to be called for in the future.

Acknowledgements

We thank the people in our laboratories for discussions. Work in our laboratories is or has been, off and on, supported by the US Department of Energy, the National Science Foundation, the US Department of Agriculture and the Arizona Agricultural Experiment Station. Visiting scientists have been supported by CONACyT (Mexico City), Agricultural Research Council (London), Rockefeller Foundation (New York), Smithsonian Institution/Carnegie-Mellon Foundation (New York), Japanese Society for the Promotion of Science (Tokyo), New Energy Development Organization (Tokyo), Max-Kade Foundation (New York), and Deutsche Forschungsgemeinschaft (Bonn). A genomics approach to finding stress tolerance determinants is funded by the US National Science Foundation (DBI-9813330).

References

Amtmann, A. and Sanders, D. (1999) Mechanisms of Na+ uptake by plant cells. *Advances in Botanical Research* 29, 75–112.
Ansell, R., Granath, K., Hohmann, S., Thevelein, J.M. and Adler, L. (1997) The two isoenzymes for yeast NAD-dependent glycerol 3-phosphate dehydrogenase enco-

ded by GPD1 and GPD2 have distinct roles in osmoadaptation and redox regulation. *EMBO Journal* 16, 2179–2187.

Apse, M.P., Aharon, G.S., Snedden, W.A. and Blumwald, E. (1999) Salt tolerance conferred by overexpression of a vacuolar Na$^+$/H$^+$-antiport in *Arabidopsis*. *Science* 285, 1256–1258.

Asada, K. (1999) The water–water cycle in chloroplasts: Scavenging of active oxygens and dissipation of excess photons. *Annual Review Plant Physiology and Plant Molecular Biology* 50, 601–639.

Berger, D. and Altmann, T. (2000) A subtilisin-like serine protease involved in the regulation of stomatal density and distribution in *Arabidopsis thaliana*. *Genes and Development* 14, 119–131.

Berná, G., Robles, P. and Micol, J.L. (1999) A mutational analysis of leaf morphogenesis in *Arabidopsis thaliana*. *Genetics* 152, 729–742.

Blum, A. (1988) *Plant Breeding for Stress Environments*. CRC Press, Boca Raton, Florida.

Bohnert, H.J. and Jensen, R.G. (1996) Metabolic engineering for increased salt tolerance – the next step. *Australian Journal of Plant Physiology* 23, 661–667.

Bohnert, H.J. and Sheveleva, E. (1998) Plant stress adaptations – making metabolism move. *Current Opinion in Plant Biology* 1, 267–274.

Bohnert, H.J., Su, H. and Shen, B. (1999) Molecular mechanisms of salinity tolerance. In: Shinozaki, K. (ed.) *Cold, Drought, Heat, and Salt Stress: Molecular Responses in Higher Plants*. RG Landes, Austin, Texas, pp. 29–60.

Boyer, J.S. (1982) Plant productivity and environment. *Science* 218, 443–448.

Bressan, R.A., Nelson, D.E., Iraki, N.M., LaRosa, P.C., Singh, N.K., Hasegawa, P.M. and Carpita, N.C. (1990) Reduced cell expansion and changes in cell walls of plant cells adapted to NaCl. In: Katterman, F. (ed.) *Environmental Injury to Plants*. Academic Press, San Diego, pp. 137–171.

Cockcroft, C.E., den Boer, B.G.W., Healy, J.M.S. and Murray, J.A.H. (2000) Cyclin control of growth rate in plants. *Nature* 405, 575–579.

Dingkuhn, M., Cruz, R.T., O'Toole, J.C. and Dorffling, K. (1989) Net photosynthesis, water-use efficiency, leaf water potential and leaf rolling as affected by water deficit in tropical upland rice. *Australian Journal of Agricultural Research* 40, 1171–1181.

Eisen, M.B., Spellman, P.T., Brown, P.O. and Botstein, D. (1998) Cluster analysis and display of genome-wide expression patterns. *Proceedings of the National Academy of Sciences USA* 95, 14863–14868.

Evans, L.T. (1993) *Crop Evolution, Adaptation and Yield*. Cambridge University Press, Cambridge.

Fassler, J.S., Gray, W.M., Malone, C.L., Tao, W., Lin, H. and Deschenes, R.J. (1997) Activated alleles of yeast SLN1 increase Mcm1-dependent reporter gene expression and diminish signaling through the Hog1 osmosensing pathway. *Journal of Biological Chemistry* 272, 13365–13371.

Flowers, T.J. and Yeo, A.R. (1995) Breeding for salinity tolerance in crop plants: Where next? *Australian Journal Plant Physiology* 22, 875–884.

Frova, C., Caffulli, A. and Pallavera, E. (1999) Mapping quantitative trait loci for tolerance to abiotic stresses in maize. *Journal of Experimental Zoology* 282, 164–170.

Giraudat, J. (1995) Abscisic acid signaling. *Current Opinion in Cell Biology* 7, 232–238.

Gustin, M.C., Albertyn, J., Alexander, M. and Davenport, K. (1998) MAP kinase pathways in the yeast *Saccharomyces cerevisiae*. *Microbiology Molecular Biology Review* 62, 1264–1294.

Halfter, U., Ishitani, M. and Zhu J.K. (2000) The *Arabidopsis* SOS2 protein kinase physically interacts with and is activated by the calcium-binding protein SOS3. *Proceedings of the National Academy of Sciences USA* 97, 3735–3740.

Hare, P.D., Cress, W.A. and van Staden, J. (1997) The involvement of cytokinins in plant responses to environmental stress. *Plant Growth Regulation* 23, 79–103.

Harmon, A.C., Gribskov, M. and Harper. J.F. (2000) CDPK's – a kinase for every Ca^{2+} signal? *Trends in Plant Science* 5, 154–159.

Hasegawa, P.M., Bressan, R.A. and Pardo, J.M. (2000a) The dawn of plant salt tolerance genetics. *Trends in Plant Science* 5, 317–319.

Hasegawa, P.M., Bressan, R.A., Zhu, J.-K. and Bohnert, H.J. (2000b) Plant cellular and molecular responses to high salinity. *Annual Review of Plant Physiology and Plant Molecular Biology* 51, 463–499.

Horton, P., Ruban, A.V. and Walters, R.G. (1996) Regulation of light harvesting in green plants. *Annual Review of Plant Physiology and Plant Molecular Biology* 47, 655–684.

Kehoe, D.M., Villand, P. and Somerville, S. (1999) DNA microarrays for studies of higher plants and other photosynthetic organisms. *Trends in Plant Sciences* 4, 38–41.

Khush, G.S. (1999) Green revolution: preparing for the 21st century. *Genome* 42, 646–655.

Kim, G-T., Tsukaya, H., Saito, Y. and Uchimiya, H. (1999) Changes in the shapes of leaves and flowers upon overexpression of cytochrome P450 in *Arabidopsis*. *Proceedings of the National Academy of Sciences USA* 96, 9433–9437.

Knight, H. (2000) Calcium signaling during abiotic stress in plants. *International Review Cytology* 195, 269–324.

Krysan, P.J., Young, J.C. and Sussman, M.R. (1999) T-DNA as an insertional mutagen in *Arabidopsis*. *Plant Cell* 11, 2283–2290.

Levitt, J. (1980) *Responses of Plants to Environmental Stress, Chilling, Freezing, and High Temperature Stresses*, 2nd edn. Academic Press, New York.

Liu, J. and Zhu, J.K. (1998) A calcium sensor homolog required for plant salt tolerance. *Science* 280, 1943–1945.

Liu, J., Ishitani, M., Halfter, U., Kim, C.S. and Zhu, J.K. (2000) The *Arabidopsis thaliana* SOS2 gene encodes a protein kinase that is required for salt tolerance. *Proceedings of the National Academy of Sciences USA* 97, 3730–3734.

Long, S.P., Humphries, S. and Falkowski, P.G. (1994) Photoinhibition of photosynthesis in nature. *Annual Review Plant Physiology and Plant Molecular Biology* 45, 633–663.

Malamy, J.E. and Benfey, P.N. (1997) Organization and cell differentiation in lateral roots of *Arabidopsis thaliana*. *Development* 124, 33–44.

Mizoguchi, T., Ichimura, K., Yoshida, R. and Shinozaki, K. (2000) MAP kinase cascades in *Arabidopsis*: their roles in stress and hormone responses. *Results and Problems in Cell Differentiation* 27, 29–38.

Munns, R. (1993) Physiological processes limiting plant-growth in saline soils – some dogmas and hypotheses. *Plant Cell and Environment* 16, 15–24.

Nelson, D.E., Shen, B. and Bohnert, H.J. (1998) Salinity tolerance – mechanisms, models, and the metabolic engineering of complex traits. In: Setlow, J.K. (ed.)

Genetic Engineering, Principles and Methods, Vol. 20. Plenum Press, New York, pp. 153–176.

Netting, G. (2000) pH, abscisic acid and the integration of metabolism in plants under stressed and non-stressed conditions: cellular responses to stress and their implication for plant water relations. *Journal of Experimental Botany* 51, 147–158.

Norlyn, J.D. (1980) Breeding salt tolerant crop plants. In: Rains, D.W., Valentine, R.C. and Hollaender, A. (eds) *Genetic Engineering of Osmoregulation*. Plenum Press, New York, pp. 293–309.

Nuccio, M.L., Rhodes, D., McNeil, S.D. and Hanson, A.D. (1999) Metabolic engineering of plants for osmotic stress resistance. *Current Opinion in Plant Biology* 2, 128–134.

Ogas, J., Kaufmann, S., Henderson, J. and Somerville, C. (1999) Pickle is a CHD3 chromatin-remodeling factor that regulates the transition from embryonic to vegetative development in *Arabidopsis*. *Proceedings of the National Academy of Sciences USA* 96, 13839–13844.

O'Leary, J.W. (1994) The agricultural use of native plants on problem soils. In: Yeo, A.R. and Flowers, T.J. (eds) *Soil Mineral Stresses: Approaches to Crop Improvement*. Springer Verlag, Berlin, pp. 127–143.

O'Toole, J.C. (1989) Breeding for drought resistance in cereals: emerging new technologies. In: Baker, F.W.G. (ed.) *Drought Resistance in Cereals*. CAB International, Wallingford, pp. 81–94.

Pical, C., Westergren, T., Dove, S.K., Larsson, C. and Sommarin, M. (1999) Salinity and hyperosmotic stress induce rapid increases in phosphatidylinositol 4,5-bisphosphate, diacylglycerol pyrophosphate, and phosphatidylcholine in *Arabidopsis thaliana* cells. *Journal of Biological Chemistry* 274, 38232–38240.

Pih, K.Y., Jang, H.J., Kang, S.G., Piao, H.L. and Hwang, I. (1997) Isolation of molecular markers for salt stress responses in *Arabidopsis thaliana*. *Molecular Cell* 7, 567–571.

Popping, B., Gibbons, T. and Watson, M.D. (1996) The *Pisum sativum* MAP kinase homologue (PsMAPK) rescues the *Saccharomyces cerevisiae hog1* deletion mutant under conditions of high osmotic stress. *Plant Molecular Biology* 31, 355–363.

Postel, S. (1999) Redesigning irrigated agriculture. In: Brown, L.R., Flavin, C. and French, H. (eds) *State of the World 2000*. W.W. Norton & Co., New York, pp. 39–58.

Quarrie, S.A., Laurie, D.A., Zhu, J., Lebreton, C., Semikhodskii, V., Steed, A., Wisenboer, H. and Calesanti, C. (1997) QTL analysis to study the association between leaf size and abscisic acid accumulation in droughted rice leaves and comparison across cereals. *Plant Molecular Biology* 35, 155–165.

Ramage, R.T. (1980) Genetic methods to breed salt tolerance in plants. In: Rains, D.W., Valentine, R.C. and Hollaender, A. (eds) *Genetic Engineering of Osmoregulation*. Plenum Press, New York, pp. 311–318.

Ratcliffe, O.J., Amaya, I., Vincent, C.A., Rothstein, S., Carpenter, R., Coen, E.S. and Bradley, D.J. (1998) A common mechanism controls the life cycle and architecture of plants. *Development* 125, 1609–1615.

Richards, R.A. (1992) Increasing salinity tolerance of grain crops: is it worthwhile? *Plant and Soil* 32, 89–98.

Richmond, T. and Somerville, S. (2000) Chasing the dream: plant EST microarrays. *Current Opinion in Plant Biology* 3, 108–116.

Roxas, V.P., Smith Jr., R.K., Allen, E.R. and Allen, R.D. (1997) Overexpression of glutathione S-transferase/glutathione peroxidase enhances the growth of transgenic tobacco seedlings during stress. *Nature Biotechnology* 15, 988–997.

Rush, P.W. and Epstein, E. (1981) Breeding and selection for salt tolerance by the incorporation of wild germplasm into a domestic tomato. *Journal of the American Society for Horticultural Sciences* 106, 699–704.

Sakakibara, H., Taniguchi, M. and Sugiyama, T. (2000) His-Asp phosphorelay signaling: a communication avenue between plants and their environment. *Plant Molecular Biology* 42, 273–278.

Sarma, R.N., Gill, B.S., Galiba, G., Sutka, J., Laurie, D.A. and Snape, J.W. (1998) Comparative mapping of the wheat chromosome 5A Vrn-A1 region with rice and its relationship to QTL for flowering time. *Theoretical and Applied Genetics* 97, 103–109.

Schroda, M., Vallon, O., Wollman, F.A. and Beck, C.F. (1999) A chloroplast-targeted heat shock protein 70 contributes to the photoprotection and repair of photosystem II during and after photoinhibition. *Plant Cell* 11, 1165–1178.

Sharp, R.E., Wu, Y., Voetberg, G.S., Saab, I.N. and LeNoble, M.E. (1994) Confirmation that abscisic acid accumulation is required for maize primary root elongation at low water potentials. *Journal of Experimental Botany* 45, 1743–1751.

Shen, B., Hohmann, S., Jensen, R.G. and Bohnert, H.J. (1999) Roles of sugar alcohols in osmotic stress adaptation. Replacement of glycerol by mannitol and sorbitol in yeast. *Plant Physiology* 121, 45–52.

Shi, H., Ishitani, M., Kim, C. and Zhu, J.-K. (2000) The *Arabidopsis thaliana* salt tolerance gene SOS1 encodes a putative Na^+/H^+ antiporter. *Proceedings of the National Academy of Sciences USA* 97, 6896–6901.

Shinozaki, K. and Yamaguchi-Shinozaki, K. (1997) Gene expression and signal transduction in water-stress response. *Plant Physiology* 115, 327–334.

Soderman, E., Hjellstrom, M., Fahleson, J. and Engstrom, P. (1999) The HD-Zip gene ATHB6 in *Arabidopsis* is expressed in developing leaves, roots and carpels and up-regulated by water deficit conditions. *Plant Molecular Biology* 40, 1073–1083.

Somerville, C. and Somerville, S. (1999) Plant functional genomics. *Science* 28, 380–383.

Turner, N.C. (1986) Adaptation to water deficits: a changing perspective. *Australian Journal of Plant Physiology* 13, 175–190.

Umeda, M., Hara, C., Matsubayashi, Y., Li, H.H., Liu, Q., Tadokoro, F., Aotsuka, S. and Uchimiya, H. (1994) Expressed sequence tags from cultured cells of rice (*Oryza sativa* L.) under stressed conditions: analysis of transcripts of genes engaged in ATP-generating pathways. *Plant Molecular Biology* 25, 469–478.

Weigel, D. and Nilsson, O. (1995) A developmental switch sufficient for flower initiation in diverse plants. *Nature* 377, 495–500.

Wysocka-Diller, J.W., Helariutta, Y., Fukaki, H., Malamy, J.E. and Benfey, P.N. (2000) Molecular analysis of SCARECROW function reveals a radial patterning mechanism common to root and shoot. *Development* 127, 595–603.

Xiong, L., Ishitani, M. and Zhu, J.K. (1999) Interaction of osmotic stress, temperature, and abscisic acid in the regulation of gene expression in *Arabidopsis*. *Plant Physiology* 119, 205–212.

Zheng, H.G., Babu, R.C., Pathan, M.S., Ali, I., Huang, N., Coutois, B. and Nguyen,

H.T. (2000) Quantitative trait loci for root-penetration ability and root thickness in rice: comparison of genetic backgrounds. *Genome* 43, 53–61.

Zhu, J.K., Hasegawa, P.M. and Bressan, R.A. (1997) Molecular aspects of osmotic stress in plants. *Critical Reviews in Plant Sciences* 16, 253–277.

Zhu, J.K., Liu, J. and Xiong, L. (1998) Genetic analysis of salt tolerance in *Arabidopsis*. Evidence for a critical role of potassium nutrition. *Plant Cell* 10, 1181–1191.

7

Plant Stress Factors: Their Impact on Productivity of Cropping Systems

U.R. Sangakkara

Faculty of Agriculture, University of Peradeniya, Peradeniya 20400, Sri Lanka

Introduction – Global Impact of Plant Stress Factors

The world is facing a food crisis. The current population of 6 billion people is estimated to reach 7 billion in 10 years (2010) and even 8–10 billion in 2025. Increased living standards are expected to generate a greater demand for food and fibre, all produced by crops, including grasslands. Hence the resources of the earth will be called upon to produce 40–50% more food and fibre by the year 2025 (Khush, 1999). More importantly, the land with its finite resources will be called upon to produce more food and fibre thus increasing the stress on factors that determine productivity and sustainability (Turner, 1998; FAO, 1999).

Today, the developed countries have ample food supplies to the extent of paying off farming land. In contrast, farmers of Africa, Asia and Latin America do face a crisis. Available agricultural land is not expected to increase in the foreseeable future. Thus, while estimates demand a doubling of food production in the next 40–50 years, available resources do not have the capacity to maintain or, more importantly, enhance production. Under these conditions, farmers, especially those in the developing countries, are striving to produce crops for consumption and/or sale with diminishing resources. This is indeed stress – both to the farmer and to the resources.

Stress Factors in the Tropical Regions

Smallholder farming systems characterize tropical crop production; some at subsistence levels while others are very intensively managed units. In most countries of the tropics, food for local populations is produced in the subsistence sector, while specialized units tend to provide commodities for export and exclusive markets. Under such conditions, the stress factors affect the productivity patterns of the smallholder farmer to a greater extent, especially as such units do not have the capacity to protect crops from adverse climatic conditions, poor soils, insufficient water and plant nutrition.

Prior to the green revolution, the sustainability of agricultural units was generally maintained. The existence of forest gardens where many perennial species were and are still grown together and traditional organic systems illustrates this, as seen in many biodynamic units in the tropics. With the green revolution, intensification of cropping and excessive use of chemicals caused both production and ecological problems. The excessive mining of soils, pollution of water and its wastage in major irrigation schemes, and lack of preserving soil organic matter have all led in many instances to the loss of sustainability.

At present, crop production in the developing nations is affected by three important abiotic factors. These are reductions in the soil organic matter and inadequate nutrients (FAO, 1999) and insufficient water (Bouwer, 2000). One also cannot negate the recent trends of research on the impact of global warming on crop production. It is postulated that this phenomenon would have a greater impact in the tropics. It is under these circumstances that farmers of the developing world are called upon to produce 60–70% more food for increasing populations.

Water – the scarce resource

Water is considered the most limiting factor in crop production in the tropics. The lack of water suitable for agriculture is a fact today with an estimated increase of 80 million people per annum. The FAO (1999) highlights that in 2030, 70% more food needs to be produced from the water resources found at present. Global warming is predicted to enhance water use and this problem could be accentuated by the increasing variability in weather, as evidenced by the recent droughts in the tropics. These trends are also compounded by the displacement of staple food crops from suitable environments in which the available water is used to produce high value crops for export, or for non-agricultural enterprises (Edmeades *et al.*, 1998). Water is fast becoming a very scarce resource for tropical agriculture, even in the wettest countries in the world such as India (Agarwal and Narain, 1999). This is a significant abiotic stress factor in crop production, especially as drought-resistant species such as cowpea (*Vigna unguiculata* L. Walp) reduce yields by over 50% when subjected to water stress (Sangakkara, 1998).

The impact of water on crop production in the tropics begins even before planting. The monsoons or irrigation determine the date of land preparation and then planting, and hold the key to successful crop establishment. Loss of plants at planting due to the lack of rainfall or irrigation could be up to

90% depending on the species. More importantly, the resultant effect of water stress is the reduction in yield, which can be as high as 80% in rice, yams and other staple food crops of the tropics. Under these circumstances, water is becoming a political and economic choice – namely, which section of the economy will give better returns per unit of water: agriculture, tourism or urban and industrial sectors (FAO, 1999). Although the situation is rather grave in the developing world, farmers have some options to utilize the water better, using available resources. This will certainly be the short-term measure, while either the state, international or local organizations develop long-term programmes. The first option that farmers would have is to utilize species, or even cultivars, that would use water more efficiently or cultivate drought-avoiding crops, which too could be susceptible to drought stress (Turner, Chapter 8, this volume). The more efficient technique would be to enhance water-holding capacities of the soil by organic matter and crop residues. This would increase crop yields significantly and also enhance nutrient uptake due to greater retention of added fertilizers (Kumar and Goh, 2000; Sangakkara and Kandapola, 2000; Table 7.1). However, much research is still needed on the use of crop residues and its management for optimizing water uses in the developing countries of the tropics. Support is required for the development of sustainable management systems for drought-avoiding and drought-resistant crops and cultivars in future research programmes.

Land and soil – scarce and finite resources

The tropics, especially in Asia and the Pacific, have reached the safe limits of horizontal expansion of agriculture. All future needs of the growing populations have to be met by intensification of existing lands. This is not an easy task, as yields in most regions are showing signs of stagnation or even decline, especially in crops grown on certain soils in the tropics (FAO, 1999).

Soil degradation, especially the loss of its capacity to produce high yields, is due to organic matter and nutrient depletion along with erosion. Declining soil productivity means less vegetative cover to soil, less return of organic matter and less microbial activity, leading to an infertile and, at times, a biologically less active soil. Crop yields in such soils decline (Berzsenyi *et al.*, 2000), thereby returns to farmers also decline, while the costs escalate and sustainability is lost. The local environment is also degraded, pollution increased and cropping systems need to be adjusted to adapt to the low level of soil productivity. The overall result due to degradation of land is the lowering of the role of crop production within the economy of the household, the region and country, with non-farming occupations becoming more prominent.

While land has become a scarce resource, the combination of soil erosion, destruction of soil structure, loss of nutrients and nutrient-holding

Table 7.1. Impact of organic matter on yields of corn and mung bean and availability of fertilizer nitrogen. T1 and T4 represent plots with and without organic matter from season 2. (From Sangakkara and Kandapola, 2000.)

Season and crop	Treatment		% N	Yield (kg ha⁻¹)	NdfF	% N recovery
Season 1						
Corn	Seed	T1	1.76	4276	13.89	23.15
		T4	1.76	4108	14.09	23.49
	Residue	T1	1.10	14610	30.24	50.40
		T4	1.02	14481	29.51	49.18
		Sx	0.04	1422	2.15	3.58
Season 2						
Mung bean	Seed	T1	4.47	2642	1.85	3.09
		T4	4.52	2104	1.29	2.15
	Residue	T1	1.12	5180	1.17	1.95
		T4	1.18	4599	0.83	1.38
		Sx	0.03	158	0.11	0.24
Season 3						
Corn	Seed	T1	1.76	4015	0.56	0.93
		T4	1.66	2699	0.33	0.56
	Residue	T1	0.78	14115	0.47	0.78
		T4	0.87	14028	0.45	0.75
		Sx	0.02	88	0.04	0.33
Season 4						
Mung bean	Seed	T1	4.54	3258	0.44	0.74
		T4	4.70	2541	0.82	0.46
	Residue	T1	2.27	4985	0.11	0.18
		T4	2.52	4014	0.18	0.31
		Sx	0.15	447	0.09	0.14

capacity, salinization or building up of toxic salts, waterlogging and acidification are leading to the degradation of production capacity of soils in the tropics (Swindale, 1998). Recently the FAO (1999) estimated that approximately 2 billion hectares of a total of 8.7 billion hectares of crop, pastoral and forest/woodlands have been degraded since 1950. Of this approximately 16% have been degraded severely and another 44% moderately degraded which can be revived through on farm initiatives. Thus crop production today, especially in the developing world, is faced with a serious stress factor of declining productivity of land and soil.

There are options that one could consider in overcoming this scenario. Intensification of good productive land by developing suitable cropping systems could be considered a suitable option. The marginally degraded lands could be improved by using well-proven management techniques such as soil conservation and protection. The degraded lands need to be rested, not by fallow but using tree species, which could also produce an economic return after a period of time. These aspects all require *in situ* studies in the developing world.

Climate – a stress factor

The climate of a region is defined as an indicator of the weather experienced in that region. Monteith and Ingram (1998) identify three aspects of climatic variations affecting crop production. These are spatial distribution of climate, seasonal changes and global climatic changes.

Farmers of the developing countries cannot mitigate the potential stress caused by climatic changes to crops. This is due to the non-availability of the technology to develop protected cropping systems. Farmers rely on past experiences and intuition to adjust cropping sequences to reduce the stress factors caused by climate. However, the rapid changes that are taking place in climates due to global warming could affect cropping in the tropics to a greater extent than in the temperate regions (Hulme and Viner, 1998).

The impact of temperature is also significant in crop production. In the temperate regions, yields generally increase with the length of the growing season. In contrast, in the tropics, vegetative yield of species generally increases with temperature and the growing period, due to the stimulation of processes such as leaf appearance rates and plant metabolism. This again is a complex interactive effect as the non-availability of soil moisture limits the response of plants to temperature, and the causal phenomena are well documented (Pessarakli, 1999).

The most recent variation causing abiotic stress in plants is global warming. This factor must be considered on the basis of crops being unconstrained and/or constrained by other stress factors. The resultant effect of stress would vary on this basis (Rozenweig and Parry, 1997). Thus, current research emphasis is placed upon the importance of increasing CO_2 concentrations in the atmosphere and associated temperature changes. The real impact of global warming on crop yields is yet to be clearly quantified, although controlled experiments, such as FACE (free air CO_2 enrichment experiments), are carried out in the developed world. These highlight changes or even increases in physiological efficiencies of crops, trees and pasture species, biomass allocation, chemical composition of plant responses to nutrients and even microbial activity of soil (Lüscher *et al.*, 1998; Champingny and Mosseau, 1999; Morilley *et al.*, 1999). In contrast, the impact of climate change on crop production in the tropics has not been studied, although simulation models predict a greater degree of stress to crops due to drier and warmer climates (Hulme and Viner, 1998). These concepts are yet to be tested and all data presented are guesses based on many simplifications and assumptions of complex biological and physical conditions.

Interactive effects of abiotic stress

Stress factors on crop growth do not occur alone and interactions between growth resources and environments are the norm in ecosystems. However, the abiotic stress factors have a greater damaging effect during certain phases of crop growth. Thus, studies are required to determine temporal variations in abiotic stress, the ability of crop plants to adapt to stress and the different interactions between the stress factors.

In all cropping situations, the abiotic components are never optimal for plant growth due to constant changes in the environment. The maximum potential of crops is seldom achieved under usual cropping conditions. The crops are subjected to fluctuations in light, moisture, high or low temperatures (both soil and ambient), soil strength, soil aeration, salinity, nutrient imbalances, global climate changes and many other environmental stress factors. More importantly most natural environments are continuously suboptimal with respect to one or more of the above features, amongst which water, soil and plant nutrition and temperatures could be considered global abiotic stress factors affecting cropping.

Generally research projects on the response of crop plants to environmental stress have focused on changes in plant growth and yields due to one specific stress or two. This is a setback in current research, although interpretation of such interactive experiments are very complicated. At times, results become meaningless due to confounding of stress factors and their interactions. The abundant availability of literature on the response of plants to different stress factors (Pessarakli, 1999) highlights this factor, and states the importance of stress caused by individual factors, but not the interactive effects.

Most agricultural ecosystems today, especially in the developing countries, are located in fragile ecosystems. This fact affects growth of crops and, more importantly, induces a greater number of stress factors affecting crop productivity. Hence greater emphasis in research on the interactive effects of stress factors, especially under field conditions, becomes important. The role of crop science in this scenario is very significant as it involves production. Most studies do not address the multiple interactive effects of stress factors. Therefore, adaptability and flexibility of crops to multiple stress factors need identification as these are the true conditions of nature where crops are grown for food and fibre.

Technological Innovations for Enhancing Stress Tolerance

Traditional farming methods prior to the intensification in the 1960s were mostly based on ecological principles. Constant utilization of all trophic levels to develop a more conducive environment for the crop was a very important concept of these systems, which were productive and sustainable. The use of fallow periods, green or animal manures, cropping and farming systems, mulching, integration of crops and animals, and matching crops to seasonal changes were all parts of these productive systems. With increasing pressure on available resources to produce greater quantities of food and fibre from existing or diminishing land and resources, traditional practices were abandoned. Artificial inputs were used to enhance productivity. This led to the under or non-utilization of natural ecological processes such as nutrient cycling, causing a greater degree of stress to crops. More importantly, the sustainability of cropping systems declined with time.

The recognition of the problem in the late 1970s led to the initiation of projects to revive the sustainability of cropping systems, especially in the tropics. New technologies, modification of old methods and integration of different systems were all evaluated in this process. Thus, current technological innovations to mitigate stress factors and maintain or possibly enhance crop productivity can be classified into three major groups:

1. Improved crop management and adoption of early systems after suitable modification to suit existing conditions.
2. Selection and development of crops to suit stress conditions.
3. Biotechnology, principally genetic engineering to develop crops suited to the environments.

Modification and adaptation of traditional practices to mitigate abiotic stress

A fundamental concept in crop production is the management of soil. Prior to the advent of chemical inputs, the nutritional status of soil was maintained by the application of organic matter, which provided a conducive physical, chemical and biological environment within the rhizosphere. The green revolution was based on an approach to provide external inputs for crop requirements and replenish losses in an easy, simplified and concentrated form (Jones *et al.*, 1997). However, in the tropics, nutrient losses from soils are generally high due to leaching and erosion. Hence, the soils do not sustain plant growth due to the inability to retain applied nutrients. This led to the placement of emphasis again on the use of organic matter and biological processes to optimize nutrient recycling, minimizing external inputs and maximizing efficiency of production systems. The final result has been the re-emphasis on maintaining soil quality in terms of adequate organic matter and development of appropriate physical, chemical and biological properties, through the use of crop residues on a scientific basis, a feature practised in the earlier times.

Improved management of soil and cropping systems on fragile ecosystems attempts to develop links towards nature and natural ecosystems (Lefroy *et al.*, 1999). This is due to the sustainability and resilience of natural

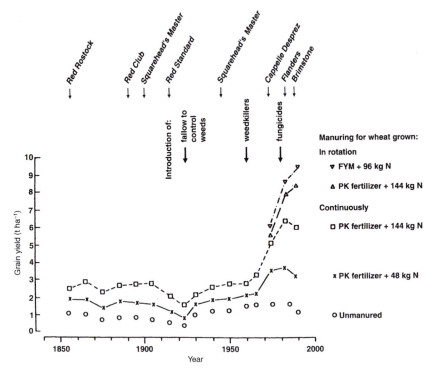

Fig. 7.1. Yields of winter wheat grown at Rothamsted with fertilizers and farmyard manure in a long-term experiment (from Johnston, 1994).

ecosystems, and the degraded nature of most intensive cropping systems lying adjacent to each other, especially in the tropics. However, long-term research does not show these effects clearly in the tropics, in contrast to that in the developed countries (Fig. 7.1). These studies clearly illustrate the benefits of maintaining balances in productivity and more importantly sustainability in cropping systems using traditional methods. Furthermore, the use of agroforestry systems which adapt traditional systems also has a significant impact in maintaining productivity. Research by Szott *et al.* (1991) highlights this concept in the tropics, where addition of loppings of trees enhances productivity of upland rice in marginal soils (Table 7.2). Hence studies on the development of crop production, especially in tropical regions, need to be based on the improvement of traditional or existing systems to maintain productivity and sustainability.

Many concepts have been developed to enhance productivity of tropical and even subtropical cropping systems. Amidst these, the use of organic matter especially in the form of manure and crop residues plays a prominent role. Studies in Africa (Fischler *et al.*, 1999) and Asia (Myers *et al.*, 1995) highlight the benefits of this method of restoring soil quality through the

Table 7.2. Grain yield of upland rice as affected by the addition of loppings of trees over four seasons. (Adapted from Szott *et al.*, 1991.)

Input applied	Addition rate per crop (kg ha^{-1})	Rice grain yields per crop (kg ha^{-1})			
		1	2	3	4
Inga edulis	6.7	1306	2235	1103	930
Erythrina spp.	6.7	2748	1718	1197	1303
Fertilizer		1844	2104	1159	1173
Control		1921	690	187	541

concept of integrated nutrient management systems. This in turn alleviates the stress factors of poor soil-water and nutrient retention, while improving soil microbial activity and thus providing a more conducive environment for crop growth.

Another aspect that is receiving attention for improved management of cropping systems, especially through the improvement of the rhizosphere, is the enhancement of the efficiency of soil microbial populations. The use of *Rhizobium* to enhance biological nitrogen fixation of legumes has been popular for many decades. More recent emphasis on the symbiotic processes of mycorrhizal fungi (Dodd, 2000) and endophytic bacteria (Sturz *et al.*, 2000) to develop sustainable cropping systems is gathering momentum. These processes enhance the efficiency of plant nutrient uptake and develop a more conducive environment for the root system by mitigating stress factors. The developments of external applications of microbial mixtures using selected species from the local ecosystem to avoid problems of contamination are also in progress. Senanayake and Sangakkara (1999) highlight the benefits of these methods in increasing crop production and environmental management through the development of soil quality and nutrient availability. The concept behind this system is the management of soil biota and this is being used to provide growth-promoting microbes, which stimulate plant growth, and suppression of disease (Swift *et al.*, 1998). Although there is a degree of concern over the use of these microbes from external environments, many commercial products do exist (Zarb *et al.*, 1999) and are being used successfully for enhancing crop growth and developing the rhizosphere. This again is a system for mitigating stress to plants.

The development of different cropping systems through techniques such as intercropping is another concept attracting attention, in order to improve productivity of tropical cropping systems (Fukai, 1993). This method of using mixtures of annual and/or perennial crops is based on the principle of utilizing different niches of an ecosystem to enhance productivity. The adoption of such systems by mimicking the structure and functions of natural ecosystems in the region helps increase productivity and also maintain biodiversity. Hence many studies highlight the value of these systems both in Asia (Trenbath, 1999) and Africa (Becker and Johnson, 1999), where productivity has been increased significantly by these systems.

While the benefits of mixed cropping of annual species received attention from the 1970s, a more recent development in this sector is agroforestry. Although research, especially in Africa, has paid a significant quantum of emphasis on this concept for over two decades, it again is an adaptation of earlier systems of slash and burn agriculture and perennial forest gardens, although on a more scientific basis. The benefits of incorporating trees, especially fast-growing legume species, lies in increasing soil fertility, obtaining complementarity in resource use, reducing environmental stress and protecting crops, thus enhancing yields (Ong and Leaky, 1999).

The crop scientists of the tropics have many options to enhance production even under the present scenario. However, there are priorities that could be developed on the basis of experience. The first would be to utilize water more efficiently, as this would be the most limiting factor for enhancing or even maintaining current yields. While research in the developed world concentrates on basic sciences, research in the developing world needs to focus on applied agronomic measures on crops, conserving soil moisture, and improving soils of the tropics. In crops, emphasis should be placed on aspects such as matching phenology with environments, especially water supply, developmental plasticity, mobilization of assimilates, leaf area maintenance and root growth patterns. In soil management, the methods could be diverse, and include options such as addition of organic matter from the surroundings or crop residues. New cropping systems which include trees as options and intensification of the more productive systems through crop combinations are also methods of enhancing productivity through modifications of existing systems, although these would be a longer-term measure as it involves detailed studies.

Selection – a criterion to mitigate abiotic stress

Species or varieties of species do survive and produce yields under a significant degree of abiotic stress. Natural vegetation also thrives in such situations. Research in plant abiotic stress has been biased towards overcoming the negative aspects (Grierson, 1999). Thus, research should consider what farmers of the traditional systems located in stressful environments do, to use the species and varieties best adapted to a given environment. This can be interpreted as the utilization of beneficial aspects of stress in developing methods to overcome stress.

Stress affects crops from seeding up to harvest, in the form of environmental changes. Thus, locating crop species, varieties or ecotypes from harsh environments and evaluating their mechanisms of stress tolerance could play a leading role in overcoming stress. The use of species such as millets and pigeon pea in the arid environments of Africa and Asia is a clear example of using the beneficial aspects of stress to improve crop productivity. Bänziger *et al.* (2001) states the usefulness of incorporating local selections in maize programmes in southern Africa to mitigate drought stress. The introduction of laws to protect natural vegetation from unwanted exploitation also emphasizes this factor that beneficial aspects of abiotic stress are important in mitigating stress factors in crops.

Biotechnology – a novel method for mitigating abiotic stress

Plant and gene modifications to improve plant productivity are cited as the next revolutions in agriculture. The numerous techniques that are currently available and are being developed have therefore given many opportunities to enhance crop productivity (Sonnewald and Herbers, Chapter 19, this volume). These methods coupled with precision agriculture could open up many possibilities to increase productivity in stressful environments, especially in the tropics (Edmeades *et al.*, Chapter 9, this volume).

Current research in biotechnology offers much hope, especially in the developing countries. However, at present this technology is based on random incorporation of genes. The studies at the International Future Harvest supported by CGIAR, especially IRRI, CIMMYT and CIAT, focus on yield improvements, phenology and stress tolerance, root systems and modifications, and many other physiological processes of staple food crops (Edmeades *et al.*, Chapter 9, this volume). However, as suggested by Blum (2000), all these technologies need testing under field conditions, especially farmers' fields of the developing world, where the need to enhance food production lies. This is a challenge faced by crop scientists of the future.

Integrated Management of Systems Under Stress Conditions

Modern trends in cropping systems research are bent on productivity and sustainability. This is due to the rapid degradation taking place in the current systems of intensive agriculture, especially that of cropping in tropical developing regions. Intensification of agriculture in these regions has often been carried out at a cost to the environment and resource base. Expansion of agriculture is on marshy lands, which are unable to sustain production, or by destroying forests which protect environments. Therefore, there is increasing pressure on farmers throughout the world, especially in the developing countries, to adopt systems that preserve the land and maintain productivity and sustainability (Turner, 1998).

Humans have little or no control over climate, therefore development of soil to sustain productivity is a very important feature in the current context of crop production. All practices adopted need to emphasize the requirements of maintaining soil productivity, and thus integrated management of abiotic resources becomes a vital factor in the current context of cropping. All practices adopted need to emphasize the benefits of integrated management.

The systems of cropping developed since the Vedic times in India and the pre-Christian era in the West adopted principles of matching climate, crops and seasons. This enabled the prolongation of the growing period, and could be considered the beginning of methodical development of cropping systems. Crop residues were incorporated thereby adding the much-required organic matter to maintain soil biological activity.

Intensification of agriculture moved away from these systems due to the development of species and varieties with high harvest indices. Current research highlights the need to adopt these ancient methods practised by farmers. Hence, emphasis is placed upon increasing biodiversity of agricultural systems and soils through the integration of either crops alone, crops and trees, crops and green manures, or even crops and animals (Pimbert, 1999). Studies on intercropping, relay cropping and even sequential cropping, along with developing systems of threshold levels in weed management have been useful in enhancing overall productivity of tropical agricultural systems. Recent research by Sharma and Ghosh (2000) illustrate that addition of 40 kg nitrogen ha^{-1} with *Sesbania aculeata* in lowland rice culture produced yields similar to those of conventional systems. Rao and Mathuva (2000) state that the addition of green manure through *Gliricidia sepium* loppings enhance productivity of maize by 27%, while intercropping with legumes enhanced productivity by 17–24%. Based on such studies, innovative farmers are adopting these systems. These are new scientific breakthroughs, although it should be stated that these are reinventions of old practices carried out over centuries by farmers in the tropics, prior to the green revolution. However, research is needed in the tropics on these themes, especially into the mechanisms of yield improvement and sustainability.

The Future – Challenges for Mitigating Abiotic Stress in Crop Production

The development of integrated systems of cropping to counteract the multiple stress factors in tropical crop production must identify the problems associated with maintaining productivity. Adopted systems must preserve water and increase its utilization efficiency. Navari-Izzo and Rascio (1999) highlight plant responses to water deficits. However, the impact of a major proportion of these studies on practical crop production is not readily reported. Emphasis is being placed on basic research for the advancement of knowledge on response of crops to stress and developing suitable systems of cropping for water stress conditions (Turner, Chapter 8, this volume). Studies on the mitigation of these factors under field conditions carried out in the form of applied research are needed, along with a reliable extension service. Studies do report the usefulness of agronomic methods such as tillage systems (Edwards *et al.*, 1999), mulching and addition of organic matter to maintain biodiversity and overcome soil-drought stress. However, systematic studies on the holistic nature of such methods are needed, both from an ecological point of view and for the new plants developed by biotechnology.

Crop production is also often limited by soil fertility. Although adequate water is made available, the inability of the soil to retain applied water and release it to crops makes all sophisticated irrigation systems a waste of capital. Hence, soil quality becomes a key factor in developing integrated systems to mitigate abiotic stress to plants. In regions where agroforestry is not feasible due to the lack of land, farmer preference and/or adaptability, systems of green manuring, organic farming and integration of animals and crops could play a successful role in mitigating soil fertility and thus help plants overcome abiotic stress. The use of suitable fertilizer systems, fertilization and other agronomic practices would help enhance productivity of marginal soils, as nutrients such as potassium help plants mitigate soil moisture and temperature stress (Sangakkara *et al.*, 1996; Johnston, 1997). The adoption of new methods such as fertigation systems especially in arid regions would ensure maximum benefits to farmers by supplying both water and nutrients in required quantities to maintain productivity and sustainability, while mitigating abiotic stress. Research does provide us with a multitude of integrated practices for producing crops under conditions of multiple stress found in all ecosystems. However, projects are just being initiated to test the validity of such programmes and methodologies for field use to maintain or possibly increase food production of cropping systems, especially in the developing world, while maintaining sustainability.

After the observation of negative effects of the green revolution on natural resources, the call arose for new systems of management. The anthropologists and sociologists studied age-old practices again as they survived the test of time, and stated that these methods have something to offer due to their sustainability. However, the recently developed systems of integrated management require large-scale testing prior to adoption. The new crops and varieties need to be evaluated under the field conditions of the developing countries. Although emphasis is today on basic research, there is a growing need to popularize and facilitate research on agronomic aspects of these crops and systems. These will include studies ranging from land preparation through seeding and propagation and management up to harvesting. It is only under such conditions that the people who produce food for increasing populations, especially in the tropical developing countries, would adopt integrated management systems of agriculture, which to date have had very little emphasis and research backing.

The future of success of crop production lies in the growth of crops with less environmental stress. Ecological models will help in the understanding of processes and indicate ways of developing suitable techniques. Thus emphasis needs to be placed on practical aspects of sophisticated techniques by crop scientists and agronomists, both in the developed and developing world.

Mitigating abiotic stress to ensure greater productivity does not mean research alone. At this time, when the world is entering into a new era of agricultural development, a call is made for cooperation, trade-offs between productivity, stability, sustainability and equality. This calls for the combination of many disciplines across natural and social sciences. Food availability is a social problem especially in the developing countries and hence cooperation in research between the developed world and the developing countries is one solution in using the available resources to the maximum in both worlds. It would also address the requirements of both farmers and the hungry, and develop appropriate research themes and agendas to optimize resources and also funds. The sociologically trained scientists of the respective countries could thereafter ensure the flow and impact of crop science research to the farming community and the world at large to produce more food under increasingly stressful conditions.

It is clear that the success of overcoming or even reducing abiotic stress lies in ecology and biotechnology. Hence, it is worthwhile looking at organic, or even low-input, agriculture as a solution to overcome stress factors. In the developing world, food requirements may not all be met with just pure organic farming, due to the fragility of the systems, which is different from that of the developed world, especially Europe. However, low-input systems could well be developed to maintain sustainability, using organic matter such as crop residues, which are available in the developing countries.

It must be accepted that biotechnology would also provide new plants with a greater potential under stressful conditions, but these plants have to be fitted into the existing ecosystems. Thus, the need arises to evaluate the plants and develop systems of cropping for the future. The modification of age old practices such as mixed and multiple cropping, agroforestry, use of organics and alternate tillage, fertilizer and water management using new concepts are all required and have a distinct role in mitigating stress to ensure production within fragile ecosystems while maintaining sustainability. It is only then that greater productivity from the new technologies developed to mitigate abiotic stress can be achieved, along with sustainability especially in fragile ecosystems of the developing countries.

Acknowledgements

Gratitude is expressed to the ETH Zurich, Switzerland, especially the Grassland Sciences and Crop Physiology group, for the facilities given for preparing this paper. The financial support granted by the ETH Zurich for a visiting fellowship is acknowledged.

References

Agarwal, A. and Narain, S. (1999) Making water management everybody's business: Water harvesting and rural development in India. *Gatekeeper Series 87.* International Institute for Environment and Development, London, pp. 1–20.

Bänziger, M., Damu, N., Chisenga, M. and Mugabe, F. (2001) Evaluating drought tolerance of some popular maize hybrids grown in sub-Saharan Africa. In: *Maize Production Technology for the Future: Challenges and Opportunities. Proceedings of the 6th Eastern and Southern Africa Regional Maize Conference.* CIMMYT and EARO, Addis Ababa.

Becker, M. and Johnson, D. (1999) The role of legume fallows in intercropped upland rice based systems in West Africa. In: Balasubramaniam, V., Ladha, J.K. and Denning, G.L. (eds) *Resource Management in Rice Systems – Nutrients.* Kluwer Academic Publishers, Dordrecht, pp. 105–120.

Berzsenyi, Z., Gyorffy, B. and Lap, D. (2000) Effect of crop rotation and fertilization on maize and wheat yields and yield stability in a long-term experiment. *European Journal of Agronomy* 13, 225–244.

Blum, A. (2000) Towards standard assays of drought resistance in crop plants. In: Ribaut, J.M. and Poland, D. (eds) *Proceedings of a Workshop on Molecular Approaches for the Genetic Improvement of Cereals for Stable Production in Water Limited Environments.* CIMMYT, Mexico.

Bouwer, H. (2000) Integrated water management – emerging issues and challenges. *Agricultural Water Management* 45 (in press).

Champingny, M.L. and Mosseau, M. (1999) Plant and crop response to trends in climatic change. In: Pessarakli, M. (ed.) *Handbook of Plant and Crop Stress.* Marcel Dekker, New York, 1087–1127.

Dodd, J.C. (2000) The role of arbuscular mycorrhizal fungi in agro and natural ecosystems. *Outlook on Agriculture* 29, 55–62.

Edmeades, G., Bolano, J., Bänziger, M., Ribaut, J.M., White, J.W., Reynolds, M.P. and Lafitte, A.R. (1998). Improving crop yields under water deficits in the tropics. In: Chopra, V.L., Singh, R.B. and Varma, A. (eds) *Crop Productivity and Sustainability. Proceedings of the 2nd International Crop Science Congress.* Oxford and IBH Publishers, New Delhi, pp. 437–451.

Edwards, P.J., Abivardi, C. and Richner, W. (1999) The effects of tillage systems on biodiversity on agroecosystems. In: Wood, D. and Lenné, J.M. (eds) *International Agrobiodiversity.* CAB International, Wallingford, pp. 305–329.

FAO (Food and Agriculture Organization) (1999) *Poverty Alleviation and Food Security in Asia.* FAO Regional Office for Asia and the Pacific, Bangkok, pp. 1–92.

Fischler, M., Wortmann, C.S. and Feil, B. (1999) *Crotolaria (C. ochroleuca* G Don) as a green manure in maize bean cropping system in Uganda. *Field Crops Research* 61, 97–102.

Fukai, S. (1993) (ed.) Intercropping – basis of productivity systems. *Field Crops Reseach (Special Issue)* 34, 239–467.

Grierson, W. (1999) Beneficial aspects of stress. In: Pessarakli, M. (ed.) *Handbook of Plant and Crop Stress.* Marcel Dekker, New York, pp. 1185–1198.

Hulme, M. and Viner, D. (1998) A climate change scenario for the tropics. *Climate Change* 39, 145–176.

Johnston, A.E. (1994) The Rothamsted classical experiments. In: Leigh, R.A. and Johnston, A.E. (eds) *Long-term Experiments in Agricultural and Ecological Sciences.* CAB International, Wallingford, pp. 9–38.

Johnston, A.E. (1997) Food security in the WANA region – the essential need for fertilizers. In: Johnston, A.E. (ed.) *Food Security in the WANA Region – The Essential Need for Balanced Fertilization.* International Potash Institute, Bern, pp. 11–30.

Jones, R.B., Snapp, S.S. and Phombeya, H.S.K. (1997) Management of leguminous leaf residues to improve nutrient use efficiency in the sub humid tropics. In: Cadish, G. and Giller, K.E. (eds) *Driven by Nature – Plant Litter Quality and Decomposition.* CAB International, Wallingford, pp. 239–250.

Khush, G.S. (1999) Green revolution, preparing for the 21st century. *Genome* 42, 646–655.

Kumar, K. and Goh, K.M. (2000) Crop residues and management practices, effects on soil quality, soil nitrogen dynamics, crop yield and nitrogen recovery. *Advances in Agronomy* 68, 197–319.

Lefroy, E.C., Hobbs, R.J., O'Conner, M.H. and Pate, J.S. (1999) What can agriculture learn from natural ecosystems? *Agroforestry Systems* 45, 423–436.

Lüscher, A., Hendry, G.R. and Nösberger, J. (1998) Long term responses to free air CO_2 enrichment of functional types and genotypes of permanent grasslands. *Oecologia* 113, 37–45.

Monteith, J.L. and Ingram, J.S.C. (1998) Climate variation and crop growth. In: Chopra, V.L., Singh, R.B. and Varma, A. (eds) *Crop Productivity and Sustainability. Proceedings of the 2nd International Crop Science Congress.* Oxford and IBH Publishers, New Delhi, pp. 57–70.

Morilley, L., Hartwig, U.A. and Aragno, M. (1999) Influence of an elevated atmospheric CO_2 content on soil and rhizosphere bacterial communities beneath *Lolium perenne* and *Trifolium repens* under field conditions. *Microbial Ecology* 38, 39–49.

Myers, R.J.K., van Noordwijk, M. and Vittyakon, P. (1997) Synchrony of nutrient release and plant demand, plant litter quality, soil environment and farmer management options. In: Cadish, G. and Giller, K.E. (eds) *Driven by Nature – Plant Litter Quality and Decomposition.* CAB International, Wallingford, pp. 215–229.

Navari-Izzo, P. and Rascio, N. (1999) Plant response to water deficit conditions. In: Pessarakli, M. (ed.) *Handbook of Plant and Crop Stress.* Marcel Dekker, New York, pp. 231–270.

Ong, C.K. and Leakey, R.R.B. (1999) Why tree crop interactions in agroforestry appear at odds with tree grass interactions in tropical savannas. *Agroforestry Systems* 45, 109–129.

Pessarakli, M. (1999) (ed.) *Handbook of Plant and Crop Stress.* Marcel Dekker, New York.

Pimbert, M. (1999) Sustaining the multiple functions of biodiversity. *Gatekeeper Series 88.* International Institute for Environment and Development, London, pp. 1–24.

Rao, M.R. and Mathuva, M.N. (2000) Legumes for improving maize yields and income in semi arid Kenya. *Agriculture Ecosystems and Environment* 78, 123–137.

Rozenweig, C. and Parry, M.C. (1997) Potential impact of climate change on world food supply. *Nature* 376, 13–138.

Sangakkara, U.R. (1998) Growth and yields of cowpea (*Vigna unguiculata* L. Walp) as influenced by seed characters, soil moisture and season of planting. *Journal of Agronomy and Crop Science* 180, 137–142.

Sangakkara, U.R. and Kandapola, C.S. (2000) Organic matter affects the retention of applied nitrogen fertilizers. In: *Proceedings of the FAO/IAEA International Symposium on Nuclear Techniques in Integrated Plant Nutrient, Water and Soil Management*. IAEA, Vienna.

Sangakkara, U.R, Hartwig, U.A. and Nösberger, J. (1996) Growth and symbiotic nitrogen fixation of *Vicia faba* and *Phaseolus vulgaris* as affected by fertilizer potassium and temperature. *Journal of the Science of Food and Agriculture* 70, 315–320.

Senanayake, Y.D.A. and Sangakkara, U.R. (1999) (eds) *Proceedings of the Fifth International Conference on Kyusei Nature Farming*, Thailand. International Nature Farming Research Center, Atami.

Sharma, A.R. and Ghosh, A. (2000) Effect of green manuring with *Sesbania aculeata* and nitrogen fertilization on the performance of direct seeded flood prone lowland rice. *Nutrient Recycling in Agroecosystems* 57, 141–153.

Sturz, A.V., Christie, B.R. and Rowak, J. (2000) Bacterial endophytes, potential role in developing sustainable systems of crop production. *Critical Reviews in Plant Sciences* 19, 1–30.

Swift, M.J., Mafongoya, P. and Ramakrishnan, P.S. (1998) Soil biodiversity – an essential foundation for sustainable soil fertility. In: Chopra, V.L., Singh, R.B. and Varma, A. (eds) *Crop Productivity and Sustainability. Proceedings of the 2nd International Crop Science Congress*. Oxford and IBH Publishers, New Delhi, pp. 321–333.

Swindale, L.D. (1998) Agroecosytems approaches to sustainable productivity. In: Chopra, V.L., Singh, R.B. and Varma, A. (eds) *Crop Productivity and Sustainability. Proceedings of the 2nd International Crop Science Congress*. Oxford and IBH Publishers, New Delhi, pp. 293–304.

Szott, L.T, Palm, C.A. and Sanchez, P.A. (1991) Agroforestry in acid soils of the humid tropics. *Advances in Agronomy* 45, 275–290.

Trenbath, B.R. (1999) Multispecies cropping in India – predictions of their productivity, stability and resilience. *Agroforestry Systems* 45, 81–107.

Turner, N.C. (1998) Managing stress – is there a productive future. In: Chopra, V.L., Singh, R.B. and Varma, A. (eds) *Crop Productivity and Sustainability. Proceedings of the 2nd International Crop Science Congress*. Oxford and IBH Publishers, New Delhi, pp. 542–544.

Zarb, J., Leifert, C. and Litterick, A. (1999) Opportunities and challenges for the use of microbial inoculants in agricultural practices. In: *Proceedings of the Sixth International Conference on Kyusei Nature Farming*, Pretoria. International Nature Farming Research Centre, Atami.

Optimizing Water Use

8

N.C. Turner

CSIRO Plant Industry, Private Bag No. 5, Wembley, Perth, WA 6913, Australia

Introduction

While water on a global scale is plentiful, 97% is saline and 2.25% is trapped in glaciers and ice, leaving only 0.75% available in freshwater aquifers, rivers and lakes. Most of this fresh water (69%) is used for agricultural production, 23% for industrial purposes and 8% for domestic purposes (Prathapar, 2000).

Water scarcity is being increasingly accepted as a major limitation for increased agricultural production and food security in the 21st century. Recent studies suggest that nearly 1.4 billion people live in regions that will experience water scarcity within the first 25 years of this century (Seckler *et al.*, 1998). More than 1 billion people live in regions that will face absolute water scarcity by 2025. These are in countries that do not have sufficient annual water resources to meet reasonable per capita water needs (Fig. 8.1). Additionally, about 350 million people live in countries that have sufficient water resources, but lack the financial resources to embark on the massive and rapid development programmes necessary to utilize them.

Of concern is that of the 1.4 billion facing water scarcity, a third live in developing countries, particularly the semiarid regions of Asia and sub-Saharan Africa that include the majority of the malnourished population of the world. The Green Revolution that significantly reduced malnutrition and poverty in the 1970s and early 1980s relied on irrigation and high inputs to increase crop yields. Much of this irrigation water has been sourced from freshwater aquifers that are now being rapidly depleted or contaminated with salt or other pollutants. Of particular concern is the decline in the water table in two of Asia's major breadbaskets, namely, the Punjab of India and the North China Plain (Seckler *et al.*, 1999). The pressure for water for

© 2001 CAB *International. Crop Science*
(eds J. Nösberger, H.H. Geiger and P.C. Struik)

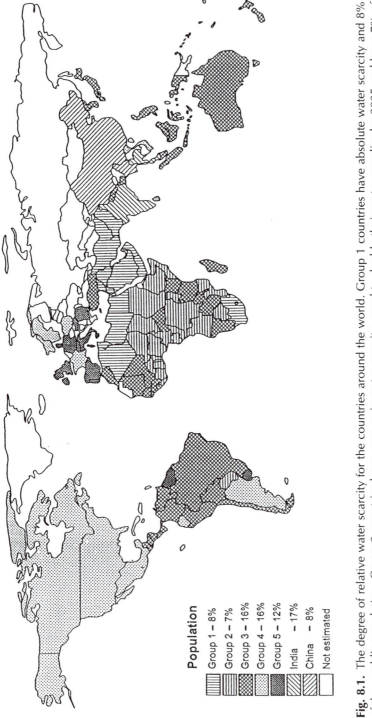

Fig. 8.1. The degree of relative water scarcity for the countries around the world. Group 1 countries have absolute water scarcity and 8% of the world's population, Group 2 countries have economic water scarcity, need to double their water supplies by 2025 and have 7% of the world's population, Group 3 countries have economic water scarcity, need to increase their water supply by 25 to 100% and constitute 16% of the world's population, Group 4 countries (16% of the world's population) and Group 5 countries (12% of the world's population) have only modest requirements (5 to 25%) for increased water. India and China are shown separately as they have areas of both high and low water scarcity. From Seckler *et al.* (1998) and Prathapar (2000).

household and industrial use and the many pressures against large irrigation schemes are likely to see less, rather than more, water available for agricultural use.

Thus, increased food production in regions of the world where rainfall is limited and crop production is marginal will need to occur in the face of limited, or even reduced, availability of water for irrigation. This will require a concerted and coordinated effort to improve the efficiency of irrigation, to improve crop water use efficiency, i.e. the biomass or economic yield per unit of evapotranspiration by the crop, and to improve the productivity of dryland (rainfed) cropping systems in water-limited environments. Improvement in water use efficiency and productivity of dryland cropping systems can be achieved by genetic improvement of the crops and also by improved management of cropping systems.

Breeding for Improved Drought Resistance and Water Use Efficiency

In water-limited environments, crop yield (CY) is a function of water use (ET), water use efficiency (WUE) and the proportion of total above-ground biomass that contributes to economic yield, i.e. the harvest index (HI) (Passioura, 1977):

$$CY = ET \times WUE \times HI \qquad (8.1)$$

Water use or evapotranspiration has two components, crop transpiration (Ec) and soil evaporation (Es). Thus crop yield is better given by:

$$CY = Ec \times TE \times HI \qquad (8.2)$$

where TE is the transpiration efficiency, the efficiency of water used by the crop itself (Fischer, 1981).

The yield of a crop in water-limited environments can, therefore, be improved by increasing water use, increasing transpiration efficiency, increasing the harvest index, or decreasing soil evaporation. Breeding for these characteristics and managing the crop to improve these characteristics have both been employed in recent years.

Empirical breeding for improved yields in water-limited environments has been slow due to the year-to-year variation in rainfall and within-season variation in rainfall distribution at dryland sites. This has led to attempts to breed for specific phenological, morphological, physiological or biochemical characteristics that putatively improve yields in water-limited environments. Turner (1997) and Turner *et al.* (2001) discuss a range of characteristics that have been identified to improve the drought resistance of crops, namely, early vigour, transpiration efficiency, stomatal control, abscisic acid accumulation, proline accumulation, osmotic adjustment, carbon remobilization, rapid grain growth, changes in hydraulic conductance, membrane stability and lethal water potential. However, few of the characteristics have been clearly demonstrated to improve yields and even fewer have been used in mainstream breeding programmes. In part this is because there is rarely one characteristic or trait that improves or maintains yield in water-limited environments and certainly not a single drought-resistance gene that can be identified and inserted into the germplasm to increase yields in all drought-prone environments. Additionally, screening for drought-resistance traits is often difficult and there are feedbacks such that an improvement in one characteristic can lead to a decline in others (Turner et al., 2001).

One example in which drought resistance has been improved by changing one characteristic is given in a subsequent chapter. Edmeades et al. (Chapter 9, this volume) show that decreasing the interval between anthesis and silking in maize increases yields in drought-prone environments and this characteristic, which can be readily screened in large populations, has been incorporated into breeding programmes for improved tropical maize.

The discovery that the isotopic signature of the carbon in the leaves of wheat could be used to identify variation among genotypes in transpiration efficiency (Farquhar and Richards, 1984) engendered considerable interest because it provided a relatively simple tool for breeders interested in selecting for improved water use efficiency. Subsequent studies showed that there was a negative relationship between transpiration efficiency and carbon isotope discrimination across a wide a range of species (Turner, 1993, 1997). Backcrossing genotypes with superior transpiration efficiency into locally adapted cultivars has produced genotypes of wheat with up to 10% higher yields at low-yielding sites with predominantly summer rainfall where the crop grows through the winter and spring primarily on stored soil moisture (Condon et al., 2002). Moreover, high transpiration efficiency showed no negative effects at higher-yielding sites and in winter-rainfall Mediterranean climates in which transpiration efficiency has no advantage (Fig. 8.2). Some of the high transpiration efficiency backcross lines are now being widely evaluated as potential new cultivars in wheat breeding programmes in the predominantly summer-rainfall regions of Australia (Condon et al., 2002).

However, selection, as opposed to evaluation, for low values of carbon isotope discrimination is not part of any mainstream breeding programme as far as this author is aware. This is because of the high cost of measuring carbon isotope discrimination in a large breeding population and because in some environments and species increased transpiration efficiency is not correlated or even negatively correlated with yield (Turner, 1993; Condon et al., 2002). Indeed, in irrigated environments reduced transpiration efficiency and high stomatal conductance may be desirable in order to maximize biomass and reduce the length of the growing season (Condon et al., 2002).

While the basis for this is clear in irrigated agriculture, it can also be important in water-limited environments. Turner et al. (1989) showed that

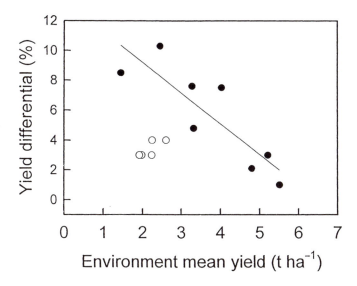

Fig. 8.2. Average yield differences between 30 backcross breeding lines selected for high transpiration efficiency (low carbon isotope discrimination) and 30 breeding lines selected for low transpiration efficiency when grown over 3 years in eight environments in eastern Australia (●) and five environments in Western Australia (○). From Condon *et al.* (2002).

increased early vigour and biomass at anthesis in water-limited Mediterranean climates can lead to increased yields and water use efficiency in wheat (Turner and Nicolas, 1998). While improved transpiration efficiency is an important characteristic in some water-limited environments, particularly those in which water saving during vegetative growth is important for survival and yield, other characteristics are important in other environments. When low temperatures and radiation limit early growth and the development of biomass, characteristics such as vigorous early growth while vapour pressure deficits are low will be more useful in boosting yields under terminal drought (Richards *et al.*, 2002).

Developments in molecular biology have provided a means whereby drought resistance traits may be more readily identified and tracked by the use of molecular markers. This methodology is now being applied with single gene traits for disease and insect resistance. Quantitative trait loci (QTLs) and amplified fragment length polymorphisms (AFLPs) for a range of drought resistance characteristics have been identified (Kleuva *et al.*, 1998; Richards *et al.*, 2002), but are not currently being widely used to speed up breeding for improved drought resistance. The impending water scarcity, especially in parts of the developing world where malnourishment is high, indicates that it is imperative that modern molecular technologies are quickly adopted in drought resistance breeding (Lipton, 1999).

Water Use in Dryland Cropping Systems

In semiarid regions, dryland crops are either grown in the rainy season(s) and/or rely on stored soil moisture. Where crops rely on incident rainfall, its distribution frequently results in both water shortage and water excess. In unimodal or bimodal rainfall patterns of the semiarid tropics, indeterminate rains result in water deficits occurring at any time during crop growth (Turner *et al.*, 2001), but high rainfall intensities result in water bypassing the root zone and crops using only a proportion of the growing-season rainfall. In Mediterranean cropping zones, crop growth frequently fails to match the seasonal rainfall distribution (Turner, 1993) resulting in terminal drought during seed filling. Taking southern Australian dryland cropping systems as an example, it can be shown that almost all the growing-season rainfall is utilized during crop growth. Nevertheless, some drainage below the root zone occurs at wetter sites and seasons, leading to recharge of the water table (Hatton and Nulsen, 1999). This deep drainage is exacerbated by large but infrequent rainfall events outside the growing season. Smettem (1998) showed that the annual deep drainage in south-west Australian agricultural systems increased from an average of about 20 mm per year at 300 mm annual rainfall to 100 mm at 1100 mm annual rainfall (Fig. 8.3). The amount of deep drainage depended on soil type, as well as rainfall, with coarse-textured sandy soils having greater deep drainage than fine-textured clay soils at a particular annual rainfall (Smettem, 1998; Asseng *et al.*, 2001b). In the ancient, flat, poorly transmissive landscapes of southern Aus-

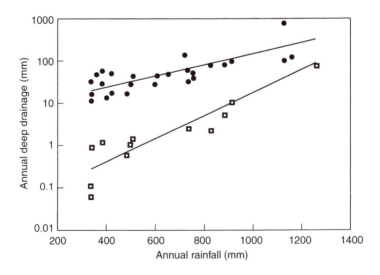

Fig. 8.3. Relationship between deep drainage (recharge) and rainfall for agricultural (●) and natural (□) systems in Western Australia. From Smettem (1998).

tralia, this loss of water below the root zone not only results in reduced yields per unit of rainfall, but also leads to rising water tables (Sadler and Turner, 1994; George *et al.*, 1997; Hatton and Nulsen, 1999).

Rising water tables result in an increased incidence of waterlogging lower in the landscape during the wet season(s). Additionally, in the regions of southern Australia with nutritionally poor, deeply weathered soil profiles the groundwater is saline, leading to deep drainage inducing secondary salinity. Recent estimates suggest that currently 10% of the cropping zone of south-western Australia is affected by salinity, that this percentage will double in the next 10–20 years, and that up to a third of the region may be affected by secondary salinity at equilibrium (George *et al.*, 1997).

Management Options for Improved Water Use and Water Use Efficiency

In annual cropping systems there is considerable scope for increasing water use efficiency, i.e. biomass or economic yield per unit of evapotranspiration, or rainfall use efficiency, i.e. biomass or economic yield per unit of growing-season rainfall. However, the scope for increasing seasonal water use is limited. Early sowing, nitrogen fertilizer use, inclusion of early-vigour genotypes and high planting density have all been suggested as methods for improving water use efficiency (Richards *et al.*, 2001), but in almost all cases the increased water use efficiency arose from increased yields not from increased water use. Recent studies by Eastham *et al.* (1999) showed that early sowing reduced water losses by soil evaporation, particularly early in the season before the crop was well established, resulting in higher rainfall use efficiency in both wheat and lupin crops. Likewise, increasing various inputs to annual pastures increased pasture yields and water use efficiency, but had only a small effect on water use (Bolger and Turner, 1999).

The season-to-season variability and the site-to-site variability make it difficult to generalize on the likely benefits of management practices on water use efficiency. However, Asseng *et al.* (2001a) used a well-tested simulation model to evaluate the impact of nitrogen fertilizer use on wheat yields and water use efficiencies in two soil types and three Mediterranean-type rainfall environments varying from 310 to 460 mm average annual rainfall. Calculated yields and water use efficiencies for up to 87 seasons were simulated using rainfall and temperature data for these seasons. The simulations showed that the increased use of nitrogen fertilizer doubled water use efficiency (wheat yield per millimetre of evapotranspiration) on average from 5 kg ha^{-1} mm^{-1} with no added nitrogen up to 10 kg ha^{-1} mm^{-1} at high nitrogen fertilizer use. The increase in water use efficiency was greater on a deep sand than on a

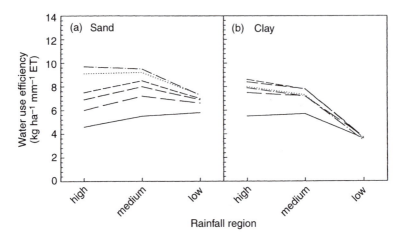

Fig. 8.4. Simulated long-term average water use efficiency for wheat crops on (a) a deep sand and (b) a clay soil, over three rainfall zones (high: 461 mm year⁻¹ average; medium: 386 mm year⁻¹ average; and low: 310 mm year⁻¹ average) with 0 (———), 30 (— —), 60 (– –), 90 (– – – –), 150 (..........) and 210 (– · – · –) kg N ha⁻¹ of applied nitrogen. From Asseng *et al.* (2001a).

clay soil and was greater at high (460 mm average rainfall) and medium (386 mm) than at low (310 mm) rainfall sites (Fig. 8.4). In the low rainfall region the water use efficiency was low (4 kg ha⁻¹ mm⁻¹) and was not affected by increased fertilizer use. The use of average values in Fig. 8.4 hides the wide range of efficiencies observed over the 87 years which varied from near 0 kg ha⁻¹ mm⁻¹ to 18 kg ha⁻¹ mm⁻¹ (Fig. 8.5). The variation was greatest at high levels of nitrogen as a result of yield reductions at high levels of applied nitrogen, especially on clay soils, in years with low rainfall. In these years, the high nitrogen regime produced high leaf areas and high water use prior to anthesis, resulting in the crops running out of water during grain filling (Asseng *et al.*, 2001a).

While management options are available to increase yields and water use efficiency of annual crops and pastures, they usually have only a small effect on growing-season evapotranspiration or on deep drainage. Figure 8.6 shows the cumulative probabilities of drainage below the root zone of wheat crops in both the same deep sand and clay soil as in Fig. 8.5. As shown by Smettem (1998), deep drainage increased with increasing rainfall and was greater in the sand than in the clay (Asseng *et al.*, 2001b). Early sowing reduced the calculated deep drainage by less than 10% in all rainfall environments and in both soil types (Table 8.1), while increasing the nitrogen fertilizer rate sevenfold doubled the yield of wheat but reduced deep drainage by less than 10% (Asseng *et al.*, 2001c). Increases in water use early in the season often resulted in decreased availability of water and

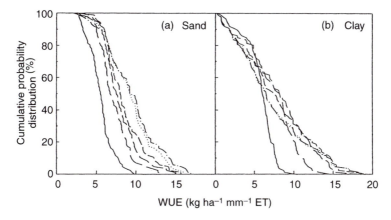

Fig. 8.5. Simulated cumulative probability distribution (based on 81 to 87 years of weather records) for water use efficiency of wheat crops on (a) a deep sand and (b) a clay soil, for 0 (———), 30 (— —), 60 (— —), 90 (– – – –), 150 (..........) and 210 (– · — · –) kg N ha^{-1} of applied nitrogen. From Asseng *et al.* (2001a).

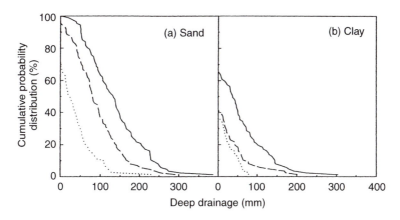

Fig. 8.6. Simulated cumulative probability distribution (based on 81 to 87 years of weather records) for annual deep drainage below a wheat crop on (a) a deep sand and (b) a clay soil, at a high (461 mm year^{-1} average, ———), medium (386 mm year^{-1} average, — —) and low (310 mm year^{-1} average, ...) rainfall zones. From Asseng *et al.* (2001b).

reduced evapotranspiration later in the season. Thus, improving crop productivity and water use efficiency will be beneficial for the farmer, but has little impact on deep drainage, rising water tables, waterlogging or secondary salinity.

Table 8.1. Simulated long-term average annual values of drainage below the root zone for wheat crops given 30 kg N ha^{-1} and sown at three different days of year (DOY) into two soil types at three zones of average annual rainfall in south-western Australia. After Asseng *et al.* (2001c).

Soil type	Rainfall (mm)	Sowing date		
		DOY 135	DOY 155	DOY 175
Sand				
	460	131	138	142
	386	86	92	97
	310	34	37	40
Clay				
	460	54	56	57
	386	23	25	26
	310	4	4	4

Role of Perennials in the Cropping System in Increasing Water Use and Decreasing Deep Drainage

To reduce deep drainage under annual cropping systems requires a significant increase in water use through a change in farming systems. The native systems in regions cleared for agriculture where secondary salinity is an increasing problem were usually woody, deep-rooted perennial shrubs and trees (Sadler and Turner, 1994; Turner, 1998; Hatton and Nulsen, 1999). These native systems have significantly lower deep drainage than agricultural systems in the same rainfall zones (Fig. 8.3). To reduce deep drainage it is now recognized that some perennial vegetation is required in at least part of the landscape (Hatton and Nulsen, 1999). To replace the agricultural systems with species that use an equivalent amount of rainfall will require reafforestation of the landscape and in high rainfall regions above 600 mm annual rainfall, where reafforestation is economically viable, this is already beginning to occur.

However, in order to maintain crop production while increasing water use, alternative systems are being evaluated and adopted. Agroforestry systems enable deep-rooted trees to be grown in parts of the landscape while pastures or crops are grown between the belts of trees. While there are several studies of water use in agroforestry systems, two will suffice here. Recent studies have shown that the fodder tree tagasaste (*Chamaecytisus proliferaa* Link var. *palmensis*) grown as single rows 30 m apart removed 140 mm of the 220 mm in the 3.7 m deep sand profile compared with 100 mm that was removed by a cereal crop (Lefroy and Stirzaker, 1999). However, the trees extracted water from only a narrow 4 m zone of soil immediately on either side of the belt of trees and had both positive and negative effects on the crops in the alleys depending on season and whether the crop was a cereal or legume (Lefroy and Stirzaker, 1999). Lefroy and Stirzaker (1999) calculated that double the numbers of rows of trees (halving the inter-row distance to 15 m) would be required to prevent water table rise at the site that has a 460 mm average annual rainfall. In another study in a similar rainfall region, an 8 m wide belt of mixed *Eucalyptus* species, each belt spaced 200 m apart, dried the soil well below 3 m and used 150 mm of water more than rainfall over an annual cycle (White *et al.*, 2001). Other evidence suggested that two of the four species accessed the groundwater in a transmissive aquifer below the trees (White *et al.*, 2000). This suggests that the trees are able to remove an additional 4 mm of rainfall over the catchment. While this is helpful, it is inadequate in removing the calculated 30 mm of average deep drainage for the rainfall zone (Smettem, 1998).

An alternate strategy that is now being widely explored is the use of perennial pasture species, such as lucerne, in the farming system. Studies in south-western Australia have shown that water use by lucerne was 50 mm more than by annual crops and pastures at a site where the annual rainfall was 480 mm (Ward *et al.*, 2001). This depleted soil water, particularly in the clay horizon below the top 0.4 m of sandy soil occupied by the majority of the roots of annual crops or pastures. Over three seasons, soil water use by lucerne in the 0.4–1.2 m depth of the soil profile was about 70 mm more than under the annual pasture. Similar results have been observed for lucerne at a wetter site (600 mm annual rainfall) in eastern Australia (Dunin *et al.*, 1999). The soil water depletion by the lucerne provided a reservoir for out-of-season rainfall that was used for lucerne regrowth and provided a buffer to reduce deep drainage in subsequent annual crops. Dunin *et al.* (1999) calculated that water use by lucerne provided about a 3-year buffer for subsequent crops. This suggests that a 3-year lucerne phase followed by a 3-year phase of annual crops will minimize deep drainage and provide a sustainable cropping system for semiarid environments.

Improving the Efficiency of Irrigation

There are two aspects to improving irrigation efficiency, namely: (i) improving the storage, delivery and uniformity of irrigation water to the crop; and (ii) improving water use by the crop itself. The first of these is beyond the scope of the present chapter, but is discussed fully in several chapters in Hoffman *et al.* (1990) and briefly in Prathapar (2000).

Improving the water use by irrigated crops requires treating water as a scarce and valuable resource. Watering to excess can lead to poor yields from waterlogging, rising water tables, off-site pollution and loss of productive land from salinization (Stirzaker, 1999). Sadler and Turner (1994) document historical examples of failed irrigation systems from salinization. Successful irrigation systems more closely balance irrigation supply to plant requirement in both time and space (Stirzaker, 1999). Indeed, many studies have shown that crops can withstand considerable soil water depletion before water deficits affect leaf area development, photosynthesis and other physiological processes (Turner, 1990a), while mild water deficits can in some circumstances increase yields (Turner, 1990b). In many crops two-thirds of the extractable soil water in the root zone can be used before evapo-transpiration and physiological activity are reduced (Turner, 1990a). As a consequence, the frequency of irrigation can be reduced until these processes are just affected. In extreme cases, a single irrigation at a critical stage near flowering and initiation of seed filling can have a profound effect on final seed yield. Deficit irrigation (English *et al.*, 1990) is now widely used in the horticultural industry, as it not only provides more efficient use of irrigation water but also reduces labour costs for pruning and often improves product quality. It has also been proposed for crop production in Mediterranean climates in which limited water is available for irrigation (Oweis *et al.*, 1998).

The monitoring of soil water deficits for irrigation scheduling is now widely used even though the location of sensors in the soil and the critical soil water status for initiation of irrigation are difficult to determine. More-over, methods for determining plant water stress are not as readily auto-mated and have, therefore, not been widely adapted. The use of infrared thermometry for sensing canopy temperature and calculation of the crop water stress index (Jackson, 1982) can be used after canopy closure, while sap flow sensors and covered leaf (stem) water potential have been suggested for accurate and site-specific irrigation scheduling for high-value woody horticultural species (Shackel *et al.*, 1997). Development of simple, reliable, automated methods of determining plant water status and irrigation need is urgently required.

Conclusions

The increasing world population, coupled with water and land scarcity, will put increasing demands on agricultural scientists to increase crop yields in marginal environments in which abiotic stresses, particularly water deficits, prevail. Breeding for improved drought resistance can demonstrably increase yields in dry environments, but has not been widely adopted, in part because drought resistance traits are difficult to screen in large populations. The use of molecular markers to select drought-resistance traits should speed up direct breeding for such characteristics, provided the focus is on traits, crops and processes that aid resource-poor farmers in marginal environments. Unless these technologies are quickly and deliberately applied to improving the nutritional requirements and drought resistance of crops, the future of the world's poor looks bleak (Lipton, 1999).

In addition to breeding for improved water use and drought resistance, it is essential to improve crop management and cropping systems to better use the limited water available, as either irrigation water or incident rainfall. There is considerable scope for reduced losses of irrigation water prior to its delivery to the crop and also for improved irrigation efficiency by matching water supply to plant requirement. Yields and water use efficiencies have been increased in water-limited environments by earlier or more timely sowing, strategic use of fertilizers and better matching of cultivars to the growing season. While yield increases have not been as dramatic in rainfed crops in semiarid environments as in temperate environments, sustained yield improvements have been demonstrated (Fig. 8.7). Yields of dryland wheat grown in semiarid regions of Australia have continued to increase consistently at about 13 kg ha^{-1} year^{-1} over the past six decades as a result of both improved genotypes and improved management in almost equal proportion (Turner and Whan, 1995). The continued application of new technologies in breeding and better understanding of changes in water use and water use efficiency induced by management will enable crop yields to be improved in marginal water-limited environments. The use of simulation models will enable better management combinations to be developed and evaluated in the future. However, in many such environments, not all the rainfall is utilized by the crop, particularly out-of-season rainfall, resulting in off-site environmental degradation. New farming systems to better use all

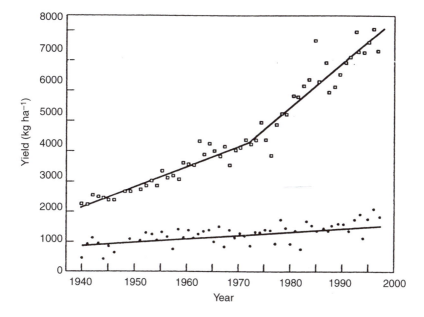

Fig. 8.7. National average wheat yields from 1940 onwards for a country with plentiful rainfall and mild climate (UK, □) and for a country with a water-limited environment (Australia, ●). Source: FAO Production Yearbooks.

the incoming rainfall are required. The development of agroforestry systems and the use of perennial crops and pastures that are deeper rooted and can utilize the out-of-season rainfall will be required.

Challenges

The challenges for the future are:

1. To apply modern breeding technologies to improve the yields and drought resistance of crops, especially minor crops grown by resource-poor farmers, under marginal water-limited environments.
2. To develop productive deep-rooted, short-term perennial crops that can utilize rainfall outside the main growing season(s).
3. To develop inexpensive and relatively simple irrigation monitoring systems based on crop performance rather than soil water monitoring to enable irrigation water to be more efficiently utilized.
4. To develop methodologies to understand the impact of management options on water use, water use efficiency and crop yields in very variable water-limited environments.

Acknowledgements

Mr M. Poole and Drs J.A. Palta and S. Asseng are thanked for helpful comments on the manuscript, and Drs S. Asseng, A.G. Condon, R.A. Richards, P.R. Ward and D.A. White are thanked for access to papers prior to publication.

References

Asseng, S., Turner, N.C. and Keating, B.A. (2001a) Analysis of water- and nitrogen-use efficiency of wheat in a Mediterranean climate. *Plant and Soil* 233, 127–143.
Asseng, S., Fillery, I.R.P., Dunin, F.X., Keating, B.A. and Meinke, H. (2001b) Potential deep drainage under wheat crops in a Mediterranean climate. I. Temporal and spatial variability. *Australian Journal of Agricultural Research* 52, 45–56.
Asseng, S., Dunin, F.X., Fillery, I.R.P., Tennant, D. and Keating, B.A. (2001c) Potential deep drainage under wheat crops in a Mediterranean climate. II. Management opportunities to control drainage. *Australian Journal of Agricultural Research* 52, 57–66.
Bolger, T.P. and Turner, N.C. (1999) Water use efficiency and water use of Mediterranean pastures in southern Australia. *Australian Journal of Agricultural Research* 50, 1035–1046.
Condon, A.G., Richards, R.A., Rebetzke, G.J. and Farquhar, G.D. (2002) Improving intrinsic water-use efficiency and crop yield. *Crop Science* (in press).
Dunin, F.X., Williams, J., Verburg, K. and Keating, B.A. (1999) Can agricultural

management emulate natural ecosystems in recharge control in south eastern Australia? *Agroforestry Systems* 45, 343–364.

Eastham, J., Gregory, P.J., Williamson, D.R. and Watson, G.D. (1999) The influence of early sowing of wheat and lupin crops on evapotranspiration and evaporation from the soil surface in a Mediterranean climate. *Agricultural Water Management* 42, 205–218.

English, M.J., Musick, J.T. and Murty, V.V.N. (1990) Deficit irrigation. In: Hoffman, G.J., Howell, T.A. and Solomon, K.H. (eds) *Management of Farm Irrigation Systems*. American Society of Agricultural Engineers, St Joseph, pp. 631–663.

Farquhar, G.D. and Richards, R.A. (1984) Isotopic composition of plant carbon correlates with water-use efficiency of wheat genotypes. *Australian Journal of Plant Physiology* 11, 539–552.

Fischer, R.A. (1981) Optimizing the use of water and nitrogen through breeding of crops. *Plant and Soil* 58, 249–278.

George, R., McFarlane, D. and Nulsen, B. (1997) Salinity threatens the viability of agriculture and ecosystems in Western Australia. *Hydrogeology Journal* 5, 6–21.

Hatton, T.J. and Nulsen, R.A. (1999) Towards achieving functional ecosystem mimicry with respect to water cycling in southern Australian agriculture. *Agroforestry Systems* 45, 203–214.

Hoffman, G.J., Howell, T.A. and Solomon, K.H. (eds) (1990) *Management of Farm Irrigation Systems*. American Society of Agricultural Engineers, St Joseph.

Jackson, R.D. (1982) Canopy temperature and crop water stress. *Advances in Irrigation* 1, 43–85.

Klueva, N.Y., Zhang, J. and Nguyen, H.T. (1998) Molecular strategies for managing environmental stress. In: Chopra, V.L., Singh, R.B. and Varma, A. (eds) *Crop Productivity and Sustainability: Shaping the Future. Proceedings of the Second Crop Science Congress*. Oxford and IBH, New Delhi, pp. 501–524.

Lefroy, E.C. and Stirzaker, R.J. (1999) Agroforestry for water management in the cropping zone of southern Australia. *Agroforestry Systems* 45, 277–302.

Lipton, M. (1999) *Reviving Global Poverty Reduction: What Role for Genetically Modified Plants? 1999 Sir John Crawford Memorial Lecture*. Consultative Group on International Agricultural Research, Washington, DC.

Oweis, T., Pala, M. and Ryan, J. (1998) Stabilizing rainfed wheat yields with supplemental irrigation and nitrogen in a Mediterranean climate. *Agronomy Journal* 90, 672–681.

Passioura, J.B. (1977) Grain yield, harvest index and water use of wheat. *Journal of the Australian Institute of Agricultural Science* 43, 117–120.

Prathapar, S.A. (2000) Water shortages in the 21st century. In: Cadman, H. (ed.) *The Food and Environment Tightrope*. Australian Centre for International Agricultural Research, Canberra, pp. 125–133.

Richards, R.A., Rebetzke, G.J., Condon, A.G. and van Herwaarden, A.F. (2002) Breeding opportunities for increasing efficiency of water use and crop yield in temperate cereals. *Crop Science* (in press).

Sadler, E.J. and Turner, N.C. (1994) Water relationships in a sustainable agriculture system. In: Hatfield, J.L. and Karlen, D.L. (eds) *Sustainable Agriculture Systems*. Lewis Publisher, Boca Raton, Florida, pp. 21–46.

Seckler, D.W., Amarasinghe, U., Molden, D., de Silva, R. and Barker, R. (1998) World water demand and supply, 1990 to 2025: scenarios and issues. *Research Report No 19*. International Water Management Institute, Colombo.

Seckler, D.W., Barker, R. and Amarasinghe, U. (1999) Water scarcity in the twenty-first century. *International Journal of Water Resources Development* 15, 29–43.

Shackel, K.A., Ahamadi, H., Biasi, W., Buchner, R., Goldhamer, D., Gurusinghe, S., Hasey, J., Kester, D., Krueger, B., Lampinen, B., McGourty, G., Micke, W., Mitcham, E., Olson, B., Pelletrau, K., Philips, H., Ramos, D., Schwankl, L., Sibbert, S., Snyder, R., Southwick, S., Stevenson, M., Thorpe, M., Weinbaum, S. and Yeager, J. (1997) Plant water status as an index of irrigation need in deciduous fruit trees. *HortTechnology* 7, 23–29.

Smettem, K.R.J. (1998) *Deep Drainage and Nitrate Losses under Native Vegetation and Agriculture Systems in the Mediterranean Climate Region of Australia.* Land and Water Resources Research and Development Corporation, Canberra.

Stirzaker, R.J. (1999) The problem of irrigated horticulture: matching the biophysical efficiency with the economic efficiency. *Agroforestry Systems* 45, 187–202.

Turner, N.C. (1990a) Plant water relations and irrigation management. *Agricultural Water Management* 17, 59–73.

Turner, N.C. (1990b) The benefits of water deficits. In: Sinha, S.K., Sane, P.V., Bargava, S.C. and Agarwal, P.L. (eds) *Proceedings of the International Congress of Plant Physiology, New Delhi, India, February 1988,* Vol. 2. Society of Plant Physiology and Biochemistry, New Delhi, pp. 806–815.

Turner, N.C. (1993) Water use efficiency of crop plants: potential for improvement. In: Buxton, D.R., Shibles, R., Forsberg, R.A., Blad, B.L., Asay, K.H., Paulsen, G.M. and Wilson, R.F. (eds) *International Crop Science 1.* Crop Science Society of America, Madison, Wisconsin, pp. 75–82.

Turner, N.C. (1997) Further progress in crop water relations. *Advances in Agronomy* 58, 293–338.

Turner, N.C. (1998) Managing stress: Is there a productive future? In: Chopra, V.L., Singh, R.B. and Varma, A. (eds) *Crop Productivity and Sustainability: Shaping the Future. Proceedings of the Second International Crop Science Congress.* Oxford & IBH, New Delhi, pp. 541–544.

Turner, N.C. and Nicolas, M.E. (1998) Early vigour: a yield-positive characteristic for wheat in drought-prone Mediterranean-type environments. In: Behl, R.K., Singh, D.P. and Lodhi, G.P. (eds) *Crop Improvement for Stress Tolerance.* CCS Haryana Agricultural University, Hisar and Max Mueller Bhawan, New Delhi, pp. 47–62.

Turner, N.C. and Whan, B.R. (1995) Strategies for increasing productivity from water-limited areas through genetic means. In: Sharma, B., Kulshreshtha, V.P., Gupta, N. and Mishra, S.K. (eds) *Genetic Research and Education: Current Trends and the Next Fifty Years.* Indian Society of Genetics and Plant Breeding, New Delhi, pp. 545–557.

Turner, N.C., Nicolas, M.E., Hubick, K.T. and Farquhar, G.D. (1989) Evaluation of traits for improvement of water use efficiency and harvest index. In: Baker, F.W.G. (ed.) *Drought Resistance in Cereals.* ICSU Press, Paris, pp. 177–189.

Turner, N.C. Wright, G.C. and Siddique, K.H.M. (2001) Adaptation of grain legumes (pulses) to water-limited environments. *Advances in Agronomy* 71, 193–231.

Ward, P.R., Dunin, F.X. and Micin, S.F. (2001) Water uptake by annual and perennial pastures from duplex soils in a mediterranean environment. *Australian Journal of Agricultural Research* 52, 203–209.

White, D.A., Turner, N.C. and Galbraith, J.H. (2000) Leaf water relations and stomatal behaviour of four allopatric *Eucalyptus* species planted in Mediterranean south-western Australia. *Tree Physiology* 20, 1157–1165.

White, D.A., Dunin, F.X., Turner, N.C., Ward, B.H. and Galbraith, J.H. (2001) Water use by contour-planted tree belts comprised of four *Eucalyptus* species. *Agricultural Water Management* (in press).

Abiotic Stresses and Staple Crops

9

G.O. Edmeades[1], M. Cooper[1,2], R. Lafitte[3], C. Zinselmeier[1], J.-M. Ribaut[4], J.E. Habben[1], C. Löffler[1] and M. Bänziger[5]

[1]*Pioneer Hi-Bred Int. Inc., 7250 NW 62nd Avenue, Johnston, IA 50131, USA;* [2]*School of Land and Food Sciences, University of Queensland, Brisbane, Qld 4072, Australia;* [3]*IRRI, PO Box 3127 MCPO, 1271 Makati City, Philippines;* [4]*CIMMYT, Apdo Postal 6-641, Mexico 06600, Mexico;* [5]*CIMMYT, PO Box MP163, Harare, Zimbabwe*

The Extent of the Problem

Per capita food production globally is influenced by two major forces affecting supply and demand: declining availability of agricultural land and water, and an annual population increase of around 90 million people. The reduction in land resources is due to loss of area to urbanization, erosion, declining fertility, salinization, compaction, and acidification. The destabilizing effects of global warming are predicted to increase variability in temperatures and rainfall. By 2020, it is estimated that 29% more staple cereal production will be needed from land and water resources similar to or less than those available today.

Quantitative estimates of yield losses due to individual abiotic stresses are essential for setting research priorities. Most estimates are based on extrapolation from trial data, the frequency of occurrence of the stress, and/or expert experience. Empirical estimates for maize assign the greatest losses to water and nutrient deficits (Table 9.1). Water deficit is also the most important abiotic stress affecting rice production, and even under irrigated conditions annual losses due to shortages of water are estimated to be of 134 kg ha^{-1} (9.9 million tons of grain annually) in Asia alone (Dey and Upadhyaya, 1996). These are expected to increase in the future, as water scarcity in Asia becomes more severe. Stresses often occur in combination: coastal areas often suffer from both salinity and submergence, and acid soils are often P-deficient as well.

Table 9.1. Empirical estimates of percentage yield losses due to abiotic and biotic stress factors in maize grown in less developed tropical versus developed temperate environments. (Sources: Cardwell, 1982; Duvick, 1992; Qiao *et al.*, 1996; CIMMYT, 1997; Heisey and Edmeades, 1999.)

Factor	Tropical areas, less developed (%)	Temperate areas, developed (%)
Drought	15	12
N deficiency	12	3
P deficiency	5	3
K deficiency	0	2
Soil acidity	4	2
Low plant density	4	5
Weeds, intercrops	8	3
Cold temperature	2	10
Heat	4	5
Compaction	4	6
Waterlogging	4	7
Salinity	2	1
Total, abiotic stresses	64	59
Biotic stresses	16	12
Losses to all stresses	80	71
Realized yield (t ha^{-1})	2.7	6.9
Potential yield (t ha^{-1})	13.5	23.8

Techniques now available can be used to obtain far more accurate estimates of loss by stress and region. For much of the temperate world, weather data can be obtained in real time. Climate surfaces and soils data can serve as inputs to crop models that predict crop yields in the absence/presence of stresses, and geographic information systems (GIS) software can be used to organize and display model output and crop distribution spatially (Hartkamp *et al.*, 1999). Pioneer Hi-Bred scientists have shown that the output from a water-constrained version of the CERES-Maize model was highly correlated ($r^2 = 0.82$; $n = 27$) with actual performance of 16 hybrids in the American Midwest, suggesting that even in this well-favoured production region water availability was a major constraint. Crop models and GIS have shown their usefulness as well in the Third World through risk assessments for drought (e.g. Hodson *et al.*, 1999) or predictions of seasonal crop production (e.g. USAID Famine Early Warning System; www.fews.org). The precision of crop models is rapidly improving, as are computing capacity and the availability of global weather data, providing powerful tools in the future for setting research priorities in marginal production environments.

Can stressful and low-yielding environments simply be avoided and resources focused on well-favoured locations? Heisey and Edmeades (1999) concluded that agricultural research for marginal environments in the developing world will continue to be justified because of the added income and security it brings to those who live there, and because it spares the use of even more marginal land.

Agronomic Interventions

Innovative crop management can alleviate the effects of most abiotic stresses. Environmental concerns, economic return and physical availability may limit usage of inputs such as fertilizer, manure, irrigation water and pesticides. Irrigated land area has increased by only 1.5% per year over the last decade, compared with 2.7% in the 1970s (FAO, 2000), and adding new irrigated areas is becoming increasingly difficult.

Promising new management technologies are becoming available. Precision farming, or site-specific management within a field through variable application of input(s) in response to yield or soil texture, promises to improve efficiency of input use, provided the causes of yield variation have been correctly identified. Other maturing technologies are:

- Irrigation and fertigation by precise application of water and nutrients through subterranean drip tubing on a demand basis.
- Compaction control through minimum tillage and in-field traffic control.
- Tropical soil fertility improvement using combined organic (legumes, agroforestry, manure, compost) and inorganic sources.
- The use of crop simulation models to guide decisions on planting date, crop, variety and fertilizer application.

It should not be overlooked, however, that in many developing countries fundamentals such as fertilizer use, liming, irrigation, adequate weed control and good grain storage remain the keys to increased and sustained production (Table 9.1). There is also a strong interaction between agronomic practices and changes in the stress tolerance of germplasm with time (Duvick and Cassman, 1999). Ideally agronomic *and* genetic improvements occur concomitantly so that genotype × management interactions can be fully exploited. Seed-based technology is inherently easier to transfer to farmers than more complex, knowledge-based agronomic practices. For this reason the focus now moves to genetic improvement of stress tolerance, focusing on staple cereal crops.

Genetic Improvement of Abiotic Stress Tolerance

Targeted processes and traits

There is extensive literature reviewing the physiology and genetics of traits associated with tolerance to specific abiotic stresses (e.g. Shannon, 1998; de la Fuente-Martinez and Herrera-Estrella, 1999; Saini and Westgate, 2000). General mechanisms by which crops obtain high stable yields under stress are emphasized here.

Yield potential

Selection in temperate maize has increased stress tolerance, capacity to recover from stress and yield potential over the past 70 years (Tollenaar and Wu, 1999). Heterosis is also a source of stress tolerance (Blum, 1997), and hybrids usually outyield varieties over a wide range of stress intensities. Richards (1992) concluded that yield potential is the most important trait when breeding for performance under saline conditions. Generally when stressed yields are <50–60% of potential it is more efficient to select for improved yields under the stress itself since stress-adaptive traits become more important than yield potential in determining performance (Bänziger *et al.*, 1997).

Phenology, escape and tolerance

Stable grain production depends on a good match between the occurrence of rainfall, adequate temperatures and nutrient supply, the timing of stress-susceptible developmental phases and the length of the crop cycle. Since earliness reduces yield potential, a combination of greater crop duration and increased stress tolerance should stabilize and increase yield simultaneously.

Tolerance to stress during reproductive growth

Grain yields in cereals under severe stress are closely associated ($r > 0.8$) with variation in kernel number per plant (Bolaños and Edmeades, 1996). Since almost 60% of global crop product depends on successful kernel set, this process is critical to success (Edmeades *et al.*, 1998). In most crops any stress that lowers photosynthesis per plant at flowering will reduce kernel set. This can be due to a loss of pollen viability, poor panicle exsertion, and spikelet sterility (Saini and Lalonde, 1998). Sap exudation rate from de-topped rice tillers under mild drought stress is a trait that has been associated with the maintenance of spikelet fertility (H.R. Lafitte, Philippines, 2000, unpublished data). Carbohydrate supply to the developing pollen grains and embryos is a major determinant of reproductive success under water stress (Saini and Lalonde, 1998). Inhibition of invertase plays a key role in stress-induced pollen sterility in wheat (Saini and Westgate, 2000).

Key determinants of reproductive success in maize under drought stress are events occurring in the ovary, namely a large flux of carbon as sucrose, and adequate levels of N and invertase (Below *et al.*, 2000; Zinselmeier *et al.*, 2000). Water deficits and shading both deplete starch reserves in the ovary thought to maintain ovary growth rates. An enhanced level of cytokinins caused by up-regulation of iso-pentenyl transferase or down-regulation of cytokinin oxidase may reduce kernel abortion, while ethylene and ABA may increase it (Jones and Setter, 2000). The risk of reproductive failure can be reduced by increasing spikelet growth rate. When under stress, maize ear growth rate slows more rapidly than that of the tassel, the anthesis–silking interval (ASI) increases and is strongly associated with grain yield ($r = -0.4$ to -0.8). Selection in tropical maize populations under drought has resulted in grain yield increases of around $100 \, \mathrm{kg \, ha^{-1} \, year^{-1}}$, reductions in ASI and spikelet number on the upper ear (Table 9.2), and an increase in harvest index across all water regimes. Drought-tolerant selections were also more tolerant of low N conditions (Bänziger *et al.*, 1999), perhaps because both stresses reduce assimilate flux to the ovary. Thus, ASI is a simple trait to use when selecting for improved maize kernel set under stress.

Plant reserves

The supply of assimilate to the filling grain can be maintained from stem reserves when the plant is stressed, though reserves are much less effective than concurrent photosynthesis at maintaining kernel set at flowering (Blum, 1997).

Roots

In drying soils an even distribution of roots with depth without an increase in root biomass would be desirable. There is little evidence that a further increase in rooting intensity over $1 \, \mathrm{cm} \, \mathrm{cm^{-3}}$ will improve water uptake (Ludlow and Muchow, 1990), though maximum uptake of N and P may

Table 9.2. Effects of selection for drought tolerance on gains per selection cycle in four maize populations when evaluated at 3–6 water-stressed (SS) sites, at 5–8 well-watered (WW) sites, or at two low N sites. Locations were in Mexico (Mex.) or outside (Int.). **, ns: rate of change per selection cycle significant at $P < 0.05$ or $P > 0.05$ (Beck *et al.*, 1996).

Population	Yield (kg ha⁻¹) SS	WW	Low N	Anthesis WW (days)	ASI[a] SS (days)	Ears plant⁻¹ SS
Evaluation 1988/91						
Tuxpeño Seq. (Mex.)	100**	125**		−0.40**		
Tuxpeño Seq. (Int.)	52ns	101**		−0.24**		
Evaluation 1992/4						
La Posta Seq. (Mex.)	229**	53ns	233	−0.52**	−1.18**	0.07**
Pool 26 Seq. (Mex.)	288**	177**	207	−0.93**	−1.50**	0.08**
Tuxpeño Seq. (Mex.)	80**	38**	86	−0.32**	−0.44**	0.02**
Pool 18 Seq. (Mex.)	146**	126**	190		−2.13**	0.05**

[a] Anthesis–silking interval.

call for root length densities of 2–5 cm cm^{-3}. Under anoxic soil conditions, however, facultative formation of aerenchyma and a high root length density in the surface soil are essential in cereals, as exemplified by rice.

Osmotic adjustment (OA)

OA maintains meristematic activity during exposure of the plant to water deficits or salinity by increasing soil water uptake, maintaining turgor, and delaying foliar senescence. In crops such as wheat, rice and sorghum, whose OA exceeds 1 MPa (Ludlow and Muchow, 1990; Nguyen *et al.*, 1997), a focus on this trait seems justified. OA has been associated with stable grain yields of maize hybrids across moderately dry environments in Argentina (Lemcoff *et al.*, 1998) despite discouraging results under severe stress (Bolaños and Edmeades, 1991).

Antioxidants

Tolerance to abiotic stresses is associated with an increase in antioxidant enzyme activity that protects cell membranes from oxidative damage. Overexpression of the antioxidant superoxide dismutase (SOD) increases tolerance to chilling in maize and to drought in wheat, and further research on the formation and metabolism of antioxidants ascorbate and glutathione is a high priority (Noctor and Foyer, 1998).

Protectants

Genetic overexpression of glycinebetaine has enhanced tolerance to cold and drought, and the exogenous application of glycinebetaine has similar but transient effects (Makela *et al.*, 1996). Late embryogenesis abundant (LEA) proteins increase sharply in tissues during desiccation or chilling, are upregulated by ABA, and are thought to stabilize plasma membrane structure.

Abscisic acid

This phytohormone increases under stress, leading to increased root growth, stomatal closure, leaf shedding, dormancy and possibly abortion of tip kernels in maize (Jones and Setter, 2000), thus favouring plant survival over production. Increased tolerance to drought in maize was associated with lowered leaf ABA concentration (Mugo *et al.*, 2000), suggesting that improved yield under stress is associated with reduced ABA production.

Staygreen

Greenness *per se* does not guarantee continued C assimilation (Thomas and Howarth, 2000), but is usually associated with improved performance under drought (Borrell, 2000) and N stress (Bänziger and Lafitte, 1997).

Blum (2000) noted the critical need for field validation of stress-tolerance mechanisms when stating that 'any claim for a genetic modification of stress resistance that is presumed to impact whole crop performance in agriculture will remain on paper unless proven with whole-plant testing systems and under field conditions'. For breeders of staple crops such modifications must demonstrate improved or stabilized grain yields under farmers' conditions before they merit adoption.

Selecting for the target environment

The value of varieties developed from selection depends on the environmental conditions encountered during selection. The intelligent use of representative testing environments can increase rates of genetic progress (Cooper and Podlich, 1999). There are two main approaches used in breeding for abiotic stress tolerance. The first involves dividing all possible target environments into mega-environments (ME), where an ME is a large crop area with similar production constraints, characterized by a small $G \times E$ interaction for yield within, and a large $G \times E$ interaction among MEs (Rajaram *et al.*, 1996). For example, in the upland rice ME, mild to severe water stress can occur at any point in the season, whereas in the rainfed lowland ME, drought occurs mainly at either end of the cropping season. Typically, data from all testing locations occurring within the ME, are weighted equally. This conventional testing strategy has produced many of the varieties in use today, but alone it does not target stress-sensitive growth stages and is wasteful of testing resources.

A second approach relies on defining the target population of environments (TPE) to describe the range of environmental conditions encountered by the crop over time (Comstock, 1977) so that the role of specific stresses in generating important $G \times E$ interactions is explicit (Cooper and Podlich, 1999; Chapman *et al.*, 2000). Frequency of occurrence of environmental types can be determined with crop simulation models using 20 to 100 years of weather data as inputs, and testing sites can be selected to best represent the TPE. Data from any given site can be weighted according to the frequency with which its specific environment occurred within the TPE. Combining this with an understanding of the genetic architecture of traits also allows breeders to evaluate alternative breeding strategies (Podlich and Cooper, 1999).

The use of well-characterized managed stress environments utilizing stresses that are common in the TPE complements both these approaches (e.g. Bänziger *et al.*, 1997; Fukai *et al.*, 1999). Managed stress environments have an advantage when testing is limited by land or seed supply (e.g. early generation testing), where the stress occurs unpredictably in the TPE, and where it can be severe. Benefits are repeatability and reliability of a stress at a level adjusted to optimize differences in genetic tolerance, and a focus on growth stages known to destabilize grain yield, e.g. stress at flowering. Data can be combined with that from an ME, but the TPE approach is preferred because data can be weighted appropriately.

Variation in stress levels encountered by a single variety within a heterogeneous field is often overlooked in crop improvement, and may be as large as the variation among test sites in the ME or TPE, especially in the tropics. Bouma *et al.* (1997) reported a within-field variation in millet yields of 360%, and a tenfold variation in infiltration on crusty soils of variable slope in the Sahel. Within-field variation in topsoil thickness (and hence water holding capacity) has been described for fields in the USA (Kitchen *et al.*, 1999), and yields of maize, soybeans and sorghum were positively correlated with topsoil depth (r^2 = 0.10–0.88, average 0.40**) and hence crop-available water. The important implication is that a single variety must be able to yield well across a wide range of stress levels, and any gain in stress tolerance must not be at the cost of unstressed yields.

The conventional breeding approach

Selection gain in a target environment (R_t) is determined by the genetic variance (σ^2_g), selection intensity (i), the genetic correlation (r_g) between the selection and target environments, and heritability (h^2_s) in the selection environment (Falconer, 1982). Thus:

$$R_t = \sigma_g i h_s r_g \qquad (9.1)$$

Selection response is maximized when the trait is highly heritable and genetically variable, a small proportion of progenies is selected, the test environments are representative of the TPE, and when test sites are numerous and on uniform soil.

Proper choice of breeding material is critical to success. Since alleles conferring tolerance to abiotic stress are found at low frequencies in most elite germplasm, progress with less risk of yield drag can often be made by screening elite germplasm for stress tolerance, even though σ^2_g may appear small. Unimproved sources may be used if yield drag from unwanted alleles can be reduced through the use of molecular markers or genetic engineering.

When different approaches to crop improvement are compared, it must be borne in mind that resources used in conventional breeding programmes are formidable. For example, each year Pioneer Hi-Bred maize breeders create over 5500 new breeding populations, generating approximately 900 new inbred lines and 100,000 new experimental hybrids for testing. Of these only 40–60 become commercial hybrids, and result in an average genetic gain of 1% per year (Duvick, 1992). Varietal development in the public IARC centres is similar, though less testing-intensive.

Unstressed versus stressed environments

When a crop is stressed σ^2_g falls and heritability declines, though the decline may not be important until yields are < 20% of potential (Bolaños and Edmeades, 1996). Experiments conducted under stress often have a high coefficient of variation and are frequently discarded, thus biasing selections towards high yielding sites. Spatial variation in soil becomes increasingly obvious as drought, excess water, low fertility and salinity reduce yields, and improved statistical designs often increase selection efficiency. Bänziger *et al.* (1997) showed that the use of an incomplete versus a randomized complete block design increased heritability for maize grain yield under low soil nitrogen by 20%. The TPE concept and the use of standardized means for across-site analysis allow data from low yielding sites to influence selection effectively.

Early testing of genotypes under stress usually requires larger managed stress locations, but allows the breeder to exploit the full range of genetic variation for tolerance among progenies. Cooper *et al.* (1999) demonstrated that progress made in rice for tolerance to rainfed conditions by early generation testing of many breeding lines across a range of environments was more cost-effective than delaying testing until fewer lines with a reduced σ^2_g remained.

Using secondary traits

Grain yield under stressed conditions is normally the primary trait for selection. A suitable secondary trait is one that is: (i) genetically associated with grain yield under drought; (ii) highly heritable; (iii) cheap and fast to measure; (iv) stable over the measurement period; (v) observed at or before flowering; and (vi) not associated with yield loss under unstressed conditions (Edmeades *et al.*, 1998). Very few proposed secondary traits meet these criteria. Using selection theory, Bänziger and Lafitte (1997) showed that the use of secondary traits plus yield improved selection gains for maize yield under low N by 20% versus selection for yield alone, and that this benefit increased as N deficiency intensified.

Planned selection for tolerance to a specific abiotic stress

One of the few successful examples involves selection for drought tolerance in tropical maize populations at CIMMYT, Mexico, under managed drought stress levels ranging from severe stress at flowering to a well-watered control (Beck *et al.*, 1996; Edmeades *et al.*, 2000). Data on stressed and unstressed yield, ASI, barrenness and staygreen under stress are being combined in a selection index used to identify superior genotypes. Improvements in yield under drought have averaged around 100 kg ha^{-1} year^{-1} (Table 9.2). Similar gains were observed under low N as well, suggesting a common mechanism of tolerance to stress at flowering (Bänziger *et al.*, 1999). These improved sources also showed their tolerance to drought when transferred to southern Africa. This, in combination with screening under managed stress levels, has recently resulted in maize hybrids with superior and stable performance in the harsh environments of southern and eastern Africa (M. Bänziger, Zimbabwe, 2000, personal communication).

Marker-assisted breeding

DNA markers

Unlike morphological characters, the detection of molecular markers is not influenced by the environment or growth stage of the crop. They have contributed greatly to genome analysis via mapping, gene tagging and quantitative trait loci (QTL) identification, and can be used for marker-assisted selection (MAS). DNA markers are either based on restriction fragment length polymorphisms (RFLPs), or on the polymerase chain reaction (PCR), where common types are random amplified polymorphic DNA (RAPDs), simple sequence repeats (SSRs), amplified fragment length polymorphisms (AFLPs) and single nucleotide polymorphisms (SNPs). Extremely abundant SNP markers show promise of a major improvement in efficiency over the current gel-based systems. Rice because it has a small genome (0.43×10^9 bp per haploid genome), and maize because it is highly polymorphic, have been more thoroughly characterized with DNA markers than wheat with its large (16×10^9 bp per haploid genome) complex hexaploid genome.

Genetic mapping of target traits

In this process a segregating population formed from a cross between two lines contrasting for the target trait is grown under stress. Genes or QTL are identified by associating markers with observed phenotypic expression, and markers can be used to direct further allelic manipulation. When several genes govern target trait(s), construction of a complete linkage map is desirable.

In major crops QTL have been identified for yield components, and for morphological and physiological characters associated with stress tolerance (e.g. Ribaut and Poland, 2000). Yield is a logical first target, though QTLs for yield typically are experiment-dependent, account for a small percentage of the phenotypic variance, and a high level of QTL × E interaction makes their use in selection questionable. It is therefore important to identify QTL for secondary traits that are correlated with yield under stress and whose variance increases under abiotic stress. Examples are ASI for maize, stay-green for sorghum, OA in rice roots, and canopy temperature for wheat. Use of an index of QTL for several traits appears to be the most successful MAS strategy for improving and stabilizing yield across stress levels.

The precision of QTL characterization depends on the accuracy of phenotypic data, and progress in incomplete block designs and spatial analyses has greatly improved field data quality. The efficiency of MAS depends also on sample size, since small samples of segregating progenies can give rise to false positives for QTL (Melchinger *et al.*, 1998). Newer, more robust mapping methods, such as composite interval mapping that considers QTL × E interactions, are improving the characterization of QTL. The evaluation of QTL × E interactions remains a challenge to every MAS scheme (Beavis and Keim, 1996), unless they can be associated directly with a specific factor in the environment. There is growing confidence that useful QTL effects can be detected in germplasm evaluated across several environments subject to a common abiotic stress because of successes in conventional breeding for drought tolerance in diverse maize germplasm.

MAS through QTL manipulation

Few MAS experiments using QTL for polygenic traits have been completed to date. Ribaut *et al.* (1999) attempted to transfer five genomic target regions from a drought tolerant donor, Ac7643, to CML247, an elite but drought-susceptible inbred line. After two backcrosses and two self-pollinations, the best genotype was fixed for the five target regions covering 12% of the genome. The 70 best lines were identified, crossed with two tester inbreds and evaluated in 1997 to 1999 under several water regimes. The mean of the 70 selected lines outperformed the CML247 checks, and the best line performed two to four times better than the control under drought. No yield reduction was observed under well-watered conditions.

Molecular marker applications in plant breeding

Markers should significantly improve the efficiency of incorporating stress tolerance traits with low heritability that are difficult or costly to measure. At present QTL identification is required whenever additional germplasm is used. Although successful, the backcross-MAS experiment described above cannot be considered as routine because of the cost of mapping, though QTL identification is helping validate candidate genes associated with stress tolerance. To improve polygenic traits efficiently, new selection schemes should be considered, where, for example, the MAS step is conducted only once at an early stage of recombination to fix favourable alleles at selected genomic regions in large segregating populations derived from crosses between elite lines with high allelic complementarity (Ribaut and Betrán, 1999). Combining information from functional genomics with new MAS schemes and breeding approaches should eventually allow the mapping step to be bypassed, and when target gene/pathways have been identified, individuals could be selected using an index of favourable alleles.

Genomics and novel sources of genetic variation

The blend of computing power, biotechnology, markers and traditional field selection techniques (Fig. 9.1) provides a powerful set of tools in the search for new genetic variation for abiotic stress tolerance.

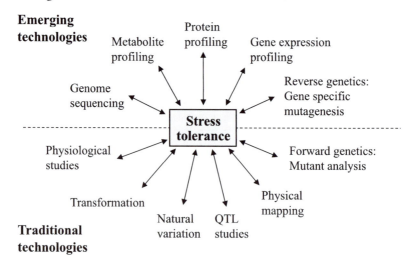

Fig. 9.1. Experimental tools available for trait discovery and manipulation.

A key to understanding gene function is the analysis of gene expression patterns. Northern Blot analysis is limited to a relatively small number of genes at one time. RNA profiling, on the other hand, quantifies thousands of transcripts from individual genes in response to abiotic stress (Brown and Botstein, 1999). This technique has the ability to detect differences among tissues, genotypes and/or environments. It can identify gene clusters that respond coordinately to a specific environmental challenge, and thus suggest critical pathways associated with stress tolerance. Stress-responsive changes in specific proteins, metabolite fluxes and pool sizes can also be profiled using the emerging tools of proteomics and metabolomics. Together these techniques are greatly aiding gene and pathway discovery.

Candidate genes conferring stress tolerance can be identified in several ways. One means is by reverse genetics, where a known gene sequence is disrupted by transposable elements and accompanying changes in the plant phenotype are observed. Using this technique Bensen *et al.* (1995) have fully characterized the *an1* gene in maize. Candidate genes can also be identified through synteny, a procedure that relies on the remarkably consistent conservation of gene order among cereal crops (Devos and Gale, 1997). Transformation, where candidate genes are integrated into the genomes of crops, provides another indispensable tool in determining gene function and effectiveness.

The initiation of large-scale expressed sequence tag (EST) sequencing projects has occurred in parallel with transformation, and the private sector has isolated and sequenced >400,000 ESTs in maize alone, thought to cover 70–80% of all maize genes. The public sector has catalogued 50,000 maize ESTs. These are readily available to researchers (http://www.zmdb.ias-tate.edu/), and sequences covering 85% of the rice genome will soon become publicly available (http://www.monsanto.com/monsanto/mediacenter/background/00apr03_rice.html). As a result the number of genes available for transformation has increased by several orders of magnitude over the past 5 years.

Effective transformation requires the use of an appropriate promoter that controls the expression of the target structural gene. Specific promoters can direct expression of genes to particular tissues and/or times in development. For example, the *sag12* promoter from *Arabidopsis* limits expression to senescing leaves, while the isopentenyl transferase (*ipt*) gene from *Agrobacterium* catalyses the rate-limiting step in cytokinin biosynthesis. When Gan and Amasimo (1995) fused these two and inserted them into the tobacco genome, cytokinin levels were elevated in maturing leaves of transformants and foliar senescence was delayed. More sophisticated control elements will be necessary in the future, perhaps targeting specific organs such as kernels, roots, meristems, the aleurone layer, stomates or transfer cells.

Recent research has demonstrated the functionality of chemically inducible promoters (McNellis *et al.*, 1998). In this case, *avrRpt2*, an avirulence gene that induces hypersensitive cell death, was ligated to a glucocorticoid-inducible promoter and inserted in the *Arabidopsis* genome. Dexamethasone (the inducer) was applied, resulting in localized expression of *avrRpt2* and a concentrated hypersensitive cell death. Such studies suggest the possibility of managed gene expression for tolerance and escape, implemented only when a stress such as drought or heat is considered likely to occur.

A new generation of transgenic crop plants, now in the gene discovery stage, is focusing on traits that are more quantitative in nature, such as improvement in drought, cold, or aluminium tolerance of cereals. Transgenes may bridge gaps among common species that differ markedly in stress tolerance. In rice, for example, it may be feasible to reduce cuticular water loss that is two to four times greater than that of sorghum and maize by introducing genes from these species (Nguyen *et al.*, 1997). In a similar manner root penetration of rice in drying soils could be increased using genes from other species or near relatives (Liu *et al.*, 2000). The use of truly isogenic lines (with or without the transgene) provides scientists with a powerful tool to test the relative importance of specific genes. These are currently being used to test drought tolerance mechanisms at IRRI and Pioneer Hi-Bred, while functional genomics are being used to identify genes and gene products conferring tolerance to anthesis-stage water deficits in rice.

Finally, transgenes will have to pass the same tests as conventional genetic variation: increased but stable yield under the targeted stress, and no yield penalty in its absence. Each must be evaluated in the TPE, alone, and when stacked with other transgenes. Lastly, each will also have to pass the tests of food, feed, environmental safety and public acceptance before it can be commercialized.

Challenges

Population growth, poor agronomic practices and global climate change are driving changes in production environments that pose the following challenges to crop scientists:

- Quantification and prioritization of stresses on a regional or even global scale to establish research goals, using the maturing resources of GIS, climate change forecasting, crop modelling driven by quality data, and market information.
- Development of agronomic techniques that increase input use efficiency and complement unique features of new cultivars, while remaining accessible and affordable.
- Refinement of the definition of TPEs so that breeding and agronomic research can be precisely focused on needs of existing and new crop production areas.
- Field-based screening techniques, supported by laboratory and statistical methods, that efficiently identify genetic variation for mechanisms that stabilize and increase production under stress at no cost to unstressed performance, followed by the optimized use of such secondary traits in selection.
- Efficient combination of conventional breeding methods with gene discovery through expression methodology, genomics and molecular markers to identify, validate and integrate key alleles and genes suited to specific environmental challenges.
- Development of tissue- and timing-specific promoters and switchable controlling elements that may eventually allow stress tolerance to be managed according to need.

Finally, the greatest challenge may not be technical. It is to place these new resource-intensive technologies and intellectual property in the hands of those who can least afford to pay for them – the resource-poor farmers of the developing world whose crop production environments are rapidly deteriorating. The way forward will call for dedication, a willingness to put the needs of the poor ahead of short-term profitability, and imaginative cooperation among private and public researchers, environmentalists and regulators to create a win–win situation for all concerned.

References

Bänziger, M. and Lafitte, H.R. (1997) Efficiency of secondary traits for improving maize for low-nitrogen target environments. *Crop Science* 37, 1110–1117.

Bänziger, M., Betrán, F.J. and Lafitte, H.R. (1997) Efficiency of high-nitrogen selection environments for improving maize for low-nitrogen target environments. *Crop Science* 37, 1103–1109.

Bänziger, M., Edmeades, G.O. and Lafitte, H.R. (1999) Selection for drought tolerance increases maize yields over a range of nitrogen levels. *Crop Science* 39, 1035–1040.

Beavis, W.D. and Keim, P. (1996) Identification of quantitative trait loci that are affected by environment. In: Kang, M.S. and Gauch, H.G. (eds) *Genotype-by-Environment Interaction*. CRC Press, Boca Raton, Florida, pp. 123–149.

Beck, D., Betrán, J., Bänziger, M., Edmeades, G., Ribaut, J.-M., Willcox, M., Vasal, S.K. and Ortega A. (1996) Progress in developing drought and low soil nitrogen tolerance in maize. In: Wilkinson, D. (ed.) *Proceedings of 51st Annual Corn and Sorghum Research Conference*. ASTA, Washington DC, pp. 85–111.

Below, F.E., Cazetta, J.O. and Seebauer, J.R. (2000) Carbon/nitrogen interactions during ear and kernel development of maize. In: Westgate, M.E. and Boote, K.J. (eds) *Physiology and Modeling Kernel Set in Maize*. CSSA Special Publication No. 29. CSSA, Madison, Wisconsin, pp.15–24.

Bensen, R.J., Johal, G.S., Crane, V.C., Tossberg, J.T., Schnable, P.S., Meeley, R.B. and Briggs, S.P. (1995) Cloning and characterization of the maize *An1* gene. *Plant Cell* 7, 75–84.

Blum, A. (1997) Constitutive traits affecting plant performance under stress. In: Edmeades, G.O., Bänziger, M., Mickelson, H.R. and Peña-Valdivia, C.B. (eds) *Developing Drought and Low-N Tolerant Maize*. CIMMYT, El Batan, pp.131–135.

Blum, A. (2000) Towards standard assays of drought resistance in crop plants. In: Ribaut, J.-M. and Poland, D. (eds) *Proceedings of Workshop on Molecular Approaches for the Genetic Improvement of Cereals for Stable Production in Water-Limited Environments*. CIMMYT, El Batan, pp. 29–35.

Bolaños, J. and Edmeades, G.O. (1991) Value of selection for osmotic potential in tropical maize. *Agronomy Journal* 83, 948–956.

Bolaños, J. and Edmeades, G.O. (1996) The importance of the anthesis–silking interval in breeding for drought tolerance in tropical maize. *Field Crops Research* 48, 65–80.

Borrell, A.K. (2000) Physiological basis, QTL and MAS of the staygreen drought resistance trait in sorghum. In: Ribaut, J.-M. and Poland, D. (eds) *Proceedings of Workshop on Molecular Approaches for the Genetic Improvement of Cereals for Stable Production in Water-Limited Environments*. CIMMYT, El Batan, pp. 142–146.

Bouma, J., Verhagen, J., Bouwer, J. and Powell, J.M. (1997) Using systems approaches for targeting site-specific management on field level. In: Kropff, M.J., Teng, P.S., Agrawal, P.K., Bouma, J., Bouman, B.A.M., Jones, J.W. and van Laar, H.H. (eds) *Applications of Systems Approaches at the Field Level*, Vol. 2. Kluwer Academic Publishers, Dordrecht, pp. 25–36.

Brown, P.O. and Botstein, D. (1999) Exploring the new world of the genome with DNA microarrays. *Nature Genetics* (Suppl.) 21, 33–37.

Cardwell, V. (1982) Fifty years of Minnesota corn production: sources of yield increase. *Agronomy Journal* 74, 984–990.

Chapman, S.C., Cooper, M., Hammer, G.L. and Butler, D.G. (2000) Genotype by environment interactions affecting grain sorghum. II. Frequencies of different seasonal patterns of drought stress are related to location effects on hybrid yields. *Australian Journal of Agricultural Research* 51, 209–221.

CIMMYT (1997) Development of new stress-resistant maize genetic resources (UNDP Project GLO/90/003). *Maize Program Special Report.* CIMMYT.

Comstock, R.E. (1977) Quantitative genetics and the design of breeding programs. In: Pollak, E., Kempthorne, O. and Bailey, T.B. (eds) *Proceedings of the International Conference on Quantitative Genetics, August.* Iowa State University Press, Ames, pp. 705-718.

Cooper, M. and Podlich, D.W. (1999) Genotype × environment interactions, selection response, and heterosis. In: Coors, J.G. and Pandey, S. (eds) *The Genetics and Exploitation of Heterosis in Crops.* ASA-CSSA-SSSA, Madison, Wisconsin, pp. 81–92.

Cooper, M., Rajatasereekul, S., Somrith, B., Sriwisut, S., Immark, S., Boonwite, C., Suwanwongse, A.S., Ruangsook, A., Hanviriyapant, P., Romyen, P., Pornurais-anit, P., Skulkhu, E., Fukai, S., Basnayake, J. and Podlich, D.W. (1999) Rainfed lowland rice breeding strategies for Northeast Thailand. II. Comparison of intrastation and interstation selection. *Field Crops Research* 64, 153–176.

de la Fuente-Martinez, J.M. and Herrera-Estrella, L. (1999) Advances in the understanding of aluminium toxicity and the development of aluminium-tolerant transgenic plants. *Advances in Agronomy* 66, 103–120.

Devos, K.M. and Gale, M.D. (1997) Comparative genome analysis in the grass family. In: Tsaftaris, A.S. (ed.) *Genetics, Biotechnology and Breeding of Maize and Sorghum.* Royal Society of Chemistry, London, pp. 3–9.

Dey, M.M. and Upadhyaya, H.K. (1996) Yield loss due to drought, cold and submergence in Asia. In: Evenson, R.E., Herdt, R.W. and Hossain, M. (eds) *Rice Research in Asia: Progress and Priorities.* CAB International, Wallingford, pp. 291–303.

Duvick, D.N. (1992) Genetic contributions to advances in yield of US maize. *Maydica* 37, 69–79.

Duvick, D.N. and Cassman, K.G. (1999) Post-green revolution trends in yield potential of temperate maize in the North-Central United States. *Crop Science* 39, 1622–1630.

Edmeades, G.O., Bolaños, J., Bänziger, M., Ribaut, J.-M., White, J.W., Reynolds, M.P. and Lafitte, H.R. (1998) Improving crop yields under water deficits in the tropics. In: Chopra, V.L., Singh, R.B. and Varma, A. (eds) *Crop Productivity and Sustainability – Shaping the Future. Proceedings of the Second International Crop Science Congress.* Oxford and IBH, New Delhi, pp. 437–451.

Edmeades, G.O., Bolaños, J., Elings, A., Ribaut, J.-M., Bänziger, M. and Westgate, M.E. (2000) The role and regulation of the anthesis–silking interval in maize. In: Westgate, M.E. and Boote, K.J. (eds) *Physiology and Modeling Kernel Set in Maize.* CSSA Special Publication No. 29, CSSA, Madison, Wisconsin, pp. 43–73.

Falconer, D.S. (1982) *Introduction to Quantitative Genetics,* 2nd edn. John Wiley & Sons, New York.

FAO (Food and Agriculture Organization) (2000) FAOStat database – agricultural production. <http://apps.fao.org/>

Fukai, S., Pantuwan, G., Jongdee, B. and Cooper, M. (1999). Screening for drought resistance in rainfed lowland rice. *Field Crops Research* 64, 61–74.

Gan, S. and Amasimo, R.M. (1995) Inhibition of leaf senescence by autoregulated production of cytokinin. *Science* 270, 1986–1988.

Hartkamp, A.D., White, J.W. and Hoogenboom, G. (1999) Interfacing geographic information systems with agronomic modeling: a review. *Agronomy Journal* 91, 761–772.

Heisey, P.W. and Edmeades, G.O. (1999) Maize production in drought stressed environments: technical options and research resource allocation. Part 1. In: CIMMYT (eds) *World Maize Facts and Trends 1997/98*. CIMMYT, pp. 1–36.

Hodson, D.P., Rodríguez, A., White, J.W., Corbett, J.D., O'Brien, R.F. and Bänziger, M. (1999) *Africa Maize Research Atlas (v. 2.0)*. CIMMYT.

Jones, R.J. and Setter, T.L. (2000) Hormonal regulation of early kernel development. In: Westgate, M.E. and Boote, K.J. (eds) *Physiology and Modeling Kernel Set in Maize*. CSSA Special Publication No. 29. CSSA, Madison, Wisconsin, pp. 25–42.

Kitchen, N.R., Sudduth, K.A. and Drummond, S.T. (1999) Soil electrical conductivity as a crop productivity measure for claypan soils. *Journal of Production Agriculture* 12, 607–617.

Lemcoff, J.H., Chimenti, C.A. and Davezac, T.A.E. (1998) Osmotic adjustment in maize (*Zea mays* L.): changes with ontogeny and its relationship with phenotypic stability. *Journal of Agronomy and Crop Science* 180, 241–247.

Liu, L., Lafitte, R. and Dimayuga, G. (2000) Genetic resources to improve rice performance under water deficit. *International Rice Research Notes* (in press).

Ludlow, M.M. and Muchow, R.C. (1990) A critical evaluation of traits for improving crop yields in water-limited environments. *Advances in Agronomy* 43, 107–153.

Makela, P., Peltonen-Sainio, P., Jokinen, K., Pehu, E., Setala, H., Hinkkanen, R. and Somersalo, S. (1996) Uptake and translocation of foliar-applied glycinebetaine in crop plants. *Plant Science Shannon* 121, 221–230.

McNellis, T.W., Mudgett, M.B., Li, K., Aoyama, T., Horvath, D., Chua, N.H. and Staskawicz, B.J. (1998) Glucocorticoid-inducible expression of a bacterial avirulence gene in transgenic *Arabidopsis* induces hypersensitive cell death. *Plant Journal* 14, 247–257.

Melchinger, A.E., Utz, H.F. and Schon C.C. (1998) Quantitative trait locus (QTL) mapping using different testers and independent population samples in maize reveals low power of QTL detection and large bias in estimates of QTL effects. *Genetics* 149, 383–403.

Mugo, S.N., Bänziger, M. and Edmeades, G.O. (2000) Prospects of using ABA in selection for drought tolerance in cereal crops. In: Ribaut, J.-M. and Poland, D. (eds) *Proceedings of Workshop on Molecular Approaches for the Genetic Improvement of Cereals for Stable Production in Water-Limited Environments*. CIMMYT, El Batan, pp. 73–78.

Nguyen, H.T., Babu, R.C. and Blum, A. (1997) Breeding for drought resistance in rice: physiology and molecular genetics considerations. *Crop Science* 37, 1426–1434.

Noctor, G. and Foyer, C.H. (1998) Ascorbate and glutathione: keeping active oxygen under control. *Annual Review of Plant Physiology and Plant Molecular Biology* 49, 249–279.

Podlich, D.W. and Cooper, M. (1999) Modeling plant breeding programs as search strategies on a complex response surface. *Lecture Notes in Computer Science* 1585, 171–178.

Qiao, G.C., Wang, Y.J., Guo, H.A., Chen, X.J., Liu, J.Y. and Li, S.Q. (1996) A review of advances in maize production in Jilin Province during 1974–1993. *Field Crops Research* 47, 65–75.

Rajaram, S., Braun, H.-J. and van Ginkel, M. (1996) CIMMYT's approach to breed for drought tolerance. *Euphytica* 92, 147–153.

Ribaut, J.–M. and Betrán, J. (1999) Single large-scale marker-assisted selection (SLS–MAS). *Molecular Breeding* 5, 531–541.

Ribaut, J.-M. and Poland, D. (eds) (2000) *Proceedings of Workshop on Molecular Approaches for the Genetic Improvement of Cereals for Stable Production in Water-Limited Environments.* CIMMYT, El Batan, Mexico.

Ribaut, J.-M., Edmeades, G.O., Betrán, F.J., Jiang, C. and Bänziger, M. (1999) Marker-assisted selection for improving drought tolerance in tropical maize. In: O'Toole, J. and Hardy, B. (eds) *Proceeding of the International Workshop on Genetic Improvement for Water-Limited Environments.* IRRI, Los Baños, pp. 193–209.

Richards, R.A. (1992) Increasing salinity tolerance of grain crops – is it worthwhile? *Plant and Soil* 146, 89–98.

Saini, S.H. and Lalonde, S. (1998) Injuries to reproductive development under water stress, and their consequences for crop productivity. *Journal of Crop Production* 1, 223–248.

Saini, H.S. and Westgate, M.E. (2000) Reproductive development in grain crops during drought. *Advances in Agronomy* 68, 59–96.

Shannon, M.C. (1998) Adaptation of plants to salinity. *Advances in Agronomy* 60, 75–120.

Thomas, H. and Howarth, C.J. (2000) Five ways to stay green. *Journal of Experimental Botany* 51, 329–337.

Tollenaar, M. and Wu, J. (1999) Yield improvement in temperate maize is attributable to greater stress tolerance. *Crop Science* 39, 1597–1604.

Zinselmeier, C., Habben, J.E., Westgate, M.E. and Boyer, J.S. (2000) Carbohydrate metabolism in setting and aborting maize ovaries. In: Westgate, M.E. and Boote, K.J. (eds) *Physiology and Modeling Kernel Set in Maize.* CSSA Special Publication No. 29. CSSA, Madison, Wisconsin, pp. 1–13.

Biotic Stresses in Crops

10

R. Nelson

International Potato Center, PO Box 1558, Lima 12, Peru

Introduction

Agricultural production has changed rapidly over recent decades, under pressure to keep pace with growing human populations. Agricultural intensification has, in turn, led to increasing pest pressures. Pesticide use is now the dominant global approach to management of biotic stresses, in spite of promising developments in resistance breeding and integrated pest management (IPM) and a rising awareness of the negative effects of pesticides. Even with massive use of pesticides, crop losses have increased as a consequence of intensification (Oerke *et al.*, 1994; Pimentel and Greiner, 1997).

Available estimates suggest that there is more scope for increasing food supplies by reducing the impact of biotic stresses than by increasing yield potential. Crop losses are significant, particularly for resource-poor farmers. Pests and diseases contribute to food insecurity despite the availability of many IPM strategies and tactics, because farmers lack access to available knowledge, skills and technologies and/or because pressures and incentives favour the use of pesticides. New avenues of research, such as for biotechnological innovations in plant breeding, have led to new information and insights, but have had little impact on pest problems in developing countries. If existing knowledge and technology cannot reach small, resource-poor farmers, it is doubtful that future advances will help them, unless investment is made in increasing farmers' access and local capacity for innovation.

This chapter attempts an overview of two broad areas of flux concerning the management of biotic stresses in crops. The first pertains to the flow of resistance genes from crop germplasm to farmers' fields. This section discusses recent developments in plant breeding, particularly in the area of biotechnology. The second concerns the flow of information and technology from research institutions to farmers and policy-makers. This section addresses trends in the way pest management problems are conceptualized, and changes in the roles of researchers, extensionists, farmers and policy-makers. These processes will determine the practical impact of innovations in crop science. The issues are considered primarily in the context of developing countries, drawing on case studies from rice and potato.

Crop losses due to biotic stresses

It is difficult to assess precisely the importance of biotic stresses in crop production, particularly in the context of developing countries. Statistics are scant and dubious. Yield losses may be overestimated for several reasons, such as the separate estimation of losses caused by multiple pests and diseases. Pimentel *et al.* (1991) cite an instance in which >100% damage was estimated when separate estimates of losses due to insects, disease and weeds were added up. It is likely that negative yields would be obtained for most crops if all the losses claimed by pathologists, entomologists and nematologists were combined.

Even if taken with a grain of salt, the existing global crop-loss estimates are impressive. The Natural Resources Institute (1992) has estimated that 30–40% of potential global food, fibre and feed is lost to insects, nematodes, diseases and weeds. These losses are valued at US\$300 billion year^{-1}. Estimated losses in the developing world are as high as 60–70% of potential production, when post-harvest losses are included. It has been estimated that if pests went uncontrolled, 70% of global production would be lost (Oerke *et al.*, 1994); it is interesting that this figure corresponds roughly to current loss estimates for developing countries.

A case study: attempts to estimate yield losses due to potato late blight

Potato is a crop for which yields are more likely to increase through reduction of losses than through increases in yield potential, particularly in developing countries. Although the crop can produce impressive amounts of food per hectare (yields of 88–104 t ha^{-1} have been achieved in the US and New Zealand, with a dry matter content of around 22%; Oerke *et al.*, 1994), it is vulnerable to many pests and diseases. While cereal breeding programmes emphasize yield potential, potato breeding programmes focus mostly on quality and pest resistance.

The most important biotic constraint to potato production is late blight, caused by the oomycete *Phytophthora infestans*. This disease, which gained infamy for its role in the Irish Potato Famine, has recently become increasingly difficult to manage, due to global migrations of pathogen strains with broad virulence and resistance to the systemic fungicide metalaxyl (Fry *et al.*, 1993). By the mid-1990s, anecdotal evidence suggested that late blight epidemics were increasingly severe worldwide, yet few data were available on the extent of the problem, particularly in developing countries.

Three approaches are being taken by researchers at the International Potato Center (CIP) to analyse yield losses due to late blight. A survey of expert estimates indicated that annual direct losses due to the disease could be valued at approximately US$2.75 billion, and that farmers spend an additional US$740 million on fungicides (T. Walker, personal communication). Because late blight is strongly driven by weather conditions, losses could also be estimated through the use of a late blight forecasting programme linked to a geographic information systems database fed with geo-referenced weather data (Hijmans *et al.*, 2000). Using the number of sprays needed for a susceptible variety as a proxy for late blight risk, a world map was produced to indicate risk zones. This approach is currently being used to evaluate the potential impact of interventions such as the use of potato varieties with particular levels of quantitative resistance.

To complement approaches based on surveys and simulation, field-level estimates were obtained for key sites. A set of baseline studies was conducted for six countries for which intensive interventions were to be undertaken (these are described in the last case study). Although resistant varieties were available in some of the countries, the majority of farmers used susceptible varieties. In the Andes, the weather strongly favoured the disease, and late blight was shown to have a strong effect on yield (r^2 for the damage–yield relationship in Bolivia was 0.82; PROINPA, 1998, personal communication). The disease often drove yields to zero, and 17% of the plots studied in Peru were completely destroyed by the disease (Ortiz *et al.*, 1999).

By combining survey information, simulation and field-level data collection, blight researchers are developing a clearer idea of where the disease is most important, while simultaneously refining their strategies for disease management.

Breeding for Resistance to Biotic Crop Stresses

International public sector plant breeding programmes have a mixed record with regard to biotic stresses. On the one hand, the first products of the Green Revolution are widely known for the pesticide dependence that they initially fostered, which led to crises in pest management (see case study below on rice IPM in South-East Asia). Resistance breeding has subsequently produced many varieties with useful levels of resistance to key pests, although the successes are interspersed with sobering instances in which resistance has been rapidly defeated.

Overall, resistance breeding has proven to be an excellent investment. In some cases, single disease or pest resistance genes have had enormous impacts. For instance, the *Lr34* and *Sr2* genes for resistance to leaf and stem rust in wheat, respectively, have been widely deployed and have provided durable resistance. The 1B/1R translocation in wheat, which carries multiple resistances, is present in more than half the wheat area in developing countries, or almost 40 million hectares. The rice blast resistance in IR64, which has been grown on millions of hectares in South-East Asia, remains useful. The *Xa4* gene in rice, which was once important in controlling epidemics of bacterial blight, no longer provides qualitative resistance, but still conditions quantitative resistance to the disease.

In spite of excellent returns to investment in conventional plant breeding, little glamour has been attributed to it in recent years. The limelight has shone instead on biotechnology. Relatively large amounts of resources have been invested in crop biotechnologies over the past two decades. Much has been promised and much expected from both marker-assisted selection and genetic engineering. Most recently, a set of high-throughput technologies ('genomics') has once again enchanted the research community with a new set of promising possibilities. From the researchers' point of view, biotechnologies have become steadily more powerful and the potential for positive impact remains tantalizingly promising.

Among the various constraints to crop productivity, biotic stresses provide some of the more obvious targets for molecular genetic analysis, genetic improvement through marker-assisted breeding, and genetic engineering. The application of molecular methods has greatly enhanced our understanding of the structure and function of plant genomes and genes. For each of the major food crops, a series of both major genes and minor genes has been identified and mapped to the plant chromosomes. The number of cloned genes has risen steadily and now is suddenly skyrocketing with the launching of massive cDNA sequencing projects.

Monogenic resistance

Monogenic types of resistance are the easiest to handle both by conventional breeding and by molecular techniques. Unfortunately, monogenic resistance has generally been shown to be non-durable (with notable exceptions), particularly when pest pressure is high. For instance, monogenic resistance for rice blast disease has been rapidly overcome by new pathogen strains in temperate Asia, where environmental conditions are favourable to the disease, while this type of resistance has been very useful for management of rice blast in South-East Asia, where environmental conditions are less conducive to disease development. For potato late blight, major genes are notoriously short-lived. Despite this patchy record, many attempts at direct gene transfer have focused on monogenic resistance. Some of the early transgenes were derived from microbes, such as the toxin genes of *Bacillus thuringiensis* (Bt) and virus coat protein genes. More recently, cloned plant resistance genes have become available. For instance, efforts are being made in China to deploy transgenic rice carrying the cloned *Xa21* gene for resistance to bacterial blight of rice (Zhai *et al.*, 2000), although bacterial populations virulent to this gene have been documented in several Asian countries (e.g. Lee *et al.*, 1999; Ochiai *et al.*, 2000).

Whether incorporated into crops by conventional breeding or genetic engineering direct gene transfer, resistance mediated by major genes is unlikely to be durable in many cases. To compensate for the inherent vulnerability of host resistance in the face of pathogen evolution, researchers have proposed a range of deployment strategies (Mundt, 1998). To manage conventional resistance, efforts have focused on using diverse sources of genes, on accumulating multiple minor factors conditioning resistance, on combining major genes with complementary resistance spectra, and employing population-level genetic diversity in time and space. Population-level strategies include the sequential deployment of lines carrying different genes (temporal diversity) and spatial patches of genetically distinct genotypes (with mixtures or multilines representing the small-patch extreme of spatial diversity).

Researchers working with transgenic resistance borrow deployment strategies at the population level, and are able to refine deployment strategies at the gene level. Careful selection of gene and promoter sequences can allow researchers to specify a type of resistance, a level of gene expression, and a developmental and tissue-specific pattern of expression. The potential or demonstrated breakdown of Bt-derived resistance has heightened awareness about the need to manage pest resistance genes as precious genetic resources. When it approved the first Bt crops in 1995, the US Environmental Protection Agency required the companies making the releases to implement plans for managing resistance in the pest populations. The most widely used strategies involve high levels of expression of toxin genes (to kill insects heterozygous for resistance to the toxin) and non-transgenic refuge plants (to promote the survival of susceptible insects). The weakness of genetic engineering strategies to date is the heavy reliance on single genes. It is anticipated that complexes of transgenes will be used in the future, to generate novel forms of quantitative resistance.

The cloning and sequencing of plant resistance genes has been a landmark success for plant molecular biology (for a review, see Hammond-Kosack and Jones, 1997). The resulting increase in knowledge of the molecular basis of host–pathogen interactions is of high scientific merit. In some cases, the use of resistance genes is likely to have major practical value. This may be true for potato viruses, where several R-genes have shown durable resistance. Wider use of virus resistance would allow farmers to maintain seed quality for longer periods, enabling them to re-use their seed for a greater number of seasons before purchasing fresh stocks. This may not be in the interests of the seed industry, but impressive advances have been made in the public sector, particularly at the UK's John Innes Centre.

For the moment at least, public opinion does not appear to support the invention of new genes and novel routes of gene flow. Consumers are more likely to accept the use of crop biotechnology to make more efficient use of natural variation within crop gene pools. Molecular techniques can be of service in various ways, from providing insights into the genetic structure of germplasm collections, to marker-assisted selection or introgression of desired alleles, to direct gene transfer of desired alleles among plants.

Marker-assisted analysis of quantitative resistance

There is a consensus that quantitatively inherited resistance is more durable than major-gene resistance for several pest systems. That is, resistance controlled by multiple genes, each of which contributes to the phenotype in a small way, is the most difficult for pathogens to overcome through the processes of mutation and selection.

For several diseases and insect pests, molecular markers have been used to locate quantitative trait loci (QTL) conditioning partial resistance. These studies have provided new insights into the numbers and locations of chromosomal segments involved in various types of resistance. But unfortunately, QTL mapping allows only a very imprecise and incomplete localization of the genes involved in conditioning a quantitative trait. The level of resolution and certainty achieved in most QTL mapping studies is so limited that it does not effectively set the stage either for marker-assisted selection or for map-based gene cloning.

Over the past few years, however, gene-mapping technologies have increasingly been brought together with the products of studies of trait physiology, biochemistry and molecular biology. Using the 'candidate gene approach', genes known or suspected to play a role in biochemical pathways involved in a trait are used as markers in genetic studies. Candidate genes that are co-localized with QTL effects are considered as likely suspects to actually control the trait. If this hypothesis can be confirmed on a substantial scale, the problems of precision for marker-assisted selection and cloning could be overcome. Defence-related genes have been co-localized with QTLs for quantitative resistance to several diseases and insect pest systems (e.g. Byrne *et al.*, 1996; Faris *et al.*, 1999; Geffroy *et al.*, 2000). To determine whether the candidate genes actually account for the QTLs in question, the alleles of interest will have to be cloned and transferred into susceptible genotypes.

Because considerable investigation has been done on the physiology and molecular biology of plant defence, many candidate genes can be identified from the literature for disease resistance. There are many fewer candidate genes for insect resistance. New developments in large-scale cloning and sequencing of cDNAs have made it possible to identify whole suites of genes that are induced by a particular challenge, such as infection or pest attack. Microarray technologies, in which the expression of staggering numbers of genes can be assayed in a single experiment, can reveal whole-genome pictures of patterns of gene expression under different conditions. By combining gene mapping with genomics approaches that allow more efficient and complete identification of genes involved in conditioning resistance, it should soon be possible to clone QTLs and to achieve efficient gene flow through direct gene transfer. The potential impact of this promising research area will depend on the availability of funds for research, and public acceptance of transgenic resistance using natural, plant-derived genes.

Breeding and biotechnology: a long engagement

Although much has been said about the potential power of integrating breeding and biotechnology, researchers in these two areas have been working more in parallel than together. Both have been productive areas of research, but their products are different. Breeders produce new and improved varieties, and biotechnologists produce better tools, biological insights and promises to transform breeding.

Many of the anticipated pay-offs of investments in plant biotechnology have been elusive to date. Marker-assisted selection is not a practical reality for most breeding programmes in the public sector. This is the case for even wealthy institutions, and obviously more so for resource-poor research organizations in developing countries. Young (1999) summarizes the history of marker-assisted genetic analysis and selection, highlighting the relatively short history of this research area, the tremendous challenge involved, and the complexity of the analytical procedures. He cites the publication of 400 papers containing the key words 'marker-assisted breeding' or 'marker-assisted selection' (MAS), noting that few (if any) of these studies report the use of MAS leading to release of germplasm or varieties. Young (1999) summarizes the reasons why marker-assisted selection has been difficult to realize in practical breeding programmes. The key limitation is that most agriculturally important traits are quantitatively inherited, and it has proven difficult to map QTLs accurately with the relatively small populations and phenotypic data sets that most laboratories are prepared to handle. Imprecise and inaccurate mapping results do not provide a sufficiently sound basis for MAS.

Impact has been seen in some cases, however, and more may become apparent in the near future. For the soybean cyst nematode, molecular markers associated with single resistance genes have been very useful in breeding programmes because of difficulties in scoring the resistance phenotype by conventional means. For both bacterial blight and blast of rice, major genes for disease resistance have been combined using marker-assisted selection and back-cross breeding, which in some instances would be difficult to achieve using conventional breeding because the presence of one broad-spectrum resistance gene can mask the presence of other resistance genes (e.g. Yoshimura *et al.*, 1995). With the support of donors such as the Rockefeller Foundation and research institutions such as the International Rice Research Institute, national programmes are applying these techniques to incorporate resistance genes for major diseases and pests of rice into genotypes of local importance.

Although the practical implementation of marker-assisted selection has been disappointing to date, our knowledge of the genes controlling diverse types of resistance continues to improve, and will be radically enhanced through the application of new genomics approaches. These research investments will eventually make it possible to greatly enhance the flow of useful alleles from gene-banks to varieties. It is not yet clear who will control the results of the costly genomics research, and what patterns of investment in biotechnology are now appropriate for small laboratories and resource-poor national programmes, to allow them to take advantage of the anticipated developments.

A case study: improving resistance to potato late blight

Strong arguments could be made for the genetic engineering of potato. A great deal of pesticide is used for potato in rich and poor countries alike. Effective natural enemies are not available for many tuber-infesting pests, nor are they an option for diseases such as late blight. Neither pesticides nor compelling levels of genetic resistance are available for bacterial wilt (caused by *Ralstonia solanacearum*), an important bacterial disease of the crop.

Conventional breeding is relatively challenging for potatoes. Potato is tetraploid and highly heterozygous, so it is difficult to determine the genetic basis of a given trait. The crop is subject to inbreeding depression, so backcrossing desirable genes into valued genotypes is not an option. Potato breeding is a lengthy process, often taking 20 years between cross and release. Even when promising new varieties are available, varietal change is slow in many countries, particularly where a substantial proportion of the crop is grown for processing. Markets are often inflexible with regard to the desired characteristics of a potato. Markets in developing countries are more open to varietal change than markets in industrialized countries, but adoption of a new variety can be difficult. While potato breeding is difficult, genetic engineering is relatively feasible. The crop can be transformed by simple methods such as *Agrobacterium*-mediated transformation. Potatoes were among the first transgenic crops.

Potato germplasm is rich in resistance to late blight; many wild relatives of potato show quantitative and/or qualitative resistance to the disease. There is an urgent need to make better use of this resistance, either by conventional breeding and/or by direct gene transfer. With the goal of producing genotypes with durable resistance, a recurrent selection programme has been in progress for nearly two decades at the International Potato Center. The levels of resistance in the breeding populations have steadily

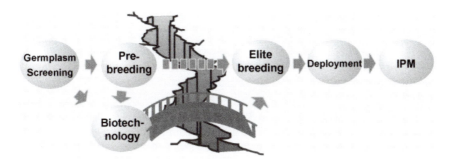

Fig. 10.1. Schematic view of the flow of disease resistance alleles from germplasm collections, through a breeding programme, and to farmers in the context of an IPM effort. The barrier to uptake of alleles by an elite breeding programme may be partially overcome through the use of biotechnology (marker-assisted selection and/or direct gene transfer).

increased, presumably through the accumulation of minor genes for resistance. Materials produced by CIP's breeding programme are selected for use in many developing countries.

Genetic engineering could help keep a strong flow of genes from germplasm into farmers' fields (Fig. 10.1). Even when market limitations do not pose insurmountable obstacles to the adoption of newly bred genotypes, potato breeders are justifiably reluctant to cross their elite materials with wild species because it is so difficult to recover a desired genotype. It would be extremely desirable to be able to transfer genes among potato genotypes without having to scramble the genotypes in the way that occurs in every cross.

Genetic maps have been produced for potato, and for several crosses the genes conditioning quantitative resistance have been located to segments of the potato chromosomes (e.g. Leonards-Schippers *et al.*, 1994). Several large-insert libraries have been prepared, with the intention of cloning of resistance genes. Initial interest has focused on major genes, because of the relatively precise genetic localization that can be achieved for this type of gene. With the identification of candidate genes associated with QTLs, it may soon be possible to clone the genes that control quantitative resistance as well. It remains to be seen whether the exploitation of quantitative resistance through direct transfer of potato genes among potatoes, which could allow for substantial decreases in pesticide use, will be acceptable to the public.

Trends in Integrated Pest Management (IPM)

In the present era, pests and diseases are to a large extent managed through the use of pesticides. Farmers use approximately 2.5 million tons of pesticides each year, spending US$20 billion (*Pesticide News* 1990, cited by Pimentel *et al.*, 1992). Pesticides are perceived to provide good returns to investment. Farmers are estimated to gain from US$3 to US$5 for every dollar they invest in pesticides. These estimates, however, do not include indirect costs associated with pesticide use, such as damage to human and environmental health, secondary pest problems caused by loss of natural enemies, and the evolution of pesticide resistance. Public awareness of these problems is rising and these costs are increasingly taken into account in cost–benefit analyses of pesticides (Crissman *et al.*, 1998).

Many of the problems associated with pesticide use have been appreciated for a long time. There is increasing documentation of problems of morbidity and mortality among pesticide applicators (Pingali *et al.*, 1994; Antle *et al.*, 1998), consumer health issues related to pesticide residues, pest resurgence and wider environmental problems. The World Health Organization recently reported that 3 million people suffer severe pesticide poisonings each year, resulting in 220,000 deaths. Although only 20% of the world's pesticides are used in developing countries, more than half of the poisonings and deaths occur in the developing world (Pimentel and Greiner, 1997).

The concept of IPM is evolving as a framework for bringing together a range of tactics for pest management. Definitions of IPM are diverse, and increasingly recognize the range of strategies employed and the desired outcomes. IPM strategies may entail implicit or explicit aims of rationalizing, reducing and/or eliminating the use of pesticides. Chemical-based IPM strategies aim at improving the efficiency and effectiveness of pesticide use through scouting and forecasting, as well as through the improvements in application technology. At the other extreme is bio-intensive IPM, which relies on natural processes and avoids chemical inputs entirely. IPM can be considered at a range of levels of ecological complexity, from single crop–pest combinations, to the pest complexes relating to a given crop, to the cropping system. Increasing attention is now being given to the higher orders of complexity.

Commonly considered IPM components include the use of rotations, host plant resistance, natural enemies, antagonistic microbes, alternative pesticides such as plant extracts, and conventional chemical control. These elements may interact in complex ways that are difficult to predict. Genetic resistance may interact not only with a pest, but also with the natural enemies that contribute to controlling populations of the pest. This interaction may be positive (additive or synergistic), or it may be negative and actually compromise the efficiency of the resistance (Thomas, 1999).

Crop rotation is an IPM strategy that has been used for millennia to control pests and maintain soil health. Although rotations are known to break pest cycles and/or reduce levels of infestation, reducing the need for pesticide application, they are not necessarily used. Traditional Andean agriculture involves complex rotations and long fallows to manage nematodes. Contemporary farmers in the Andes are facing increasing nematode problems as they intensify production. In the USA, rotation is recognized to reduce pest pressure and allow reduced pesticide use in maize and cotton, but 25% of US maize crops and 60% of cotton crops were grown without rotation for 1991–1993 (Agricultural Resources and Environmental Indicators, 1994).

The research community is steadily building a body of knowledge that relates to the complexities of agroecosystems and the ways in which this information can be used to design increasingly sustainable pest management systems. This body of information can only be useful if it can be utilized by farmers. Unless farmers are able to become knowledgeable and ecologically responsible decision-makers, IPM must remain a largely academic concept. The next section considers the factors that affect farmers' access to information related to the management of biotic stresses.

Information Flows in Pest and Disease Management

The ecological interactions upon which IPM is based are complex, and successful IPM requires substantial knowledge. In addition, impact at the farm level can only be realized if public policies make IPM economically and practically feasible. Key policy issues include support for IPM research and extension; subsidies, bans and other factors affecting pesticide prices; support for incentive and training programmes for farmers; and programmes that enhance farmer and consumer awareness of the health effects of pesticide use. For these reasons, it is crucial that appropriate research results be generated, that policy-makers create appropriate conditions, and that farmers understand and implement IPM.

Farmers cannot recover losses that have already occurred, so they make decisions regarding pest management in order to prevent the crop losses that they fear or anticipate. To make sound predictions and rational decisions, farmers need basic agroecological knowledge and access to risk-reducing technologies such as resistant varieties. Integrated management of crop diseases poses particular problems for poor farmers, because of the invisible nature of the microbes that cause plant disease. Several studies dealing with farmers' knowledge of plant disease in developing countries have found that while farmers are often quite knowledgeable about the environmental factors that affect disease development, they are unaware of the microbial nature of the causal agent, and they are poor at differentiating symptoms and diseases (Bentley and Thiele, 1999; Ortiz *et al.*, 1999). This lack of understanding hampers farmers' ability to manage crop diseases.

There are tremendous weaknesses in the systems that enable flow of information among researchers, extensionists and farmers. In developing countries, the agricultural research sector is often critically under-funded, and in many countries, no formal extension system exists. Communications between research and extension organizations are often minimal. The following case studies illustrate examples in which efforts at increased communication among researchers, extensionists and farmers have improved farmers' ability to avoid crop losses due to biotic stresses. In these and many other cases, the roles of farmers, extension organizations and researchers in technology development, implementation and integration is being considered creatively. Farmers are increasingly seen as partners in the development and local adaptation of both seed-based and information-based technologies.

A case study: insect pest management for rice in South-East Asia

Management of the brown planthopper (*Nilaparvata lugens*) in South-East Asia offers a particularly spectacular example of the consequences of knowledge flows among researchers, extension organizations, farmers and governments. During the 1950s and early 1960s, Indonesia was the world's largest rice importer and food was scarce. In response to turbulence and food insecurity, the Indonesian government instituted a series of policies aimed at achieving self-sufficiency in rice production. High-yielding rice varieties, bred at the International Rice Research Institute (IRRI), were introduced into South-East Asian rice production systems beginning in the late 1960s. Government extension programmes in Indonesia and the Philippines aggressively promoted the use of these varieties, together with both fertilizers and pesticides.

The new technology led to increased but destabilized rice production. Rice yields in Indonesia doubled between the late 1960s and the mid-1980s, and Indonesia became self-sufficient in rice production in 1984. But while the newly bred rice varieties were responsive to fertilizer inputs, they were susceptible to diseases and insect pests, and massive pest outbreaks occurred in spite of the huge increases in insecticide use, as brown planthopper damage occurred in 1975–1977, and again in 1985–1986 (van de Fliert, 1993). Research conducted in the late 1970s indicated that brown planthopper outbreaks were caused by pesticide-induced mortality of the pests' natural enemies (Kenmore, 1980). By the mid-1980s, it was apparent that insecticide use was not cost-effective in rice.

In Indonesia, the government responded by banning 57 broad-spectrum pesticides, gradually eliminating pesticide subsidies, reducing pesticide production, and implementing a national IPM programme. Rice varieties resistant to the brown planthopper were available from IRRI by that time, and were widely accepted in Indonesia. The government declared IPM to be its national policy, and a large-scale farmer-training programme was implemented to enhance ecologically based pest management. These actions were taken in consultation and collaboration with a team of FAO researchers and educators, which developed an innovative adult-education approach known as the Farmer Field School (FFS).

Beginning in 1986, farmers (in groups of approximately 25) learned about rice-field ecology as a basis for improved decision making in rice pest management. With the support of the FAO programme and national and local extension organizations, millions of farmers were trained through participation in weekly FFS sessions throughout a complete cropping cycle. The IPM programme was widely recognized for effecting decreases in pesticide use without increases in crop losses. The pesticide bans were crucial for providing an environment conducive to the success of the programme.

In the Philippines, the national response to the pesticide crisis was much slower. In spite of concerns over the health and environmental effects of massive pesticide use, and growing evidence of negative impacts, the Philippine government did not ban the most hazardous pesticides until 1992. A national IPM programme involving large-scale farmer training through FFS was initiated only in 1993, although the FAO's regional IPM programme had its headquarters in the Philippines. The delay in the government response to the inappropriate use of pesticides in rice, particularly in comparison with the response in neighbouring Indonesia, has been attributed to political environments that valued the interests of agrochemical companies over farmers' health (Loevinsohn and Rola, 1998).

There is consensus among scientists and economists that pesticide use is unnecessary and indeed counterproductive for pest management in rice production in South-East Asia. Natural biological control is extremely efficient in that system, perhaps because of the exceptional richness of rice-field biodiversity. In a field school on rice, farmers can learn to respect the power of natural enemies and the ability of the rice plant to compensate for reasonable levels of leaf damage. These natural processes may not be so effective in every pest system, however; attentive non-intervention will not always lead to a good harvest. The following case study deals with such a system.

A case study: disease management for potato in the Andes

As noted earlier, potato farmers around the world are struggling with increasing epidemics due to late blight. Although Andean farmers have been growing potatoes for thousands of years, they are unable to cope with the disease, in part because recent pathogen migrations present farmers with a new and dangerous situation. Moderate levels of genetic resistance are present in varieties, which may or may not be available to resource-poor potato farmers. In any event, available levels of resistance are not sufficient to protect crops from the disease, and crop losses are common.

On a worldwide basis, the disease is primarily controlled through the use of fungicides. For resource-poor farmers, however, fungicides are not a good solution. Fungicides are often used unsafely and ineffectually, and the chemicals may be unaffordable when they are needed. Varietal resistance offers the only powerful alternative to fungicide use, as most agronomic and biological control tactics are of minimal utility for this disease. Desperate farmers are willing to adopt new disease-resistant varieties. The situation in poor countries is the reverse of that in rich countries, where precise and timely use of fungicides is relatively feasible and cost-effective, while there is little perceived scope for varietal change. Most research effort on late blight in industrialized countries has dealt with the use of computer-based forecasting methods for improving the efficiency and effectiveness of fungicide use.

Studies on farmer knowledge, opinions and practices were conducted for key potato production areas in developing countries. Interviews confirmed that late blight was farmers' top concern in the Andes (Ecuador, Peru and Bolivia) as well as in China, while bacterial wilt (caused by *Ralstonia solanacearum*) was equally important in sub-Saharan Africa (Uganda) (O. Ortiz, CIP, personal communication). Although farmers in the Peruvian Andes consider themselves potato growers and potato eaters, they are often unwilling to plant potatoes in the rainy season because of the high probability of enormous crop losses due to late blight. In the dry season, potato yields are limited by lack of water. Farmers spray when they see a late blight problem, at which point it is often too late to avoid heavy losses.

In part inspired by the success of Farmer Field Schools for rice in Asia, training programmes have been initiated on a pilot scale for potato farmers in seven countries. In each of the countries, season-long potato field schools are being developed through collaborations involving local extension organizations, local research organizations, and the International Potato Center. Farmers use mini-microscopes to observe the pathogen and disposable plastic boxes to conduct experiments on disease aetiology. Understanding the concept of latent period helps farmers avoid spraying fungicides too late to save the crop. Knowledge of the pathogen allows farmers to understand the biological basis for a number of important disease-management tactics, such as high hilling and elimination of diseased foliage before harvest to avoid infection of the tubers. In addition to the hands-on agroecological training activities, the potato farmers conduct field experiments to test promising new breeding lines. The farmer groups also learn about and investigate a range of other pest and crop management issues. Farmers take data on their study plots, recording their observations on posters to share within their communities, with other communities during 'field days', and with researchers (Fig. 10.2).

In Peru, pilot-scale field schools have been conducted for 3 years. The field schools have focused on providing farmers with useful knowledge relating to disease management, and access to resistant varieties and breeding lines. The participating communities have rapidly adopted the best-performing new varieties tested, aided by credit provided by the non-

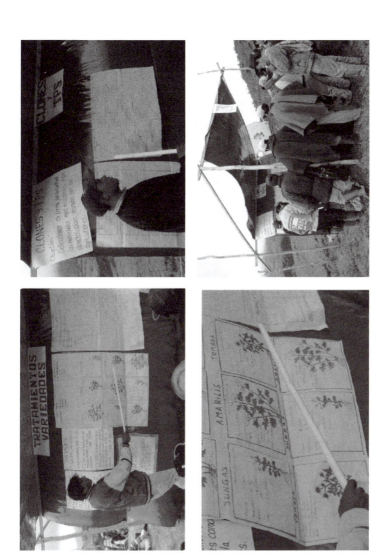

Fig. 10.2. Farmers at a 'field day' in Cajamarca, Peru, presenting the results of experiments conducted as part of FPR-FFS (farmer field schools incorporating farmer-participatory research). The FPR-FFS were organized through a collaboration between the non-governmental organization CARE-Peru and the International Potato Center.

governmental organization CARE, which is implementing the field schools. In Bolivia, pilot field schools focused on improving the efficiency of fungicide use. This was shown to be extremely profitable to the farmer participants. Returns of $2000 per hectare were documented in a year with strong late blight pressures. A range of farmer training methods was utilized and the more intensive, participatory approaches were found to produce the most impact (Torrez *et al.*, 1999).

Policies relating to fungicides are extremely important in determining farmers' approach to disease management. Some of the commonly used fungicides are suspected human carcinogens, but considering the quantities of fungicide that are applied to potato crops, surprisingly little data are available on the health effects of these chemicals. Andean farmers apply fungicides with backpack sprayers, often without protective clothing, and complain of feeling bad after spraying. Similarly, little is known about the effects of fungicide use on insect natural enemies, or on beneficial soil microbes. If no health and/or environmental cost is attributed to fungicides, farmers in most developing countries would be well advised to spray more fungicides. On the other hand, if these chemicals were known to cause significant health impacts, it would be much easier to leverage adoption of resistant varieties.

Challenges

Impressive research advances have been made in breeding, biotechnology and IPM. Public awareness of the negative impacts of pesticides is increasing, as the effects on human health and environment are better understood. Yet the pesticide paradigm retains a strong hold in the management of biotic stresses of crops. Wide-scale change will require continuing investment in resistance breeding and agroecological research. Unless this investment is made in the public sector, the results are unlikely to benefit the world's poor. In addition to research, it is essential that efforts be made to ensure that poor farmers have access to appropriate knowledge, skills and technology. Economic and/or legislative pressures will also be important in leveraging a shift from the pesticide paradigm to more sustainable approaches to pest and crop management.

It remains to be seen whether plant biotechnology will contribute significantly to meeting humanity's future food needs. This is an important question not only for the major food crops, which have received substantial research investments in breeding and biotechnology, but also for the 'orphan crops'. There are many crops with tremendous local nutritional and cultural importance, for which specific pest problems have not been resolved. For instance, integrated pest management has not successfully controlled damage by the sweet potato weevil in sub-Saharan Africa, and transgenic approaches are considered to provide much-needed 'last resort' management options. Similarly, weevils have nearly driven the Andean tuber crop oca

(*Oxalis tuberosa*) out of cultivation. Oca is important in marginal mountain areas where there are few other adapted crops. Transgenic approaches could be of great importance in preserving an important crop and thus sustaining local biodiversity.

This chapter has attempted to summarize trends that affect the flow of genes through breeding programmes, and the flow of information to those who need it. With developments in genomics and internet access, there is scope for dramatic change in both of these flow rates. It may become possible to transfer valuable crop resistance alleles more efficiently in the future, tremendously increasing the utility of crop genetic resources. As noted above, access to information is a limiting factor for resource-poor farmers. Within a few years, the internet may be a mechanism to allow greater sharing of information among farmers, researchers and extension providers. The research community should exert all efforts to ensure that exciting new developments benefit the rural poor.

References

Agricultural Resources and Environmental Indicators (1994) *Agricultural Handbook No. 705.* USDA Economic Research Service.

Antle, J., Cole, D. and Crissman, C. (1998) Further evidence on pesticides, productivity and farmer health: potato production in Ecuador. *Agricultural Economics* 18, 199–207.

Bentley, J. and Thiele, G. (1999) Bibliography of farmer knowledge and management of crop disease. *Agriculture and Human Values* 16, 75–81.

Byrne, P.F., McMullen, M.D., Snook, M.E., Musket, T.A., Theuri, M.J., Widstrom, N.W., Wiseman, B.R. and Coe, E.H. (1996) Quantitative trait loci and metabolic pathways: genetic control of the concentration of maysin, a corn earworm resistance factor, in maize silks. *Proceedings of the National Academy of Sciences USA* 93, 8820–8825.

Crissman, C., Antle, J. and Capalbo, S. (1998) *Economic, Environmental and Health Tradeoffs in Agriculture: Pesticides and the Sustainability of Andean Potato Production.* Kluwer Academic Publishers, Dordrecht.

Faris, J.D., Li, W.L., Liu, D.J., Chen, P.D. and Gill, B.S. (1999) Candidate gene analysis of quantitative disease resistance in wheat. *Theoretical Applied Genetics* 98, 219–225.

Fry, W.E., Goodwin, S.B., Dyer A.T., Matuszak, J.M., Drenth, A., Tooley, P.W., Sujkowski, L.S., Koh Y.J., Cohen, B.A., Spielman, L.J., Deahl, K.L. and Inglis, D.A. (1993) Historical and recent migrations of *Phytophthora infestans*: chronology, pathways and implications. *Plant Disease* 77, 653–661.

Geffroy, V., Sevignac, M., Oliveira, J., Fouilloux, G., Skroch, P., Thoquet, P., Gepts, P., Langin , T. and Dron, M. (2000) Inheritance of partial resistance against *Colletotrichum lindemuthianum* in *Phaseolus vulgaris* and co-localization of quantitative trait loci with genes involved in specific resistance. *Molecular Plant–Microbe Interaction* 13, 287–296.

Hammond-Kosack, K. and Jones, J.D.G. (1997) Plant disease resistance genes. *Annual Review of Plant Physiology and Plant Molecular Biology* 48, 575–607.

Hijmans, R.J., Forbes, G.A. and Walker, T.S. (2000) Estimating the global severity of potato late blight with GIS-linked disease forecast models. *Plant Pathology*, 49, 697–705.

Kenmore, P. (1980) Ecology and outbreaks of a tropical insect pest of the Green Revolution, the Brown Planthopper, *Nilaparvata lugnes* (Stal). PhD thesis, University of California, Berkeley, USA.

Lee, S.W., Choi, S.H., Han, S.S., Lee, D.G. and Lee, B.Y. (1999) Distribution of *Xanthomonas oryzae* pv. *oryzae* strains virulent to *Xa21* in Korea. *Phytopathology* 89, 928–933.

Leonards-Schippers, C., Gieffers, W., Schaeffer-Pregel, R., Ritter, E., Knapp, S.J., Salamini, F. and Gebhardt, C. (1994) Quantitative resistance to *P. infestans* in potato: a case study for QTL mapping in an allogamous plant species. *Genetics* 137, 67–77.

Loevinsohn, M. and Rola, A. (1998) Linking research and policy on natural resource management: the case of pesticides and pest management in the Philippines. In: Tabor, S. and Faber, D. (eds) *Closing the Loop: From Research on Natural Resources to Policy Change. Policy Management, Report No. 8*. Maastricht: European Centre for Policy Management, pp. 88–113.

Mundt, C.C. (1998) Deployment of disease resistance genes. In: Hasmid, A.A., Yeang, L.K. and Sadi, T. (eds) *Proceedings of the Rice Integrated Pest Management Conference*, Malaysian Agricultural Research and Development Institute, Perpustakaan Negara, pp. 108–113.

Natural Resources Institute (1992) *A Synopsis of Integrated Pest Management in Developing Countries in the Tropics*. Natural Resources Institute, Chatham.

Ochiai, H., Horino, O., Miyajima, K. and Kaku, H. (2000) Genetic diversity of *Xanthomonas oryzae* pv. *oryzae* strains from Sri Lanka. *Phytopathology* 90, 415–421.

Oerke, E.-C., Dehne, H-W., Schonbeckand, F., Weber, A. and Oerke, E.-C. (1994) *Crop Production and Crop Protection: Estimated Losses in Major Food and Cash Crops*. Elsevier, Amsterdam.

Ortiz, O., Winters, P. and Fano, H. (1999) La percepción de los agricultures sobre el problem del tizón tardío o rancha (*Phytophthora infestans*) y su manejo: estudio de casos in Cajamarca, Peru. *Revista Latinoamericana de la Papa* 11, 97–120.

Pimentel, D. and Greiner, A. (1997) Environmental and socio-economic costs of pesticide use. In: Pimentel, D. (ed.) *Techniques for Reducing Pesticide Use: Economic and Environmental Benefits*. John Wiley & Sons, Chichester, pp. 51–78.

Pimentel, D., McLaughlin, L., Zepp, A., Lakitan, B., Kraus, T., Kleinman, P., Vancini, F., Roach, W., Graap, E., Keeton, W. and Selig, G. (1991) Environmental and economic effects of reducing pesticide use. *BioScience* 41, 402–409.

Pimentel, D., Acquay, H., Biltonen, M., Rice, P., Silva, M., Nelson, J., Lipner, V., Giordano, S., Horowitz, A. and D'Amore, M. (1992) Environmental and economic costs of pesticide use. *BioScience* 42, 750–760.

Pingali, P.L., Marquez, C.B., Palis, F.G. and Rola, A.C. (1994) Impact of pesticides on farmer health: a medical and economic analysis. In: Teng, P.S., Heong, K.L. and Moody, K. (eds) *Rice Pest Science and Management*. International Rice Research Institute, Manila, Philippines, pp. 277–289.

Thomas, M. (1999) Ecological approaches and the development of 'truly integrated' pest management. *Proceedings of the National Academy of Sciences USA* 96, 5944–5951.

Torrez, R., Tenorio, J., Valencia, C., Orrego, R., Nelson, R. and Thiele, G. (1999)

Implementing IPM for late blight in the Andes. In: *Impact on a Changing World. Program Report 1997–1998*. CIP, Lima, pp. 91–99.

van de Fliert, E. (1993) Integrated pest management: farmer field schools generate sustainable practices: a case study in Central Java evaluating IPM training. PhD thesis, Wageningen Agricultural University, The Netherlands.

Yoshimura, S., Yoshimura, A., McCouch, S.R., Baraoidan, M.R., Mew, T.W., Iwata, N. and Nelson, R.J. (1995) Tagging and combining bacterial blight resistance genes in rice using RAPD and RFLP markers. *Molecular Breeding* 1, 375–387.

Young, N. (1999) A cautiously optimistic vision for marker-assisted selection. *Molecular Breeding* 5, 505–510.

Zhai, W., Li, X., Tian, W., Cao, S., Zhao, X., Zhu, L., Zhou, Y., Zhang, Q. and Pan, X. (2000) Introduction of a rice blight resistance gene, *Xa21*, into five Chinese rice varieties through an *Agrobacterium*-mediated system. *Science in China* 43, 361–368.

Management of Complex Interactions for Growth Resources and of Biotic Stresses in Agroforestry

11

C.K. Ong[1] and M.R. Rao[2]

[1]*ICRAF, United Nations Avenue, Gigiri, PO Box 30677, Nairobi, Kenya;* [2]*11, ICRISAT Colony-I, Akbar Road, Cantonment, Secunderabad-500 009, AP, India*

Introduction

Future increase in food and wood production in the tropics will have to be achieved from existing land and water resources, therefore the challenges to crop research are to improve the efficiency with which land and water are currently used. One promising option for improvements of this kind is by using agroforestry, with the ultimate aim of achieving sustainability of production and resource use (Young, 1997). Unlike the Green Revolution concept, agroforestry is perceived as a complex, 'green' but low input technology for marginal lands or fragile lands (Sanchez, 1995). In such conditions, mixtures of trees and crops have the potential to improve land management via their ability to reduce soil erosion, improve nutrient cycling and increase rainfall utilization. However, despite the demonstration that such systems can dramatically reduce soil losses and improve soil physical properties, the beneficial effects on crop yield are often unpredictable and insufficient to attract widespread adoption. In semiarid areas, crop yield increases are rare in alley cropping because fertility and microclimate improvements do not offset the large competitive effect of trees with crops for water and nutrients (Rao *et al.*, 1998). Major disappointments with alley cropping in the semiarid tropics and elsewhere led to a greater emphasis on sequential systems, such as improved fallows, which segregate trees and crops over time in order to remove the undesirable competitive effects of trees. It is only recently that agroforestry researchers have examined the trade-offs between crop yields and environmental functions (light, water, nutrients, pests) from an ecological point of view (Lefroy and Stirzaker, 1999). An ecological approach to agroforestry research might provide a more scientific basis for developing strategies to reduce the trade-offs between crop productivity and environmental functions.

A powerful incentive for promoting agroforestry in the tropics is the consideration of sustainability. Applied to land and water, sustainability means meeting the production needs of present land users whilst conserving for future generations the resources on which that production depends. Agroforestry is defined as a collective name for land use systems in which woody perennials (trees, shrubs, etc.) are grown in association with herbaceous plants (crops, pastures) or livestock, in a spatial arrangement, a rotation, or both, and in which there are usually both ecological and economic interactions between the trees and other components of the system. In simplified terms, agroforestry means combining the management of trees with productive agricultural activities. Agroforestry provides opportunities for forest conversion in the true sense of the term, replacement of natural forest with other tree-based land use systems. There are also opportunities to use agroforestry for the prevention or reversal of land degradation in the sub-humid and humid tropics.

This chapter describes the rationale and the basic physiological concepts concerning the complex interactions between trees and crops for growth resources and pests, in order to develop better strategies for management of agroforestry systems.

Rationale of Agroforestry

It is often assumed that appropriate agroforestry systems can provide the environmental functions needed to ensure sustainability and maintain microclimatic and other favourable influences, and that such benefits may outweigh their complexity (Sanchez, 1995). It is also assumed that agroforestry might be a practical way to mimic the structure and function of natural ecosystems, since components of the latter result from natural selection towards sustainability and the ability to adjust to perturbations (van Noordwijk and Ong, 1999). It is this opportunity for agroforestry to mimic the interactions between trees and other plants in natural ecosystems that led to the recent redefinition of agroforestry, in which different agroforestry practices are viewed as stages in the development of an agroecological succession akin to the dynamics of natural ecosystems (Ong and Leakey, 1999). Recent reviews of agroforestry findings have, however, highlighted several unexpected but substantial differences between intensive agroforestry systems and their natural counterparts, which would limit their adoption for solving some of the critical land use problems in the tropics (Ong and Leakey, 1999; van Noordwijk and Ong, 1999). The key differences are the low density of trees in savanna ecosystems and the high proportion of woody aboveground structure compared with foliage in mature savanna trees. The most intractable problem for agroforestry in the semiarid tropics is how to manage the complex interactions between trees and crops for water, light and nutrients.

Resource Capture Concept

The principles of resource capture have been used to examine the influence of agroforestry on ecosystem function, i.e. the capture of light, water and nutrients (Ong and Black, 1994), and to understand better the ecological basis of sustainability of tropical forests. The concept of complementary resource use is not new in ecological studies and it was proposed by de Wit (1960) and others that mixtures of species may have greater capacity to exploit growth resources and hence be more productive than monocultures. However, recent re-interpretations of published results indicate that increased yield of combinations of annual crops was not always associated with greater resource capture or utilization. Nevertheless, current ideas on agroforestry interactions continue to be rooted in the complementary resource use concept. For example, Cannell *et al.* (1996) proposed that successful agroforestry systems depend on trees capturing resources that crops cannot. The capture of growth resources by trees and crops can be grouped into three broad categories to show competitive, neutral or complementary interactions (Fig. 11.1). In the neutral or trade-off category, trees and crops

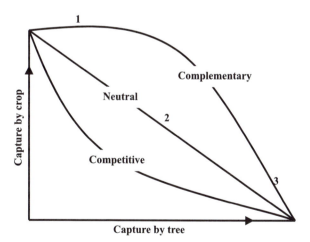

Fig. 11.1. Resource capture by tree and crop showing complementary, competitive and neutral interactions. (1) Parkland or savanna (late-stage agroecosystem); (2) boundary planting; and (3) alley cropping (early-stage agroecosystem). (From Ong and Leakey, 1999.)

exploit the same pool of resources so that increases in capture by one species result in a proportional decrease in capture by the associated species. If trees were able to tap resources unavailable to crops, then the overall capture would be increased as shown by the convex curve, i.e. complementary use of resources. In the third category, negative interactions between the associated species could result in serious reduction in the ability of one or both species to capture growth resources (concave curve). It is important to bear in mind that tree–crop interactions may change from one category to another depending on the age, size and population of the dominant species as well as the supply and accessibility of the limiting growth resources.

Interactions for Light, Water and Nutrients

Successful plant mixtures appear to be those which make 'better' use of resources, by using more of a resource, by using it more efficiently, or both. In terms of the water use of an agroforestry system a central question is, therefore: does intercropping woody and non-woody plants increase total harvestable produce by making more effective use of rainfall? It is possible, at least theoretically, that a mixture of trees and crops may improve the overall rainfall-use efficiency either directly, by more rain being used for transpiration, or indirectly by improved transpiration efficiency (e_w), i.e. more dry matter produced per unit of water transpired.

In theory, the potential of agroforestry to improve e_w is limited compared with intercropping, as the understorey crops are usually C_4 species and the overstorey trees are invariably C_3 species in agroforestry. Improvement in e_w is most likely if the understorey crop is a C_3 species, which are usually light saturated in the open, and partial shade may have little effect on assimilation with the result that e_w is improved by concurrent reduction in transpiration. This may explain why cotton yield in the Sahel was not reduced by the heavy shading of karite (*Vitellaria paradoxa*) and nere (*Parkia biglobosa*) in parklands while yields of millet and sorghum were reduced by 60% under the same trees (Kater *et al.*, 1992). The same reason may explain the observation in the South and Central American savannas that C_3 grasses are found only under trees and never grow in open grassland dominated by C_4 grasses. These examples of the parkland and savanna systems suggest that they represent the upper end of the complementary curve in Fig. 11.1.

There is also the potential for microclimate modification in agroforestry systems due to the presence of an upperstorey tree canopy. This may alter not only the radiation, but also the humidity and temperature around an understorey crop. Some evidence for this has been found where crops have been grown using trees as shelter belts, and decreases in saturation deficit have been reported for several crops. Data from an agroforestry trial in Kenya also show that the air around a maize crop growing beneath a *Grevillea robusta* stand is more humid than the free atmosphere above the trees.

In Kenya, Belsky and Amundson (1997) observed improved micro-climate along with higher soil biotic activity and N mineralization, higher infiltration rate and greater beneficial effects in more xeric environments. Plants grown in the open sites were more nutrient-limited than under the tree canopy but artificial shade generated smaller increases in understorey vegetation. However, Belsky and Amundson were unable to firmly conclude whether microclimate changes or nutrient enrichment were more important in increasing understorey productivity. It is significant that the positive tree effects on understorey vegetation are limited to certain sites and species combinations, including both nitrogen and non-nitrogen fixing trees. It is also difficult to determine precisely whether the tree–grass interactions in tropical savannas are typical of the competitive category (Fig. 11.1) as the literature evidence is primarily based on plot-level analysis. If the improvement in soil fertility is due to redistribution of nutrients in the landscape, this would be an example of the neutral category in Fig. 11.1.

Evidence from a series of shade-cloth trials on maize and beans at Machakos, Kenya, confirmed the small but beneficial effects of shading on crop temperature and crop production when rainfall is inadequate for crop production but, unlike the savanna situations, crops failed under *Grevillea robusta* trees because below-ground competition consistently outweighed the benefit of shade. In contrast, increased soil water (4–53% greater than in the open) was reported in the crop root zone beneath *Faidherbia albida* canopies in the savanna of Malawi. In theory, trees can increase soil water content underneath their canopies if the water 'saved' by their shade effect on reducing soil evaporation and rainfall redistribution, e.g. funnelling of intercepted rainfall as stem flow, exceeds that removed by the root systems beneath tree canopies (Ong and Leakey 1999). At high tree densities, the proportion of rainfall 'lost' as interception by tree canopies and used for tree transpiration would exceed that 'saved' by shading and stem flow, resulting in drier soil below the tree canopy. Van Noordwijk and Ong (1999) expressed this as the amount of water used per unit of shade. This may be one of the most important factors for the observed difference between savanna and alley cropping findings.

The potential for capture of deep water and nutrients coupled with recent innovations in instrumentation (mini-rhizotrons, sap-flow gauges) have stimulated a resurgence in root research and increased attention on spatial complementarity in rooting distribution and the potential beneficial effects of deep rooting. Agroforestry is also considered as critical for maintaining ecosystem functioning in parts of Australia, where deep-rooted perennial vegetation has been replaced by annual crops and pastures, leading to a profound change in the pattern of energy capture by vegetation, rising water tables and associated salinity (Lefroy and Stirzaker, 1999). The Australian example showed that compared with the natural ecosystem it replaced, the agricultural system is 'leaky' in terms of resource capture. Recent investigations in West Africa suggest that a similar magnitude of 'leakiness' is possible when native bush vegetation or woodland, which provides little runoff or groundwater recharge, is converted into millet fields. The expectation is that agroforestry systems will be able to reduce this leakiness because of extensive tree-root systems. Earlier research on South African savannas has shown that tree roots extend into the open grassland, providing a 'safety net' for recycling water and nutrients and accounting for 60% of the total below-ground biomass.

One of the earliest detailed studies of resource capture in agroforestry systems was that described by Ong *et al.* (1996) in semiarid India (Hyderabad) for a C_4 crop, millet (*Pennisetum americanum*), and a C_3 tree, *Leucaena leucocephala*. Total intercepted radiation during the rainy season was 40% greater in the alley crop than in sole millet, primarily because the presence of leuceana increased fractional interception during the early stages of the growing season. The sole leucaena and alley leucaena intercepted twice as much radiation again during the following long dry season when cropping was not possible. The evidence from this study shows that the main advantage of alley cropping was in extending the growing period into the dry season and increasing the annual light interception. However, interception by the more efficient C_4 crop was reduced to only half that of the sole millet. This system falls within the lower end of the complementary curve (Fig. 11.1).

The alley crop produced 7 t ha^{-1} biomass compared with 4.7 t ha^{-1} of sole millet despite the high amount of light interception because of the low photosynthetic rate or conversion coefficient of a C_3 species. The conversion coefficient (e_r), defined here as the ratio of biomass production to intercepted light per unit area, and provides a measure of the 'efficiency' with which the captured light is used to produce new biomass; the alternative term radiation use efficiency is also commonly used. This and other studies by Ong and co-workers showed that the less efficient C_3 overstorey (tree) component dominated the total light interception while the increased e_r of the understorey component was insufficient to compensate for the reduced light interception. These results are typical of many alley-cropping studies where the tree populations were so high that reduction in crop yield was inevitable since the trees captured most of the resources at the expense of the crops. Although crop yields were seriously reduced, these are examples of complementary interactions in terms of resource capture.

As for light, agroforestry offers substantial scope for spatial and temporal complementarity of water use resulting from improved exploitation of available water. However, the opportunity for significant complementarity is likely to be limited unless the species involved differ appreciably in rooting patterns or duration. Recent findings for a range of tree species at Machakos, Kenya, showed that when rainfall was low (250 mm) maize yield was linearly and negatively related to the amount of water used by the trees. This relationship broke down when rainfall exceeded 650 mm. This example illustrates that the trees were using water from the same soil profile as the maize, i.e. neutral response. Similar studies in India and Kenya confirm that there is a greater opportunity for agroforestry to increase rainfall utilization by using off-season rainfall when cropping is not possible, i.e. temporal complementarity.

Competition for Below-ground Resources: Root Structure and Function

Early studies of spatial complementarity in agroforestry began by examining the rooting architecture of trees and crops grown as pure stands. For example, Jonsson *et al.* (1988) described the vertical distribution of five tree species at Morogoro, Tanzania, and concluded that the tree root distribution and maize were similar except for *Eucalyptus camaldulensis*, which had uniform distribution to 1 m. Thus, they concluded that there was little prospect of spatial complementarity if these trees and crops were grown in combination. Recent reviews of the rooting systems of agroforestry systems by Ong *et al.* (1999) essentially supported the earlier conclusion of Jonsson *et al.* (1988).

What is the extent of complementarity in water use when there is such a considerable overlap of the two rooting systems? Results at Machakos, Kenya, consistently showed that there was no advantage in water uptake when there was little water recharge below the crop-root zone. However, when recharge occurred following heavy rainfall tree roots were still able to exploit more water below the rooting zone of the crops, even when there was a complete overlap of the root systems of trees and crops. This is another example of temporal complementarity.

Direct measurement of tree root function was facilitated in recent years by the availability of robust sap-flow gauges, which offer a unique opportunity for quantifying the amount of water extracted by tree roots from different soil layers, and hence for assessing spatial complementarity with crops. Experiments in which the lateral roots were progressively severed or the soil excavated indicated that 3-year-old trees were capable of extracting up to 80% of their water requirements from beneath the crop-rooting zone. As trees grew larger they depleted the soil water and became more and more dependent on current rainfall and severe shoot pruning was necessary to improve infiltration of soil water and below-ground complementarity.

The lack of below-ground spatial complementarity in alley cropping was advanced by Van Noordwijk and Purnomosidhi (1995), who observed that repeated prunings of trees in alley cropping had the danger of enhancing below-ground competition by promoting the proportion of superficial roots. They imposed three pruning heights on five tree species (in the sub-humid site at Lampung, Sumatra, Indonesia). Recent measurements of the long-term (3 years) effects of pruning on two species, *Senna spectabilis* and *Gliricidia sepium*, at Machakos, Kenya, confirmed the findings in Lampung and showed that rooting depths of both pruned trees and crops were almost identical in the alley-cropping treatment. The evidence so far suggests that below-ground competition for water is inevitable in alley-cropping systems where water is limiting.

The most remarkable example of temporal complementarity in water use is the unusual phenology of the Sahelian tree, *Faidherbia albida*, which retains its leaf shedding habit in the rainy season even when planted in the Deccan plateau of India, where the water table is too deep for tree roots. One of the few deliberate experiments conducted at Hyderabad, India, comparing *F. albida* with a tree having 'conventional' leaf phenology was reported by Ong *et al.* (1996). Comparison of sap flow rates of *F. albida* and a local Indian tree, *Albizia lebbek*, showed that transpiration rate of *F. albida* begins when the understorey crops have developed a full canopy. In contrast, *A. lebbeck* produced a full canopy well before the onset of the rains and shed its leaves when the *F. albida* started to develop its canopy. Crop yields beneath both trees were about the same suggesting that both tree species were utilizing water from the same soil profile. This is clearly an example of a neutral category (Fig. 11.1). Thus, phenology on its own is not adequate for complementary use of resources since temporal complementarity for light was not matched by the supply of water.

Where groundwater is accessible to tree roots there is clear evidence for spatial complementarity. For instance, measurements of stable isotopes of oxygen in plant sap, groundwater and water in the soil profile of windbreaks in the Majjia valley in Niger showed that neem trees, *Azadirachta indica*, obtained a large portion of their water from the surface layers of the soil only after rain, when water was abundant, but during the dry season tree roots extracted groundwater (6 m depth) or deep reserves of soil water. In contrast, at a site near Niamey, West Africa, where groundwater was at a depth of 35 m, it was found that both the trees and millet obtained water from the same 2–3 m of the soil throughout the year.

Interactions with Weeds and Pests

Agroforestry research hitherto has concentrated on tree–soil–crop–livestock interactions affecting the use of growth resources, productivity of component species, and soil resource base. But little attention has been given to learn how the component interactions affect herbivores and other soil biota, and how pests and diseases affect the performance of agroforestry systems. This is partly because of the overriding importance given to agroforestry to solve more urgent problems of soil fertility, soil erosion and to generate income, and partly because of the notion that trees are less susceptible to pests and diseases than crops. But agroforestry trees are attacked by a number of insect pests and diseases at all stages of their growth like any other crop species. Although trees are more resilient than crops with some innate mechanisms to overcome damage in later stages of their growth, they are particularly vulnerable to pests, diseases and weed competition in the seedling stage. Three issues of interest when discussing weeds, pests and diseases are: (i) whether agroforestry can be employed to reduce pests and weeds; (ii) whether trees increase pests of crops; and (iii) how to manage pests and diseases of the agroforestry systems being practised. There are few situations where agroforestry is solely used for pest management, but reduction of pests is often an accompanying benefit.

The principles of tree–crop interactions with respect to growth resources also apply to interactions of these components with weeds in agroforestry and controlling weeds is essentially a matter of applying those principles to favour crops and trees. Canopy and litter management for shade and soil cover is the single most important low-cost biological tool for weed management. Two other mechanisms worth mentioning are allelopathic effects of trees on weeds through root secretions and/or litter decomposition products, and trees serving as trap plants for some weeds. However, there is too little knowledge at present on these mechanisms to exploit trees for the practical purpose of controlling weeds. Weeds impose a severe biological stress to crops and trees in the early stages of establishment but they can be managed to minimize competition in the later stages. Weeds may not pose a major threat in the fully established multistrata and perennial tree–crop based systems because of their thick canopy cover and litter fall providing a continuous soil cover. However, in systems (boundary plantings, alley cropping and parkland systems) where the tree canopy does not fully cover the ground, shade tolerant and perennial weeds may persist under trees. The difficulty for normal tillage close to trees using draught animals further aids the survival of weeds.

The traditional bush fallow of the slash-and-burn systems in the humid and sub-humid tropics has reduction of weeds and pests as one of the major objectives, besides restoring soil fertility. Natural fallows reduce weed appearance and growth, and deplete weed seed pools in the soil by preventing addition of new seeds and degrading seed viability over time because of their longer land occupancy (Gallagher *et al.*, 1999). Shade-based control of *Imperata cylindrica* through planted tree fallows is an excellent example for the potential use of agroforestry to reclaim millions of hectares of abandoned land infested by *Imperata* for agricultural production in South-East Asia. Tree fallows with fast growing trees such as *Acacia auriculiformis*, *A. mangium*, *Gliricidia sepium*, *Calliandra calothyrsus* and *Gmelina arborea* can be used to eradicate the grass as well as produce poles and wood over a 5–7-year period (Macdicken *et al.*, 1997). Some management input is needed to ensure that planted tree fallows achieve the objectives of natural fallows in a relatively shorter time period. For example, trees in planted fallows need weed control in the beginning to help them establish better, particularly where tree densities used are low. Furthermore, the effectiveness of tree fallows to smother weeds can be increased by combining them with herbaceous cover crops.

The shorter-term managed fallows of 1–2 years, now widely experimented for soil fertility replenishment, reduce weeds and/or shift the composition of weed species from difficult-to-control grasses to easily controllable broad-leaved weeds (Rao *et al.*, 1998). However, because of relatively shorter land occupancy, these fallows may not reduce the weed seed bank in the soil as effectively as the longer-term natural fallows. Certain weeds may serve as improved fallow species due to their high nutrient scavenging ability in low fertility environments and their ease of establishment. Species such as *Lantana camara*, *Tithonia diversifolia*, *Euphatorium inulifolium* and *Chromolaena odorata* can be used for short-term fertility restoration. However, prospects for their long-term use over successive fallow–crop rotation cycles is not yet known as these daisy fallows do not input N into the soil through biological N_2 fixation. Short-term (1–2-year) fallows of *Sesbania sesban* and *Desmodium distortum*, besides reducing normal weeds, were found to reduce the parasitic weeds *Striga asiatica* and *S. hermonthica* that parasitize cereal crops in southern and eastern Africa, respectively (Rao and Gacheru, 1998). The leguminous shrubs act as false hosts and stimulate germination of *Striga* seeds in the soil without being parasitized. The germinated *Striga* withers out later in the absence of a host. The combined effect of reduced viable seeds and increased mineral nitrogen by *Sesbania* and *Desmodium* fallows substantially decreases *Striga* on the following maize crops. Crops can be grown subsequent to fallows for one to three seasons free from the competition of weeds and pests.

Agroforestry with multiple species and complex interactions among them provides opportunities for creating pest problems as well as for controlling them. Pest problems could arise due to exotic and/or indigenous insects or pathogens on both indigenous and exotic trees. Exotic trees are widely used in agroforestry throughout the tropics and there have been many instances of those trees severely succumbing to pests in new areas because of the absence of natural enemies. Introduction of germplasm from a narrow genetic base and promotion of few lines further increase their vulnerability to pests. The recent devastation caused by the leucaena psyllid (*Heteropsylla cubana*) to *Leucaena leucocephala*, and little leaf disease on *Gliricidia sepium* caused by a phytoplasma, are reminders to researchers and development agencies of the importance of pests and diseases in agroforestry. Indigenous pests on indigenous trees generally are less harmful, but they can assume epidemic proportions under favourable climatic conditions. *Sesbania sesban*, a shrub indigenous to Africa, rarely experiences major pests in natural stands, but it has been found to be severely damaged by defoliating beetles, *Mesoplatys ochroptera* and *Exosoma* sp., in managed fallows. The exotic cypress aphid *Cinara cupressi* affecting the indigenous African tree *Juniperus procera*, and *Celosterna scabrator* a minor insect that attacks *Acacia* spp. in India affecting eucalypts introduced long ago, are examples of new associations in agroforestry (Rao *et al.*, 2000).

Pest problems differ with the nature of agroforestry systems. In rotational systems, both natural and planted fallows increase pests on subsequent crops if fallow species serve as alternative hosts to pests until the main crop host is available later. *S. sesban* and *Tephrosia vogelii* increase root-knot nematode (*Meloidogyne* spp.) populations in the soil over a 1–2-year growing period, though they may not be seriously damaged by the nematodes. Therefore, fallows of these species pose the risk of damaging the nematode-susceptible crops grown in rotation such as bean, tobacco and cotton. However, fallows contribute to the control of pest and disease populations in a field, if the species selected are non-hosts, less suitable hosts or somehow inhibit pests and disease-causing organisms.

The pest interactions in simultaneous agroforestry systems can be broadly grouped under biological (i.e. species-related), physical (microclimate-related) and chemical (through competition for nutrients and emission/suppression of odours) interactions. Pest problems will be aggravated if trees host the same insects that attack crops; trees reduce vigour and growth of crops by competing for growth resources and create a favourable microclimate for multiplication of pests. Trees contribute to pest reduction by restricting movement of insects or pathogens, sheltering crops from unfavourable environmental stresses, masking the odours emitted by the host crops and affecting the host-searching ability of insects, and increasing parasite and predator populations by creating favourable niches for their survival and multiplication. The influence of the tree component in modifying pests in agroforestry may vary from a small area underneath the scattered trees in parkland systems to narrow strips of land close to the tree-row in shelter belts and boundary plantings, to the entire field in multistrata systems. The species and microclimate-related aspects are further described below.

Species Diversity and Pests in Agroforestry

Observations based on natural systems and mixed annual crops suggest that systems with diverse plant species have fewer pest problems than pure crops. Mixed systems encourage natural enemies to develop by providing alternative food sources in the absence of main prey and offering congenial habitats to survive unfavourable conditions. This is contrary to the observation that highly diverse systems render prey-searching more difficult for specialist predators or parasites. On a positive note, diverse systems can make host searching equally difficult for specialist insects and reduce their infestations of crops. The dilution of host plants in mixed systems is known to reduce pest infestations, especially fungal disease infection. As agroforestry systems generally are more diverse, they are expected to have less pest problems than annual crop systems. Few studies have actually quantified whether this holds true in agroforestry. Many insects and pathogens attack both taxonomically related and unrelated species, so for diversity to reduce pests the species within a system should be outside the host range of pests.

The diversity in agroforestry systems varies widely from no more than two species as in alley cropping and a few trees in parkland systems to well over 100 species in Javanese home-gardens. The multistrata systems are also highly diverse with many species fulfilling different functions. Therefore, the general rules of diversity on pests and diseases, if they exist, are unlikely to apply the same way to all agroforestry systems. Diversity by itself may not reduce pests, and some combinations of species can increase pests. Much depends on the species involved. However, planned diversity based on the knowledge of tree–herbivore interactions can help reduce pests. For example, compared with pure *Sesbania* fallow, mixed fallows of *Sesbania* and *Crotalaria* did not increase the root-knot nematode populations and yield of the nematode-susceptible bean after mixed fallow was not reduced (Desaeger and Rao, unpublished data). *Crotalaria* being antagonistic to root-knot nematodes, it helped to suppress the nematode populations. Examples of successful manipulations of biodiversity within an ecosystem for minimizing pest damage have been achieved case by case, rather than starting from a theoretical basis. This is how the indigenous systems have evolved, so in attempting to develop systems designed for fewer pests, contributions should be sought from indigenous knowledge as well as from theoretical ecology. The use of biodiversity as a strategy to reduce pests is conditioned by farmers' preferences for species and availability of markets for the chosen species.

Physical and Microclimate Effects and Pests

Trees in agroforestry systems alter the microclimate by shading and consequently decreasing soil and canopy temperature of the understorey components, reducing wind speed, increasing humidity, reducing evapotranspiration and prolonged leaf wetness. These changes may not be uniform throughout the field in all agroforestry systems, but depend on the proportion of the tree component, height and thickness of canopy, and climatic conditions. They are more marked under and close to the trees than at distances away from them. These changes may have a positive or negative impact on insects and their natural enemies (parasites, predators, entomopathogenic microorganisms) and diseases.

Hedgerows and tree-rows affect the flight direction and landings of many flying insects through the landscape by their windbreak effect. Such insects may accumulate in the calmer air in the leeward side of a windbreak or on trees from where they move on to crops. Trees may also form barriers to the movement of insects in and out of a plot due to the repelling effect of non-host species and difficulty of searching for the host in the presence of non-hosts to specialist herbivores. Many fungal diseases are wind-dispersed, so windbreaks can reduce the spread of disease propagules from already infected plantations and their entry into healthy plantations, especially protecting plants in the lower canopy.

Many hymenopteran parasites exhibit greater host-searching capacity under bright light, but moderate shade favours the parasitic wasp (*Cephalonomia stephanoderis*) of the coffee berry borer. However, lower temperatures combined with higher humidity under trees increase parasitism by the egg parasite *Trichogamma* sp. (Pu, 1978) and the effectiveness of entomopathogenic fungi (Jacques, 1983). Similarly, shade increases the persistence of the entomopathogenic fungus *Beauveria bassiana* on the coffee berry borer. The length of infective period of *Bacillus thuringiensis* was greatly reduced under direct sunlight.

Given the distinctly different effects of shade and other microclimatic parameters on different pests and natural enemies, a generalized scheme cannot be developed for using microclimate as a pest management strategy. The question of optimum shade or microclimate depends on the major pests and diseases under consideration, the local environmental conditions, and level of fertilizer use. An important principle, however, is to manage tree density and its canopy such that it protects the understorey component(s) from unfavourable environmental stresses. The shade should be uniform rather than patches of dense and sparse shade that create favourable niches for certain pests, which could become the source of inoculum for further spread later.

Integrated Pest Management

Unless the biological constraints imposed by weeds, pests and diseases are removed, the potential benefits of agroforestry in terms of increased capture and efficient use of resources cannot be translated into economic benefits. These biological constraints should be managed through integrated pest management (IPM), which is all the more important given its particular relevance to poor farmers and low value of many trees used in agroforestry in the tropics. IPM in agroforestry should encompass some or all of the following strategies: (i) substituting susceptible tree species with alternatives that provide similar products and services; (ii) exploiting host-plant resistance to insects; (iii) soil and plant management that reduces pest populations but favours natural enemies; (iv) strategic use of environmentally safe chemicals and botanical pesticides; and (v) exploiting biological control by introducing effective natural enemies (Rao *et al.*, 2000). In contrast to the traditional approach of integrated technological package to IPM, an ecological approach that incorporates local understanding of agroecosystems and how they can be managed to enhance natural enemy action into IPM fits well with agroforestry.

In a number of instances trees susceptible to pests and diseases can be replaced by alternative species that perform the same function and produce similar products. Species substitution is possible for ecological functions of trees such as nutrient cycling (*Crotalaria* instead of *Sesbania*), shade (*Gliricidia* instead of *Erythrina*), soil conservation (*Gliricidia* instead of *Senna*), and to some extent for products such as fodder (e.g. *Calliandra* instead of *Leucaena*) and firewood. However, this strategy is not feasible where the trees produce specific products such as fruit, medicine, quality timber, etc. Host-plant resistance as a tactic for pest management is particularly suitable for agroforestry trees that produce low value products and for trees used for service functions. Limited studies in this area have indicated considerable variability within some species, for example in the susceptibility of *Albizia lebbek* to psyllids and aphids, *Leucaena leucocephala* to leucaena psyllid and *Sesbania sesban* to *Mesoplatys* beetle. Emphasis should be placed on germplasm collection, and species and provenance screening for selecting tolerant material.

To exploit the biological control it is essential to know the potential predators and parasites, the ecological conditions under which they are most effective, and management practices that enhance their populations. At present there is very limited knowledge on the prospects for biological control of pests in agroforestry. There are few cases where biological control has been tried and there was limited success even in the well-known case of Leucaena psyllid. Chemical control is beyond the means of resource-poor small-scale farmers in the tropics, but it may be justified in the case of high value trees and against termites in large-scale plantations to improve stand establishment. Botanical pesticides may be encouraged because of their low cost, and safety to the users, livestock and environment, but their use is constrained by the inadequate supply of plant material. However, trees such as neem (*Azadirachta indica*), *Melia azedarach*, *Melia volkensii*, *Tephrosia vogelii* and *Euphorbia tirucalli* can be exploited both as agroforestry trees and for pest control value.

Challenges

The understanding of the hydrological, ecological and physiological processes in alley cropping and other simultaneous agroforestry systems has advanced considerably during the last few years. Despite recent attention to root research, progress in selecting suitable tree species for intercropping in drylands has been surprisingly slow. Thus, it appears that the twin goals of fast growing, preferred by farmers and researchers, and low competitiveness are mutually exclusive when nutrients and water are confined to the topsoil. The greatest challenge is how to integrate trees and crops on the same piece of land. Management practices such as crown and root pruning must be developed to improve complementarity in resource use.

Long-term studies on the effect of root pruning are needed because such information is crucial for the technology to be adopted by farmers, who are completely new to root pruning since they cannot 'see' below-ground competition and only a few appreciate the benefit of crown pruning. Few farmers in Africa are aware of below-ground competition and root pruning is considered to be too difficult and impractical. Pruning offers farmers the chance to increase yields of both crops and tree products, using tools they already own, and they could even increase the number of trees they grow. This will help reduce poverty in rural areas, especially when these approaches are combined with selecting trees for the improved value of their products.

There is little information on how tree–crop interactions modify interactions at other trophic levels such as trees–herbivores and herbivores–natural enemies. Which herbivores and under what conditions they assume pest status, and economic losses caused by them are not known for many agroforestry systems. Given the complexity of interactions and limited knowledge of pests and diseases in agroforestry, it is difficult to design systems with fewer pests. An understanding of the ecological changes that occur with the adoption of agroforestry and ecological causes for pest problems should provide a basis for developing pest management strategies. Information is urgently needed on the biology and population dynamics of pests and their natural enemies over seasons, and adequate human and financial resources should be invested in such agroforestry research in the future by the concerned institution.

References

Belsky, A.J. and Amundson, R.G. (1997) Influence of savannah trees and shrubs on understorey grasses and soils: new directions in research. In: Bergstrom, L. and Kirchmann, H. (eds) *Carbon and Nutrient Dynamics in Natural and Agricultural Tropical Ecosystems*. CAB International, Wallingford, pp. 153–171.

Cannell, M.G.R., van Noordwijk, M. and Ong, C.K. (1996) The central agroforestry hypothesis: the trees must acquire resources that the crop would not otherwise acquire. *Agroforestry Systems* 34, 27–31.

de Wit, C.T. (1960) On competition. *Verslagen van Landbouwkundige Onderzoekingen* 66, 1–82.

Gallagher, R.S., Fernandes, E.C.M. and McCallie, E.L. (1999) Weed management through short-term improved fallows in tropical agroecosystems. *Agroforestry Systems* 47, 197–221.

Jacques, R.P. (1983) The potential of pathogens for pest control. *Agriculture, Ecosystems and Environment* 10, 101–126.

Jonsson, K., Fidgeland, L., Maghembe, J.A. and Hogberg, P. (1988) The vertical root distribution of fine roots of five tree species and maize in Morogoro, Tanzania. *Agroforestry Systems* 6, 63–69.

Kater, L.J.M., Kante, S. and Budelman, A. (1992) Karite (*Vitellaria paradoxa*) and nere (*Parkia biglobosa*) associated with crops in South Mali. *Agroforestry Systems* 18, 89–105.

Lefroy, E.C. and Stirzaker, R.J. (1999) Agroforestry for water management in southern Australia. *Agroforestry Systems* 45, 277–302.

Macdicken, K.G., Hairiah, K., Otsamo, A., Duguma, B. and Majid, N.M. (1997) Shade-based control of *Imperata cylindrica*: tree fallows and cover crops. *Agroforestry Systems* 36, 131–149.

Ong, C.K. and Black, C.R. (1994) Complementarity in resource use in intercropping and agroforestry systems. In: Monteith, J.L., Scott, R.K. and Unsworth, M.H. (eds) *Resource Capture by Crops*. Nottingham University Press, Loughborough, pp. 255–278.

Ong, C.K. and Leakey, R.R.B. (1999) Why tree–crop interactions in agroforestry appear at odds with tree–grass interactions in tropical savannahs. *Agroforestry Systems* 45, 109–129.

Ong, C.K., Black, C.R., Marshall, F.M. and Corlett, J.E. (1996) Principles of resources capture and utilisation of light and water. In: Ong, C.K. and Huxley, P.A. (eds) *Tree-crop Interactions – a Physiological Approach*. CAB International, Wallingford, pp. 73–158.

Ong, C.K., Deans, J.D., Wilson, J., Mutua, J., Khan, A.A.H. and Lawson, E.M. (1999) Exploring below ground complementarity in agroforestry using sap flow and root fractal techniques. *Agroforestry Systems* 44, 87–103.

Pu, Z. (1978) *Principles and Methods for Biological Control of Insect Pests*. Beijing Academic Press, Beijing.

Rao, M.R. and Gacheru, E. (1998) Prospects of agroforestry for *Striga* management. *Agroforestry Forum* 9, 22–27.

Rao, M.R., Nair, P.K.R. and Ong, C.K. (1998) Biophysical interactions in tropical agroforestry systems. *Agroforestry Systems* 38, 3–50.

Rao, M.R., Singh, M.P. and Day, R. (2000) Insect pest problems in tropical agroforestry systems: contributory factors and strategies for management. *Agroforestry Systems* 50, 243–277.

Sanchez, P. (1995) Science in agroforestry. *Agroforestry Systems* 30, 5–55.

Van Noordwijk, M. and Ong, C.K. (1999) Can the ecosystem mimic hypotheses be applied to farms in African savannahs? *Agroforestry Systems* 45, 131–158.

Van Noordwijk, M. and Purnomosidhi, P. (1995) Root architecture in relation to tree–soil–crop interactions and shoot pruning in agroforestry. *Agroforestry Systems* 30, 161–173.

Young, A. (1997) *Agroforestry for Soil Management*. CAB International, Wallingford.

Optimizing Crop Diversification

12

D.J. Connor

Department of Crop Production, The University of Melbourne, Victoria 3010, Australia

Introduction

In agroecosystems, the control of crop diversification resides with the managers of individual farm enterprises, where in each region, climate and soils and available markets for products determine the range of economically viable commodities. Given a realistic overall objective to maximize long-term income, maintain cash flow and maintain the resource base, managers look to produce a range of crops to minimize the risk associated with low commodity prices, spread labour requirements, and also to gain the significant benefits that can accrue from interactions between distinct crop species. These benefits relate to biological processes of complementarity and competition between species that can increase yields and/or reduce the need for external inputs in the maintenance of soil, the management of its fertility, and in the control of weeds and diseases. The questions for individual producers are numerous. How many different crops to grow? How to mix or sequence them? How do the physical characteristics of their farm, e.g. size and soil types, and their financial position and attitude to risk, determine a different optimal solution to those of their neighbours?

This chapter will focus on crop diversification at the farm level. It will explain, with examples, how farmers devise and manage cropping systems that meet their objectives. It will also evaluate how the decisions of farmers contribute to or conflict with the broader requirements of society. The first step, however, is to establish the nature of cropping systems, the role of management and inputs to their productivity and sustainability, and the concept of optimization.

© 2001 CAB *International. Crop Science*
(eds J. Nösberger, H.H. Geiger and P.C. Struik)

Crop Ecology

Cropping systems are collections of plant communities that are established, managed and renewed for the production of food and industrial products. These systems exist on individual farms under individual management. They are clearly goal-oriented. The cropping pattern of a region is then the agglomeration of contiguous farms that are individually managed within comparable climatic, but often with distinct soil, economic and sociological environments.

In one ecological sense crops are disclimaxes (Clements, 1916), i.e. communities removed from their ecological equilibrium, that require physical, chemical and energetic inputs for their persistence and continuing productivity – i.e. sustainability. The design and management of those inputs have a major intellectual component. They require understanding of the same physical and biological processes that occur in natural vegetation, together with the socio-economic issues of human communities. In another ecological sense, a trophic analysis reveals that crops are natural food chains that have been simplified (shortened) to maximize the diversion of photo-assimilate to the desired product(s). Consequently, crops are less bio-diverse than most natural plant communities, including those from which they have been developed. In fact, at the global level, crop production, at least of staple commodities, is dominated by systems constructed of monocultures of individual crops. Finally, despite their relative simplicity, crop communities contain the same productive processes that operate under the same ecological rules as do natural communities, even though man is a dominant part of them.

In this chapter, the focus is on mainstream crop production of the type that is practised by relatively few people over wide areas to supply the basic food requirements of the much larger urban population. It focuses, therefore, on systems that currently use a small number of crops drawn from the cereal, legume and oilseed groups, and that integrate those activities with pasture for grazing animals.

A common misunderstanding is that the crops grown in each region are, or should be, the ones that are most closely adapted to the particular combination of soil and climate. That is, however, not the driving force in agriculture. Adaptation to the physical and biological environment serves only to establish the range of options from which farmers make economic choices. They will grow those crops that are the most profitable, in the long term, as determined by yield, product prices, and costs of inputs to maintain the productivity of the system. Selected crops must tolerate the environmental conditions reasonably well but most crop producers face significant biological as well as economic risks and must develop production strategies and tactics that avoid them or minimize their adverse effects on production, economic performance, and on the resource base itself.

Biological risk is associated with the effects of weather on yield, either directly by response of the crop itself, or indirectly through the response of competing weeds and/or consuming pests. In many cropping systems, the effects of uncertain rainfall can be minimized by irrigation/drainage but much of the world's broad-scale cropping is rainfed and therefore subject to the risk of drought and/or flood as well as extremes of temperature (and wind). Economic risk relates to the uncertainty of costs of inputs and prices for products. The latter, particularly, are subject to substantial inter-seasonal variation because yield responds to weather at the site of production while price also responds to yield of the same or competing crops grown elsewhere. Farmers make a significant investment to establish each crop, and once committed, must then evaluate the cost of each successive possible management tactic against an uncertain economic return. Here, skill, developed by training, and more frequently by long experience, becomes a critical element of success, and there is substantial variation of management and technical skill among the farmers of any region.

Optimization

A rational objective for farmers is to adopt cropping practices that maximize expected long-term income, maintain cash flow to sustain their business, and maintain their resource base. The long-term aspect embodies the requirement for sustainability. This does not, however, mean that transient periods of resource depletion or adoption of individual practices that are 'unsustainable' in the long term cannot be part of an optimal solution. Ley farming, for example, involves the sequential accumulation and loss of soil structure and fertility in pasture and crop phases. In this case, long-term sustainability requires that soil condition is recoverable following each cropping phase. A currently topical issue concerns the use of herbicides, to which, it seems, weeds inevitably develop resistance. From that perspective, herbicide use is unsustainable but that does not mean individual herbicides have no role in sustainable systems. It is rational to use herbicides sensibly and not squander them while they remain effective.

Economic survival requires maintenance of cash flow to avoid unmanageable debt, even if that means choosing less risky options that are less profitable. Risk is an important feature of cropping activities, especially of extensive, rainfed cropping operations in low-rainfall regions. There is plenty of evidence that farmers are risk averse and further that attitudes to risk depend upon financial status. Farmers with limited financial reserves are understandably the most risk averse. As wealth increases, or as financial instruments to 'smooth' variable returns become available, farmers, as other decision makers, are less influenced by the absolute degree of risk they face.

The focus of this discussion is how can crop diversification best contribute to farmers' objectives? An overriding issue is that farmers must design and manage their farms as a whole and not as a series of individual activities. This is essential because cropping systems typically contain many positive and negative interactions, at both biological and economic levels, between component activities in productivity and management. An example of the extent of these issues can be seen in Table 12.1, which lists the range of factors to be considered in the diversification of a southern Australian rainfed cereal-cropping system to include crop and pasture legumes (Pannell, 1995). In each case, optimum diversification will be that level of crop diversity that best contributes to a farmer's particular objective. Farms in a given region may be similar in their natural resource base, but the optimum diversification for neighbouring farmers may differ considerably with size of farm, stage of development, level of debt, and willingness and ability to accept risk.

To summarize then, the optimal solution for each farm depends not only on its natural resources, but also on size, the returns from alternative products, the costs of inputs including labour, the interactions between component activities, the wealth of the farmer, and attitude to risk. This means that optimum solutions, and hence the associated degree of crop diversification, vary between regions and from farm to farm within individual regions. The complexity and individuality of the problems require comprehensive models of cropping systems for their solution. At present, the most appropriate models are whole-farm linear, dynamic and discrete stochastic programming models. Some of these models contain strong biological interactions but, as yet, typical crop simulation models are not sufficiently comprehensive to deal with the range of environmental and management issues that contribute to the optimal solution of crop diversity.

Establishing and Maintaining Diversity

There are two, non-exclusive, ways of diversifying crop production. Monocultures of established cultivars can be grown in sequences (rotations), or alternatively, individual crops can themselves comprise combinations of cultivars or species. Combinations of cultivars are known as 'blends' and, in the special case of cultivars with different resistances to disease, as 'multiline' crops. Combinations of species, called intercrops, can take a range of forms, being combinations of annuals and/or perennials, that can vary greatly in spatial arrangement from random to highly stratified (Vandermeer, 1989). Any of these methods of combining species increases the biological diversity of cropping systems, including that of associated organisms.

There exists a belief and developing paradigm that the sustainable cropping systems of the future will be found only in multispecies crops designed to mimic the structure and processes seen in the natural systems that they have replaced (e.g. Soule and Piper, 1992; Ewel, 1999). Others do not subscribe to that view (Loomis and Connor, 1996). First, natural systems are themselves dynamic passing through successional stages of uncertain future in the face of inevitable climatic change. The importance is to understand

Table 12.1. Factors affecting decisions on incorporation of legumes in cereal-based cropping systems (Pannell 1995).

1.	*Short-term profit factors*	
	1.1	Legume grain yield (depends on weather, soil type, weeds, etc.)
	1.2	Legume crop stubble production and quality
	1.3	Pasture production level, quality and timing
	1.4	Yields of non-legume crops and pastures
	1.5	Input costs
	1.6	Output prices (for legume crops, livestock, livestock products and non-legume crops)
2.	*Dynamic factors (short- to medium-term)*	
	2.1	Nitrogen fixation and yield boost from other factors (e.g. disease break, soil structure)
	2.2	Pasture density
	2.3	Legume crop disease
	2.4	Stubble management
	2.5	Weed control
	2.6	Tillage method
	2.7	Carry over of fertilizer
3.	*Sustainability factors*	
	3.1	Herbicide resistance
	3.2	Soil degradation (acidification, organic matter decline, erosion, nutrient decline)
	3.3	Pasture legume persistence
	3.4	Pasture establishment costs
4.	*Risk factors*	
	4.1	Yield variability
	4.2	Price variability
	4.3	Yield/price covariance
	4.4	Flexibility of the enterprise in response to changed conditions
	4.5	The farmer's attitude to risk
5.	*Whole-farm factors*	
	5.1	Total crop area
	5.2	Machinery
		5.2.1 Total capacity
		5.2.2 Timing of requirements of different enterprises
	5.3	Total feed supply (timing and quality)
	5.4	Feed requirements of livestock on hand (timing and quality)
	5.5	Finance availability and cost
	5.6	Labour availability, quality and cost
	5.7	The farmer's objectives (profit, risk reduction, sustainability, leisure)
	5.8	The farmer's knowledge and experience

processes and how to manage them in either simple or mixed crop communities, rather than to construct mimics. Second, not all natural systems are bio-diverse. Some are effectively monocultures that none the less can compare in continuing productivity with their more bio-diverse counterparts. Third, while sustainability is a parameter that is useful to describe the continuing productivity of agricultural systems, it is not appropriate to describe the changing species composition or primary productivity of natural systems.

The emphasis here in agroecological analysis is on the processes and balance of resource supply and capture, and on the competitive and complementary relationships between the planned and unplanned (associated) biodiversity. This analysis can be achieved, and the basis for sustainable productivity of cropping systems established, without the fundamentalism of 'ecosystem mimicry'. The latter reflects a selective interpretation of the nature of natural systems and of the nature and purpose of agriculture (see Wood, 1998).

Role of Diversification in Cropping Systems

Diversification is a feature of cropping systems because it assists the achievement of cropping objectives (sustainable productivity) by allowing farmers to employ biological cycles to minimize inputs, maximize yields, conserve the resource base, and also to reduce risk due to both environmental and economic factors. Growing a range of crops suited to different sowing and harvesting times also enables farmers to manage greater areas while attending to each crop at optimal times. The benefits of biodiversity arise from differences in productivity of species, their product prices, nutritional requirements, responses to stresses, and from the biological contributions they can offer to the control of weeds, pests and diseases. Biodiversity will only reduce risk when the yields of the alternative crop choices are not positively correlated. Benefit arises when cropping systems are designed and managed to utilize these differences and establish complementarity. The advantages can be considerable, and there are many examples in the literature. At the same time, however, it must be emphasized that observed benefits do not necessarily persist under all conditions, and further that alternative non-biological interventions can achieve the same objectives, and may do so more easily or more cheaply. Crop producers cannot ignore those possibilities.

A range of functions through which biodiversity can improve the performance of cropping systems is summarized in Table 12.2. It covers a range from yield stabilization, through crop nutrition, weed, disease and pest control, to soil and water conservation. That range is bounded, on the one hand, by strategies and tactics to increase economic yield that dominate agriculture on robust agricultural land, and on the other, the conservation of the basic soil and water resources on more fragile land. This distinction

Table 12.2. Functions provided to agroecosystems by their biodiversity.

Problem	Function provided by biodiversity
Small and variable yield	Increase yield, reduce inter-seasonal variation.
Losses to weeds	Select competitive crop species, range of herbicide responses.
Losses to pests	Crops selected as disease breaks, provide biological control of disease spread with mixtures of species or cultivars (multilines).
Leaching of nutrients	Deep-rooted species to capture nutrients, planting drainage lines to recover nutrients.
Soil erosion	Suitable crops to protect soil, chosen according to soil type, slope and aspect. Possibilities for windbreaks and contour strips – intercropping.
Rising water tables, salinization	Deep rooted evergreens with high evapotranspiration, perennials on recharge areas to increase water use.

between productivity and conservation provides a useful contrast of the utility of diversification in cropping systems.

Productivity and stability

Crop rotations and intercrops can increase and stabilize harvestable and economic yield by promoting biological processes to improve crop nutrition or control weeds, pests and diseases and so reduce the need for external inputs. These issues dominate the design and management of cropping systems on robust land.

Crop rotations contribute to the management of these biological inputs in many ways. For individual fields, they allow accumulation of water and mobilize nitrogen during fallow, accumulate organic N under leguminous crops, provide disease breaks for soil-borne diseases, control weeds through the differential competitive ability of individual crops, and slow the development of tolerance to individual herbicides by allowing the use of a more diverse group of herbicides. In space, they slow the spread of aerially borne diseases because the landscape is heterogeneous with regard to crop cover. A feature of crop rotations is the dynamic effect on the soil resource, seen for example in the time-trends of soil N content. Properly managed, rotations sequentially accumulate and exploit soil structure and fertility, but maintain it at all times within recoverable levels. The same applies to soil-borne pathogens that decrease under non-host crops or to soil structure that is impaired by tillage during a cropping phase but is then recovered during a period of pasture, or under zero-tillage cropping.

In many cases, biological processes are not impaired when management seeks to improve growing conditions by using external inputs. Thus, N fertilizer can be used in legume–non-legume cropping sequences to complement that fixed by legumes. Success requires that N fertility is low at the start of the legume phase to promote active N fixation. Likewise, competitive forces that reduce the impact of weeds can be complemented by the judicious use of herbicides. In other cases, however, there is potential conflict between biological and non-biological approaches. These can be seen in the careful design that is needed for successful programmes of integrated pest, disease or weed management. Crop diversity is critical to the success of these integrated management strategies, the range being selected to maximize biological control, while allowing tactical applications of therapeutic agrochemicals as required.

In the search for greater yields and improved production efficiency, there remains the question of the role of intercrops as alternative production strategies to rotations. Intercrops are common in the agriculture of less developed (usually tropical) countries (LDC) but are almost absent from the mechanized agriculture of more developed countries (MDC). In MDCs, intercrops are restricted to inter-row cropping of annuals during the early stages of establishment of tree (often fruit) crops. In those cases, the annual plants are a secondary consideration, being planted to utilize the 'space' for as long as it remains available for growth and can be accessed by machinery. While attempts have been made to use intercrops in mechanized agriculture, and although some possible yield gains have been demonstrated, no system is yet economically viable. For example, many cultivars of field pea (*Pisum sativum* L.) are scrambling climbers. Structural support will increase growth and yield formation and facilitate harvesting. The question is, can an associated species provide that support and do it with greater overall yield compared with the two crops grown separately? Experiments with field pea combined with canola (*Brassica napus* L.) (Soetejdo *et al.*, 1998) have shown yield advantages, and the two seeds are sufficiently different in size to be separated during mechanical harvesting. Neither that combination nor any other have, however, survived the test of economic viability. Rather, plant breeding has turned to the development of short, stout pea cultivars of greater intrinsic yield. The same switch of plant type has also provided improved bush beans as a replacement to climbing beans in the maize (*Zea mays* L.)–bean (*Phaseolus vulgaris* L.) intercrops that are common in the tropics.

Intercropping in LDCs is a completely different proposition. In the first place, it applies where crops are harvested manually. Secondly, success does not arise because the combinations of crop species are able to substantially improve the capture of resources more than when grown separately. Rather, analysis identifies two characteristics of intercrops in which yield advantages have been shown to accrue. The first is that the intercrop is a combination of legume and non-legume, and the second is that the system operates at low N fertility, such that even combined yields are small. At high fertility, the previously large advantage of small N transfer during the growing season is largely lost to the competitive forces for light and other nutrients.

Multilines and isolines of single cultivars differing in disease resistance genes have found a place in mechanized agriculture because their overall physiological and phenological similarity allows common management, including harvesting. Multilines of oat (Frey *et al.*, 1977) are used in the USA and of barley in Europe (Wolfe, 1985). The oat multilines are collections of isolines, each with individual resistance genes to strains of stem rust. The notion is that the spread of the disease from unpredictable disease strains in individual crops will be slowed by resistance in one or more isolines. The result is to save the crop from serious yield loss. Under this scenario, the complex genetic structure of the seed has to be maintained for all crops, some (perhaps many) of which are likely to carry some infection. The alternative production strategies to multilines are either to use cultivars with broad multi-gene resistance to likely strains or to switch from cultivar to cultivar as single-gene resistance breaks down. It is not clear what is the best strategy. Trenbath (1984) has shown theoretically that there may be little difference over the long term between the use of multilines, the currently most-resistant component, or a rotation of lines with single-gene resistance. In any case, multilines are just one way of introducing diversity into cropping systems, the other strategies also introduce and rely on diversity, in time rather than in space.

Resource protection

Mixed cropping is a particular characteristic of areas where the resource base is vulnerable to damage, for example from soil erosion or where rising water tables cause salinity (Passioura and Ridley, 1998).

Erosion, by either wind or water, is best controlled by maintaining vegetative cover, usually requiring perennial species, which are infrequently harvestable crop species also. It is not surprising then, in such areas where crop production is also sought, that crops are grown in combination with trees, in alleys on flat land subject to wind erosion, and on contours on steep land subject to water erosion. In extreme cases the most vulnerable sections of land are taken out of crop production completely, e.g. on steep lands and along water courses, which were inadvisably cleared for agriculture in the first instance. In all cases, land-use design and management are directed to reducing erosion to acceptable if not zero levels.

Perennial species also play an important role in areas where cropping of annual plants has reduced total water use and allowed water tables to rise, with resultant salinization. In such areas, an appropriate density of trees in 'agroforestry' systems can re-establish a hydrological balance that keeps the water table and its salt content below the root zone of crops (Lefroy and Stirzaker, 1999). Agroforestry has extended rapidly in south-east Australia where, in 100 years since evergreen sclerophyll forests of eucalypts were cleared for annual crops and pasture, substantial areas have been lost to agricultural production through dryland salinization. The socio-economic dimension of the solution is complex. The technical solution identifies restoration of an appropriate hydrological balance as the basis of the solution; implementation is problematic. The complexity lies in the physical separation of recharge areas where treatment is required from discharge areas where response to treatment is sought. These infrequently occur on individual farms, are not restricted within individual watersheds, and may even be separated by hundreds of kilometres.

The major distinction, applied here, between the roles of diversity in 'productivity' and 'resource protection' serves to highlight the range of environmental challenges that confronts agriculture. In practice there is a continuum between these extremes and each situation must be treated according to the nature of the problem. The only generality is that crop diversity has a major role in most situations. The challenge is to identify what diversity can achieve, what management is required to sustain the effects, and also to assess what other methods could achieve the same objectives.

Costs of Maintaining Diversity

The monetary, labour and skill costs of maintaining multispecies crop communities are not considered here because the large requirement for manual labour in management and harvesting render them uneconomic for large-scale cropping systems. Trenbath (1999) has provided a recent analysis of many of those issues. The focus is rather on the diversity that is established using crop sequences to maximize productivity and/or the inclusion of strips of perennials for resource protection.

Equipment

The success of broad-scale cropping depends upon the availability of equipment for all cultural practices. In practice, the capacity of equipment must be large so that operations can be carried out quickly at optimal times. Equipment represents a large investment in cropping enterprises. In a typical southern Australian wheat (*Triticum aestivum* L.) cropping enterprise, for example, the total investment of Aus$2 million includes Aus$0.6 million in machinery for cultivation (including chemical measures), sowing, spraying and harvesting. Sowing and harvesting equipment is increasingly flexible for a range of grain types, in response to the demands of growers wishing to diversify crop production.

Skill

Willingness to invest the substantial cost of establishing a crop requires considerable confidence in a successful outcome. In the variable environmental and economic conditions that characterize cropping, the development of skill to justify that confidence is not gained quickly. Rather it may take a decade during which time the optimal production schedule may change substantially with the availability of new cultivars and agrochemicals.

Thus, aversion to diversification based on levels of skill and confidence should come as no surprise. Specialists abound in all sections of our society for obvious reasons. There is no reason why farming should be different. When farmers do increase their range of crops, they do so slowly, growing relatively small areas on which they adopt old and develop new skills. This is one way they minimize the risk of commencing new ventures. As they develop skill, yields and economic viability increase. Such increase in confidence and yield was seen clearly during the first two decades of the introduction of the narrow-leaf lupin (*Lupinus angustifolius* L.) into the cereal cropping systems of Western Australia (Pannell, 1995), and the return to canola cultivation in the 1990s following the release of black leg (*Leptosphaeria maculans* (Desm.) Ces. et de Not.) resistant cultivars.

Being averse to the risk of failure with new crops is not only a matter of developing the requisite production skills; there is also the possibility that the addition of a further crop may take the system beyond the optimum. That ground has to be charted by some growers before it is then possible to make the sorts of comprehensive analyses needed to explore optimum diversification. An example concerning the inclusion of narrow-leaf lupin into cereal systems is presented in a later section.

Perennials

Resource conservation is a major issue in southern Australian cereal cropping zones that has parallels in many other parts of the world. There are problems of soil loss by wind and water erosion and also by salinization that find potential solutions in the incorporation of perennials into the systems. The benefits from these activities must, however, be weighed in each case against negative interactions of resource capture and competition. It is important to understand clearly in each case what diversification is designed to achieve and what flexibility remains in management.

There has been much work on the inclusion of trees for soil conservation and control of drainage to water tables (see Lefroy and Stirzaker, 1999). Strip, often contour, plantings are the most appropriate because they allow mechanized cropping to continue. The issue in all cases is to find the optimum combination that minimizes the negative interactions of competition between hedgerows and crops. Cropping systems involving perennials are long lived and dynamic, requiring continual adjustment to management. A major limitation is their relative inflexibility. Unlike systems based on annual crops, it is not possible to make substantial changes to take advantage of favourable climatic or economic conditions. Such changes can be the basis of success in climatically variable environments (Kingwell *et al.*, 1992) where productivity is the more dominant issue to resource protection.

Crop Diversification in the Australian Wheat Zone

There has been a continuous search for optimal solutions to growing cereals in southern Australia in response to changing biophysical, environmental and socio-economic conditions. A feature of this search has been the implementation of increasingly complex crop sequences (Connor and Smith, 1987). When the industry commenced in 1870 it was characterized by continuous cultivation of wheat on parts of individual farms with the remainder used for grazing. Later (early 20th century), fallowing was introduced to improve water and nitrogen supply to crops grown every second year but it was not until after World War II, when attractive prices for wool encouraged farmers to fertilize legume-based pastures for sheep production, that a system of alternating cereals and pasture was introduced. The impact was positive for both crop and grazing enterprises and represented the first real step in diversification of crop production. Since that time, small returns from wool have encouraged further intensification of crop production. The availability of adapted cultivars of pulses and oilseeds, and markets for those products, have made that possible. These developments in adjusting cropping systems to the environmental and economic climate have also required a continual increase in farm size. Much of the cropping lands were settled in the early 20th century as 150 ha farms. Now viable farms are 1000–2000 ha depending upon rainfall and soil type, i.e. they are amalgamations of eight or more of the original farms.

The impact of these changes on crop diversification is revealed in the following analysis of an individual farm in the Victorian Wimmera since 1982 (Fig. 12.1). Then, there was a fallow–wheat rotation, each activity occupying 50% of the farm area. In 1998, fallow was just 10% of farm area and six individual crops were grown. The largest area (23%) was in canola, followed by barley (*Hordeum vulgare* L.) (22%), lentil (*Lens culinaris* L.) (18%), wheat (17%), field pea (7%) and chickpea (*Cicer arietinum* L.) (3%). A significant feature of this farm relevant to the issue of (economic) sustainability is that it gradually increased in size from 1160 ha in 1982, to 1547 ha in 1988, and to 2113 ha in 1999.

Faced with such a range of options, the challenge for individual growers to find their optimal farming strategy, and the level of crop diversity that supports it, is substantial. Most have found their way by their own experi-

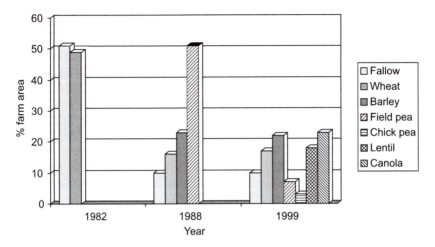

Fig. 12.1. Crop diversity (% area) on a 2000 ha Australian farm, 1982–1999.

ence and that of neighbours. Others have been guided by research that seeks to develop tools as in the following two examples.

Analysis of Optimum Diversification in Cereal Cropping Systems

The agronomic literature is replete with reports of studies of crop diversification but few are sufficiently comprehensive to address the issue of optimal diversity. The following two examples are taken from the work of Pannell and colleagues on the diversification of broad-scale cereal-based cropping systems. The important feature of the work has been the development and use of comprehensive biological and economic models of complete farm production systems. These models, MIDAS (Morrison *et al.*, 1986; Kingwell and Pannell, 1987) and MUDAS (Kingwell *et al.*, 1992), consider the effects of positive and negative external and internal factors on the performance of alternative strategies and tactics.

The first example assesses how attitude to risk determines the rational responses of Syrian farmers to new opportunities for diversification (Pannell and Nordblom, 1998) and the second (Pannell, 1995, 1998) the impact of soil type and product prices on the inclusion of crop and pasture legumes into wheat-based rotations in Western Australia.

Farm size and risk aversion in Syria

Syrian agriculture has been based on wheat, barley, lentil, vetch (*Vicia sativa* L.) and sheep for thousands of years. Farms are small, yields are small and

Table 12.3. Optional cropping activities for Syrian farmers.

Rotation
Lentil–wheat
Vetch–wheat
Fallow–wheat
Water melon–wheat
Medic pasture–wheat
Medic pasture–barley
Lentil–wheat without sheep
Vetch–wheat without sheep
Fallow–wheat without sheep
Watermelon–wheat without sheep

highly variable, and although there are few market instruments for risk management, farmers are now free to deviate from past strict controls requiring them to maintain specified crop mixes. The question they now face is how to best allocate land, activities, and resources to maximize expected gain, depending upon the size of their farm and their attitude to risk.

The analysis considers a range of optional activities as listed in Table 12.3, for farms of two sizes, the common size of 16 ha and a large farm of 64 ha. Of the ten possible activities, only three, lentil–wheat, watermelon (*Citrullus lanatus* (Thunb.) Matsumura & Nakai)–wheat, and medic (*Medicago* spp.)–barley rotations comprise the optimal solutions for farms of both sizes, regardless of their attitude to risk (Figs 12.2 and 12.3). In these comparisons, relative risk aversion (RRA) acknowledges that absolute risk decreases as wealth increases. Farmers who accept risk (RRA = 0) would optimally concentrate on the lentil–wheat rotation, with a small allocation to medic–barley, regardless of farm size. As risk aversion increases, however, other activities are included and differences develop depending upon farm size. For the small farm (Fig. 12.2), the optimal solution remains relatively constant and includes an area of the profitable but labour-demanding watermelon. For the large farm (Fig. 12.3), a smaller area of water melon is included but with increasing risk aversion, farmers allocate increasingly more land to medic–barley.

Crop and pasture legumes in Western Australia

The factors affecting decisions on the inclusion of legumes into cereal-based systems were presented in Table 12.1. They include technical and economic issues, some with immediate and others with long-term consequences. This analysis considers how soil type and long-term mean product prices determine optimal crop diversification. The options considered, for seven soil

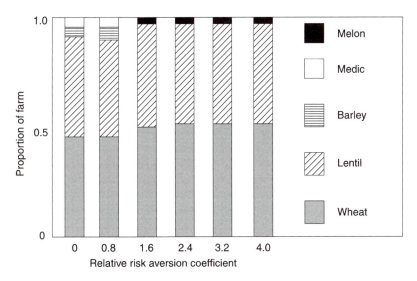

Fig. 12.2. Effect of risk aversion on optimal crop diversity of a 16 ha Syrian farm (adapted from Pannell and Nordblom, 1988).

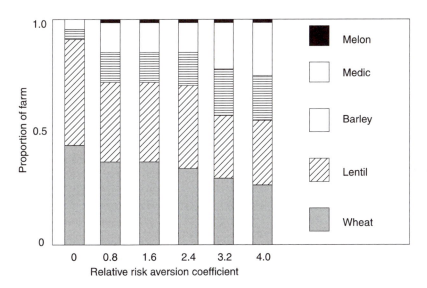

Fig. 12.3. Effect of risk aversion on optimal crop diversity of a 64 ha Syrian farm (adapted from Pannell and Nordblom, 1988).

types, are wheat and the pulse crops, field pea and narrow-leaf lupin, together with legume-based pasture to be grazed by sheep. The soils are briefly described in Table 12.4.

Table 12.4. Definition and distribution of soil types on a 2500 ha Australian farm.

Soil type	Area (ha)	Description
1. Acid sand	500	Yellow, loamy or gravelly sands, pH <5.5
2. Good sand plain	500	Deep yellow-brown loamy sands, pH 5.5–6.0
3. Gravelly sands	250	Yellow-brown gravelly sands, pH 5.5–6.0
4. Duplex	250	Grey sandy loams, loamy sands, gravelly sands and sands over yellow- or red-mottled white clay, pH 5.5–6.5
5. Medium heavy	375	Red-brown sandy loam over clay subsoil, pH 6.0–7.0
6. Heavy	500	Dark red-brown, sandy clay loams, pH >6.5
7. Heavy (friable)	125	As for soil 6 but better structured

Soil characteristics play a major role in crop selection. Gross margins are presented in Table 12.5 for the optimum, and some sub-optimal, cropping scenarios for each soil type. Profitability varies greatly between soil types, as does the optimum crop sequence. In this example, no two soils have the same ranking for the various rotations. The analysis is extended to consider the optimal allocation of land and resources in response to the price of wheat on a 'model' farm of 2500 ha with the mix of soil types described in Table 12.4. The

Table 12.5. Gross margins (Aus$ ha^{-1}) of selected rotations averaged over the length of the rotations on a 2500 ha Australian farm. (Pannell, 1998.)

Soil type[a]	1	2	3	4	5	6	7
Optimum	PPPP[b]	WL	WWL	WWL	WWF	PPPP	WWF
	($15)	($69)	($57)	($66)	($77)	($26)	($77)
Others	PPPW	WWL	WL	WWWW	PWWW	PPPW	WWWF
	($4)	($67)	($54)	($61)	($70)	($20)	($74)
	PPWW	WPPWL	WPPWL	WWWF	WWWW	PPWW	WF
	(–$5)	($54)	($49)	($58)	($67)	($17)	($64)
	WWL	PPWW	WWWW	WL	PPWW	WWWW	WWWW
	(–$12)	($43)	($46)	($58)	($66)	($3)	($62)
	WWWW	WWWW	PPWW	PPWW	WWL	WWF	PPWW
	(–$15)	($39)	($43)	($53)	($58)	($3)	($57)

[a] Soil types described in Table 12.4.
[b] F, field pea; L, lupin; P, pasture; W, wheat.

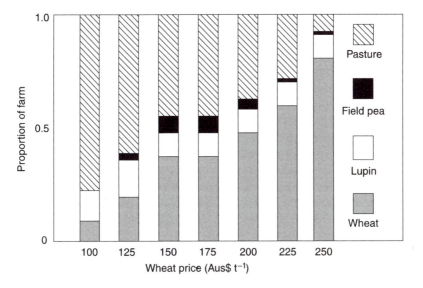

Fig. 12.4. Effect of wheat price on optimal crop diversity on a 2500 ha Australian farm (Pannell, 1995).

analysis is presented in Fig. 12.4. It reveals that all solutions include areas of pasture and lupins, but that the proportions vary greatly. Field pea becomes a part of the optimum crop diversification for mid-range prices of wheat.

Taken together, these two examples of optimum crop diversification in cereal-based systems demonstrate the complexity of the problem faced by farmers in the allocation of land and resources. Climate and soils play the role of selecting the range of crops that can be grown, but the optimal mix for individual farmers then depends upon prices for the alternative products, costs of inputs, the size of their operation, the infrastructure at their disposal, their wealth, and their attitude to risk. In any one region, the optimum solution changes from farm to farm in response to these factors, and from time to time as the economic environment changes. Farmers are not easily able to move quickly to a new optimum solution because they are constrained by existing rotations, infrastructure, and skills. In any event, the optimum is difficult to find. Intuition may keep some farmers close to the optimum, but a thorough analysis requires comprehensive models of entire farm systems.

Society's Views and Role in Crop Diversification

These views can be grouped in two classes. The first concerns the appearance of the farming landscape. The second, highly topical at this time, concerns the sustainability of mechanized broad-scale (pejoratively 'industrial') agriculture, including the maintenance of the genetic resource base. Some people, as discussed previously, propose a return to multispecies cropping systems, modelled on natural systems, as a solution to both these issues.

Cropping inevitably replaces natural vegetation with a smaller array of species whether that change is to monocultures or to intercrops. The change is dramatic but 'beauty is in the eye of the beholder' and there is no reason why a landscape of crops is intrinsically less appealing than one of natural vegetation. For crop landscapes, beauty can reside in colours and textures and their changing patterns, but also in the evident technical and/or manual industry of the participants and in the promise of a fruitful harvest. The great benefit of high-production agriculture to nature conservation is that it reduces the pressure to bring additional land into agriculture. The consequence is that more land can be retained for its many other, including aesthetic, values (Waggoner, 1994).

The genetic structure of crop production appears to have changed markedly during the past century and the current advances in biotechnology suggest more change. The landraces that developed from the self-sufficiency of individual farmers for seed have been successively replaced in MDCs by fewer open-pollinated cultivars, and more recently by even fewer hybrids. A rapid conclusion is that much genetic diversity, the potential solution to productivity gain and to future problems of disease and pest resistance, has been lost. This may not, however, be so for two reasons. First, loss of landraces and cultivars does not necessarily relate to loss of genes, as revealed by the similarity now evident in the genetic structure of even diverse organisms. Second, there is a considerable and well-organized effort at genetic conservation of crop germplasm within national and international agricultural research institutes. The collections of major crop species are large and well maintained. It is, therefore, not necessarily true that genetic diversity will be irrecoverably lost if farmers do not maintain it themselves (Fowler and Mooney, 1990), or that the hope of conservation lies with organic farmers (van Elsen, 1999).

The basis of conflict between optimal solutions to crop diversification, as seen by farmers and others, lies firmly in differences in objectives. To farmers, crop diversification is a consequence of their continuing search to maintain or increase net income and reduce its variability; for others who question the adequacy of that solution, crop diversification is an end in itself.

If society is to seek crop diversification beyond the optimal solutions established by farmers, then it must be prepared to contribute financially to that change to landscape management. Two contrasting methods that have been tried are price support and direct subsidy. Experience and analyses, including those presented here, reveal that risk aversion is a major factor in farmers' decisions to diversify cropping activities. Price support has the effect of reducing risk and so without other controls is most likely to reduce diversity. Direct subsidy, to clearly defined goals, would appear to be the better strategy.

Conclusion

Farmers seek to achieve their goals in the face of physical, biological and economic constraints. Their objective is to maximize returns, minimize costs, maintain the resource base, and avoid risky options to maintain minimum seasonal profitability for living costs and debt servicing. For this they must employ strategies and tactics suited to variable seasonal climatic conditions. Crop diversification plays an important role in the optimum solution because it impinges on most operational considerations. A diversity of crops introduces biological interactions that increase crop yields and allow reduced use of inputs in the management of soil structure, fertility, and in the control of competing weeds, diseases and pests. A diversity of crops also allows a more efficient allocation of labour and reduces economic risk.

The optimal solution for individual farmers in each region can vary greatly, independently of soils, because of differences in farm size, wealth, skill and attitude to risk. The search for each optimal solution must be made at the level of the entire farm operation and be able to adjust to changing prices for products and inputs. The search for the optimal solution is a substantial challenge that can now be assisted by whole-farm bio-economic models.

The characteristic of broad-scale cropping is the use of sequences of monocultures amenable to mechanization. In these systems diversity exists in time for individual fields and in space for individual regions. Increasing diversification has been a feature of Australian cereal-based cropping systems since the inception of the industry in late 19th century. That diversification was made possible by the gradual development of adapted cultivars of legumes and oilseeds and by the opening of markets for those products. Farmers are now able to include the advantages of greater crop diversity in their cropping enterprises and face the significant challenge of finding and continuously adapting their own optimal solution. The range of those solutions that now exist, adopted by either intuition or quantitative analysis, has produced a more diverse cropping landscape than has ever existed previously.

Increasing world population requires more products from agriculture. The response has been the conversion of more land to production and the raising of productivity by using fewer, adapted species and greater inputs of fertilizer and therapeutic chemicals. There is a widespread concern in society over the loss of natural ecosystems to agriculture. There are also more restricted concerns over the nature of modern agriculture itself, especially related to the dominance of a few crop species and the reliance on monocultures. The solution to the first problem of 'excessive' conversion of land to agriculture is most effectively addressed by increasing the productivity of agriculture on the most robust land so that more land can be preserved out of agriculture for its other values. Monocultures and high inputs are suited to high-production systems, and as shown in this chapter, the considerable diversity that is introduced by crops grown in sequence contributes greatly to increasing yield and maintaining farm businesses and the resource base.

Challenges

In looking to the immediate future, two major challenges appear in the relationships between crop diversification, sustainable food production, and the preservation of biodiversity. The first is to ensure that social and economic considerations drive farming practice along with the biological bases of crop productivity. Mankind is an evident part of 'crop ecology' and production theory must not be allowed to develop without that recognition. Second is the importance to extend and improve the efficiency of the technological, intellectual and physico-chemical inputs that are needed to maintain the productivity of agricultural systems. There is no possibility to return to low-input systems if, on the one hand, the large world population is to be fed, and on the other, a significant part of the natural environment is to be retained for its other values. The development of computer models that combine the biological and bio-economic components of agriculture should be a major part of this thrust of the application of technology to agriculture.

Acknowledgements

I am grateful to Drs Dave Pannell, David Smith, Tony Fischer, Rob Norton and Bill Malcolm for valuable discussions and suggestions on the content and structure of this chapter, but accept responsibility for the resultant document. Mr Mark Johns kindly supplied the crop diversity data from his family farm used to construct Fig. 12.1.

References

Clements, F.E. (1916) *Plant Succession: an Analysis of the Development of Vegetation.* Carnegie Institute, Washington, DC.
Connor, D.J. and Smith, D.F. (eds) (1987) *Agriculture in Victoria.* Australian Institute of Agricultural Science, Melbourne.
Ewel, J.J. (1999) Natural systems as models for the design of sustainable systems of land use. *Agroforestry Systems* 45, 1–21.
Fowler, C. and Mooney, P. (1990) *Shattering: Food, Politics and Loss of Genetic Diversity.* University of Arizona Press, Tucson, Arizona.
Frey, K.J., Browning, J.A. and Simons, M.D. (1977) Management systems for host genes to control disease. *Annals of the New York Academy of Science* 287, 255–274.
Kingwell, R.S. and Pannell, D.J. (eds) (1987) *MIDAS, a Bioeconomic Model of a Dryland Farm System.* PUDOC, Wageningen.

Kingwell, R.S., Pannell, D.J. and Robinson, S.D. (1992) Tactical responses to seasonal conditions in whole-farm planning in Western Australia. *Agricultural Economics* 8, 211–226.

Lefroy, E.C. and Stirzaker, R.J. (1999) Agroforestry for water management in the cropping zone of southern Australia. *Agroforestry Systems* 45, 277–302.

Loomis, R.S. and Connor, D.J. (1996) *Crop Ecology; Productivity and Management in Agricultural Systems*, 2nd edn. Cambridge University Press, Cambridge.

Morrison, D.A., Kingwell, R.S., Pannell, D.J. and Ewing, M.A. (1986) A mathematical programming model of a crop–livestock farm system. *Agricultural Systems* 20, 243–268.

Pannell, D.J. (1987) Crop–livestock interactions and rotation selection. In: Kingwell, R.S. and Pannell, D.J. (eds) *MIDAS, a Bioeconomic Model of a Dryland Farm System*. PUDOC, Wageningen, pp. 64–73.

Pannell, D.J. (1995) Economic aspects of legume management and legume research in dryland farming systems of southern Australia. *Agricultural Systems* 49, 217–236.

Pannell, D.J. (1998) Economic assessment of the role and value of lupins in the farming system. In: Gladstones, J.S., Atkins, C.A. and Hamblin, J. (eds) *Lupins as Crop Plants: Biology, Production and Utilization*. CAB International, Wallingford, pp. 339–351.

Pannell, D.J. and Nordblom, T.L. (1998) Impact of risk aversion on whole farm management in Syria. *Australian Journal of Agricultural and Resource Economics* 42, 227–247.

Passioura, J.B. and Ridley, A.M. (1998) Managing soil water and nitrogen to minimize land degradation. In: Michalk, D.L. and Pratley, J.E. (eds) *Proceedings of the Ninth Australian Agronomy Conference, Charles Sturt University, Wagga Wagga, Australia*. Australian Society of Agronomy, pp. 99–106.

Soetedjo, P., Martin, L.D. and Tennant, D. (1998) Productivity and water use of intercrops of field pea and canola. In: Michalk, D.L. and Pratley, J.E. (eds) *Proceedings of the Ninth Australian Agronomy Conference, Charles Sturt University, Wagga Wagga, Australia*. Australian Society of Agronomy, pp. 469–472.

Soule, J.D. and Piper, J.K. (1992) *Farming in Nature's Image. An Ecological Approach to Agriculture*. Island Press, Corelo, California.

Trenbath, B.R. (1984) Gene introduction strategies for the control of crop diseases. In: Conway, G.R. (ed.) *Pest and Pathogen Control; Strategic, Tactical and Policy Models*. John Wiley & Sons, Chichester, pp. 142–168.

Trenbath, B.R. (1999) Multispecies cropping systems in India. Predictions of their productivity, stability, resilience and ecological sustainability. *Agroforestry Systems* 45, 81–107.

Vandermeer, J. (1989) *The Ecology of Intercropping*. Cambridge University Press.

van Elsen, T. (1999) Species diversity as a task for organic agriculture in Europe. *Agriculture, Ecosystems and Environment* 77, 101–109.

Waggoner, P.E. (1994) *How Much Land Can Ten Billion People Spare for Nature? Task Force Report No. 121*, Council for Agricultural Science and Technology, Ames, Iowa.

Wolfe, M.S. (1985) The current status and prospects of multiline cultivars and variety mixtures for disease control. *Annual Review of Phytopathology* 23, 251–273.

Wood, D. (1998) Ecological principles in agricultural policy: but which principles? *Food Policy* 23, 371–381.

Biodiversity of Agroecosystems: Past, Present and Uncertain Future

13

P.J. Edwards and A. Hilbeck

Geobotanical Institute, ETH Zürich, Zürichbergstrasse 38, CH–8044 Zürich, Switzerland

Introduction

The accelerating loss of biological diversity throughout the world is an issue of growing scientific and political concern. In some regions which have experienced a great intensification of agricultural production in recent years, this loss has been very rapid (Aebischer, 1991), and has provoked a scientific debate about the importance of biodiversity for the sustainability of agriculture (Matson *et al.*, 1997). In parts of Europe, major efforts are now under way to prevent further loss of biodiversity, and even to restore diversity in those agricultural areas which have been most impoverished. However, large-scale programmes of this kind are expensive, and it is often doubtful whether the money used for these purposes is well invested. In this chapter, we argue that effective conservation and restoration of biodiversity must be based upon a sound understanding of the ecological status of the biota of agroecosystems. We describe here the origins and development of this biota, and discuss how biological diversity is affected by the introduction of modern intensive farming systems. We pay particular attention to evolutionary processes and the factors affecting the genetic diversity of species in agricultural systems.

The Components of Agrobiodiversity

On an evolutionary time scale, all agricultural landscapes are new, and were occupied until recently by natural ecosystems. While in some farming systems, notably shifting cultivation in the tropics, large parts of the natural ecosystem remain, in many others the original vegetation has disappeared almost without trace. For example, the wheat belt of North America is an almost entirely new landscape, replacing the former prairies which once occupied two-thirds of the area of the USA and now account for a few square kilometres, mainly in protected areas. Between these extremes, there are many agricultural areas, for example in central Europe, in which the original vegetation persists as fragments, sometimes heavily modified or disturbed, and ranging in size from a few square metres to many square kilometres. These fragments of more or less natural habitats, such as forest or grassland, contribute greatly to the total species diversity of these landscapes.

However, the most prominent component of biodiversity in most agro-ecosystems is that which is determined directly by humans. This is the 'planned diversity' which includes the crop plants and livestock (Matson *et al.*, 1997). Humans influence to a great extent the genetic composition, i.e. diversity, of these organisms. Since the advent of agriculture over 10,000 years ago, there has been selection, both deliberate and unconscious, for phenotypes with desirable characteristics such as high yield, good flavour and resistance to disease. Early in the history of agriculture, this process of selection must have produced a wide range of local cultivars or landraces which were adapted to the prevailing environmental conditions and suited the needs of the farmer. More recently, our ability to determine the genetic constitution of plants and animals has increased dramatically, both through the techniques of plant and animal breeding, and subsequently through genetic engineering.

Agroecosystems also include an 'unplanned diversity', consisting of the many other organisms which live in areas used for agriculture. From the farmer's point of view the unplanned diversity can be beneficial or harmful. 'Resource biota' (Swift and Anderson, 1996) and 'functional biodiversity' (Altieri, 1999) are terms used to describe those organisms which contribute positively to agroecosystems by providing ecosystem services, such as recycling of nutrients, regulation of the abundance of undesirable organisms, regulation of local hydrological processes and many more. If these functions are lost, they have to be substituted for by the farmer. In contrast, the 'destructive biota' includes weeds, animal pests and microbial pathogens, and management is aimed at reducing the abundance of this component.

This classification of agrobiodiversity into the biota of natural and semi-natural habitats, the planned biota and the unplanned biota, is inevitably somewhat arbitrary. For example, in traditional agricultural systems the planned biota often includes plant species which come from the surrounding natural forest vegetation, while many animal species are not confined either to areas used for agriculture or to natural habitats, but move between them. Nevertheless, this classification helps us to focus on the strongly contrasting processes affecting different groups of organisms in the agricultural landscape and we shall therefore use it as the framework for the rest of this chapter.

The biota of natural and semi-natural areas

The species diversity of natural ecosystems is extremely variable. Whereas tropical rainforest may contain as many as 280 tree species per hectare, the boreal coniferous forests contain only a tiny fraction of this diversity. This is not the place to describe all the factors which influence species diversity in natural ecosystems, but a few remarks are needed. Firstly, the number of species that can be supported is positively related to the area of habitat available; this relationship can be described mathematically and used to predict, sometimes with surprising accuracy, the number of species in a particular area (Rosenzweig, 1995). However, it is important to realize that the species–area curves are typically different for areas of continuous habitat and for fragmented habitats or islands. In general, the number of species per unit area which can be supported is lower in fragmented habitats than in continuous areas. Secondly, high species diversity tends to be associated with heterogeneity of habitat conditions. Such heterogeneity can be due to local variation in factors such as soil parent material or topography, or be produced by the dynamics of the vegetation itself. Natural forest vegetation, for example, is usually composed of a mosaic of patches which represent different stages of regeneration of the dominant tree species (Remmert, 1991). The various phases of this regeneration cycle provide distinct habitats for many subordinate organisms, and their life histories may be finely tuned to the spatial and temporal dynamics of the mosaic cycle.

When an area is used for agriculture, the natural vegetation becomes fragmented to some degree. The extent and spatial arrangement of the remaining fragments vary and are affected by many cultural, historical and topographic factors. Often, natural vegetation only persists in the most infertile areas or on sites which are too steep or otherwise inaccessible for agriculture. In other cases, as in the traditional woodlands in many parts of Europe, forest ecosystems are preserved in a modified form and managed as a source of wood for fuel and building materials. In yet other cases, for example in parts of India, the only natural vegetation to persist is associated with sacred sites and temples.

The mathematics of species–area relationships implies that any significant reduction in the extent of natural vegetation is likely to lead to the loss of some species, at least locally (Rosenzweig, 1995). Why should this be so? One reason is simply that the remaining area of natural habitat is not large enough to include all of the species which occurred in the region. For example, some tropical tree species occur as isolated individuals dispersed over vast areas of forest. The minimal area (a measure of the area needed to include the majority of species) for tree species in a tropical rainforest is many hectares. Another reason is that the fragmented habitat does not provide the conditions needed for the continued survival of all species. An obvious example of doomed species is the isolated large forest trees which are often left standing in areas cleared for agriculture in savanna and rainforest regions. The mature trees can survive for decades, but the conditions required for establishment of their seedlings no longer exist and so regeneration is impossible. In fact, environmental conditions in fragments of natural vegetation differ from more extensive areas in many ways which can affect species survival. Some bird species will not nest close to a woodland edge, and are therefore dependent upon larger tracts of habitat. Small fragments can also be strongly affected by the spray drift and fertilizer originating from neighbouring agricultural land. They may also be overrun by light-demanding, forest-edge species such as, for example, the exotic invasive species kudzu (*Pueraria montana* var. *lobata*) in the south-eastern USA. Kudzu was introduced from Japan into the USA in 1876 at the Philadelphia Centennial Exposition, where it was promoted as a forage crop and an ornamental plant. From 1935 to the mid-1950s, farmers in the south were encouraged to plant kudzu to reduce soil erosion, and Franklin D. Roosevelt's Civilian Conservation Corps planted it widely for many years. Seventy-five years after its intentional introduction, kudzu was recognized as a pest weed by the US Department of Agriculture and, in 1953, was removed from its list of permissible cover plants. Kudzu kills or degrades other plants by smothering them under a solid blanket of leaves, by girdling woody stems and tree trunks, and by breaking branches or uprooting entire trees and shrubs through the sheer force of its weight. Once established, kudzu plants grow rapidly, extending as much as 30 m per season at a rate of about 30 cm per day (http://www.nps.gov/plants/alien/fact/pulo1.htm).

Small populations in fragmented habitats are vulnerable to the effects of environmental variation, demographic stochasticity and loss of genetic variability. These three effects act together, so that a decline in population due to one factor makes the population more vulnerable to the other two factors (Primack, 1998). Recent work points in particular to the loss of genetic variation in small populations as a key factor in their demise. One reason for this loss of variability is that processes such as pollination and seed dispersal may be reduced or prevented because the mutualist organisms are absent or scarce. In addition, there may be a gradual loss of genetic variation as a result of inbreeding and genetic drift. The consequences of these effects are well illustrated by the work of Kéry *et al.* (2000) who investigated reproduction and offspring performance in relation to population size in the self-incompatible perennials, *Primula veris* and *Gentiana lutea*. In both species, plants in small populations tended to produce fewer seeds per plant. Reproduction was strongly depressed in *P. veris* when the populations were smaller than about 200 plants, and in *G. lutea* in populations of less than 500 plants. Plants from small populations of *P. veris* also exhibited lower phenotypic plasticity than plants from larger populations. This study, and similar work by other authors, suggest that populations of many threatened species, both plants and animals, are doomed for demographic reasons, even though they may not disappear for many decades (Primack, 1998).

The planned biota

The range of both plant and animal taxa selected for domestication was determined at a relatively early stage (<10,000 years ago) and has not been greatly extended in more recent times. In most cases, only a limited part of the diversity of the ancestral species was tapped, perhaps through a single domestication event, or through separate domestications in different parts of the species' range (e.g. for *Capsicum* spp.; Pickersgill, 1989). The speed with which cultivated varieties were developed was sometimes remarkably rapid (Wood and Lenné, 1999). For example, the wild einkorn and emmer wheats and wild barley may have been domesticated within only two centuries (Hillman, 1990), while the morphological evolution of maize occurred over a few hundred years (estimates range from 300 to 1000 years) and was largely controlled by one gene (Wang *et al.*, 1999). The strong divergence from the ancestral form which is seen in many domesticated species reflects the continued selection for desirable traits. This has often been associated with the evolution of barriers to gene flow between the domesticated and natural populations, for example through polyploid formation, selection for predominantly self-pollinated genotypes and the predominance of vegetative reproduction. Thus, domestication has captured and maintained only a tiny proportion of the genetic diversity present within the wild biota.

The diversity of species used in different production systems varies greatly (Swift *et al.*, 1996). At one extreme are some traditional agricultural systems of the tropics which resemble natural ecosystems in the diversity of species they support. Home-gardens, which used to occur throughout the tropics but are disappearing rapidly in many areas, are typically small plots of 0.5–2 ha associated with habitation. They include a remarkably wide range of plant species, mainly perennial, from which the farmer obtains a variety of food products, firewood, building materials, medicines, spices and ornamental flowers. Associated with the home-gardens, there is often traditional animal husbandry, for example of poultry and pigs, which contributes further to the diversity of these systems. Traditional mixed arable and livestock farming as practised in northern temperate regions also supported a wide range of planned biota. The traditional arable rotations ensured a diversity of crop species, and most farms were self-sufficient in fruit and vegetables, as well as producing products for local markets. In the post-World War II period much of this diversity has been lost through the introduction of modern farming systems reliant upon chemical inputs for the maintenance of soil fertility and the control of pests and diseases. The most extreme reduction of planned diversity is seen in the extensive monoculture of crops such as maize, wheat or grapes which dominate the landscape in some of the most productive regions of the temperate zone.

The reduction in genetic diversity associated with the introduction of modern farming systems has been even more dramatic than the loss of species diversity. A high level of genetic diversity is a feature of most traditional agricultural systems, and farmers have preserved this diversity through recognizing and maintaining a wide range of cultural varieties of different crops. An example of the amazingly wide knowledge base of some traditional agriculturists is provided by the Hanunoo, a mountain tribe of the Philippines, who could distinguish over 1500 useful plants, including 430 cultigens (Conklin, 1954; Thurston *et al.*, 1999). In the Ticino of Switzerland, where the chestnut *Castanea sativa* has been cultivated for almost 2000 years, well over 100 cultivars have been recognized and 94 different chestnut variety names are known (Conedera *et al.*, 1994). Moreover, even within traditional cultivars there is usually a high level of genetic diversity. The combination of distinctiveness yet variability which characterizes many landraces was well described by Harlan (1975):

> Landraces have a certain genetic integrity. They are recognizable
> morphologically; farmers have names for them and different landraces are
> understood to differ in adaptations to soil type, time of seeding, date of
> maturity, height, nutritive value, use and other properties. Most important,
> they are genetically diverse. Such balanced populations – variable, in
> equilibrium with both the environment and pathogens, and genetically
> dynamic – are our heritage from past generations and cultivators.

In contrast, modern cultivars and varieties are sometimes, though not always, very uniform genetically, especially in self-pollinated species. As Cox and Wood (1999) remark:

> A so-called 'pure-line' cultivar of wheat (*Triticum aestivum* L.) – a
> self-pollinated species – may be just that if it was developed in a European
> country with licensing requirements that require strict genetic uniformity.
> Conversely, a wheat cultivar developed for the central or northern Great
> Plains of the USA – where uniformity requirements are looser . . . may
> harbour even more genetic diversity than a wheat landrace that is descended
> from a single, highly homozygous plant selected and propagated by a farmer.

Genetic erosion in crop diversity has been dramatic in the past 50 years. According to the FAO, approximately 75% of the global genetic diversity of our crop plants has been lost during the last century (Hammer, 1998). Fifty years ago, 2000 rice varieties were grown in Sri Lanka, but only five varieties are in widespread cultivation today. The situation in India appears even more extreme: today, only ten varieties account for 75% of Indian rice production, where there were formerly 30,000 in cultivation (UNEP, 1993). Within a decade or less, a comparatively limited number of new, high-yielding varieties or hybrid crops have replaced almost entirely the traditional, locally bred varieties derived from and consisting of a multifold of landraces (Chrispeels and Sadava, 1994).

Some scientists worry that genetic engineering of crop plants, predominantly carried out in the private sector, may accelerate the loss of genetic diversity in crop plants. This could occur if plant breeders focus on engineering a limited number of high yielding varieties and uniform hybrids, adding individual foreign genes or traits at the expense of significant genetic variability. Others argue that there will be an increase in diversity at the gene level, especially when intellectual property rights expire and transgenes enter into the public domain and can be used to transform any variety. Certain transgenes will then tend to become universal in the crop (Witcombe, 1999). However, it is important to recognize that such diversity will consist of a limited number of interchangeably re-arranged alleles coding for potent, novel traits in a small number of otherwise genetically uniform elite lines grown in monocultures. The consequences of such technological developments for crop genetic diversity need to be monitored carefully. Already, the adoption rates of a small number of transgenic maize and cotton cultivars in US agriculture equals or surpasses that of hybrids, confirming the expected further trend away from growing a diversity of crop varieties (Witcombe, 1999; James, 2000).

There has been a similar decline in the genetic diversity of livestock. In the USA, Holstein cows represent almost 91% of the dairy stock. Similarly, over 60 chicken breeds have been abandoned, and only five breeds supply most chicken meat and brown eggs in the USA. White eggs come almost exclusively from a single breed, white leghorns (Raloff, 1997). According to the FAO, at least 1500 of roughly 5000 domesticated livestock breeds throughout the world are now rare, i.e. represented by less than 20 breeding males on the planet or less than 1000 breeding females (Raloff, 1997).

The unplanned biota of agricultural land

Like natural ecosystems, agricultural landscapes consist of a mosaic of distinct habitat elements (Edwards *et al.*, 1999). These include the various crop types and non-crop elements such as hedgerows, uncultivated marginal strips and trackways. These different elements provide contrasting abiotic conditions and are usually disturbed (e.g. by ploughing, harvesting, herbicides, burning, etc.) with different frequencies and at different times of the year. An important difference compared with natural ecosystems is that the dynamics of these habitat patches, and also to some extent the abiotic conditions, are strongly influenced by the farmer who decides which crop to plant, when to plough and harvest, and what kinds of external inputs to use. The unplanned diversity of these habitats is drawn from a wide range of natural ecosystems. Very often, we do not even know from which natural habitats the plants and animals inhabiting agroecosystems originated. Probably many arable weeds are natives of regularly disturbed habitats such as scree slopes, river terraces and coastal beaches (Ellenberg, 1988). These organisms could only become established on agricultural land because their life histories were compatible with the habitat conditions in the new environment; in other words, the species of the unplanned biota are 'pre-adapted' to the conditions they encounter in agroecosystems (Harper, 1982).

In general, agricultural areas provide favourable conditions for growth – the farmer ensures that there are adequate levels of water and nutrients – but the level of disturbance is very high. If this disturbance is regular, the life history and phenology of organisms can be closely attuned to that rhythm. For example, some traditional hay meadows in Europe have had a remarkably regular management for centuries, with the hay crop being cut at the same time each year. Such grasslands typically contain tall, early flowering species which reproduce freely from seed such as *Fritillaria mele-agris*, *Tragopogon pratensis* and *Silene flos-cuculi* (Harper, 1977; Ellenberg, 1988). The species composition contrasts strongly with grasslands which have been managed by grazing or by more frequent mowing, and which tend to contain perennial grasses such as *Lolium perenne* and *Cynosurus cris-tatus* and low-growing, clonally reproducing dicotyledonous herbs such as *Bellis perennis* and *Ranunculus repens*. If the disturbance is frequent yet irregu-lar, there is likely to be selection for organisms with a short lifespan, high reproductive output and high dispersal ability. Weed species which persist in agroecosystems where the pattern of disturbance is irregular also tend to show variability in their germination requirements, patterns of dormancy and growth phenology. This provides a kind of 'risk spreading' in an unpre-dictable environment. In particular, such species often produce long-persisting seed banks or seeds which are polymorphic in their germination behaviour (e.g. *Chenopodium album*; Williams and Harper, 1965).

The spatial organization of habitat patches is important for the persist-ence of many species in the agricultural landscape for various reasons. Firstly, many mobile animals, such as birds, require a variety of habitat elements (Tucker, 1997). For example, predatory birds such as the barn owl may nest in woodland fragments, and hunt their prey by moving along hedgerows (de Bruijn, 1979). Similarly, the skylark may nest in the crop but search for food in marginal vegetation where there are more insects to be found. Secondly, for more mobile species such as birds and mammals, undisturbed habitat elements represent refuges to which they can flee when the habitat they occupy is damaged or destroyed. Particularly important in this respect are permanent or semi-permanent structures such as hedgerows and fragments of grassland (Wratten and Thomas, 1990). Finally, the undis-turbed elements represent a source of individuals for recolonization of habi-tats which have been disturbed. Many studies have shown that populations of insects such as carabid beetles in cereal fields are greatly enhanced if there are strips of grassland or wildflowers, or hedgerows adjacent to a field (e.g. Dennis and Fry, 1992).

Because the unplanned biota accumulates gradually as new species are introduced, diversity tends to increase with time. For example, Mermod (2000) lists 60 plant species which occurred as arable weeds in a Neolithic settlement (*c.* 2500 BC) on the shores of Lake Neuchatel in Switzerland. With very few exceptions, these species still occur as arable weeds in the region, but the present arable flora includes many other species introduced more recently. Not surprisingly, old agroecosystems tend to be richer in unplanned diversity than more recent ones. The agricultural tradition in Europe is at least 1000 years older than in North America and as a consequence the number of invasive European weeds in North America is much higher than vice versa. The rice paddy system in many Asian countries (Heckman, 1979; Edwards *et al.*, 1999) developed over thousands of years. Schoenly *et al.* (1996a) found a cumulative total of some 687 arthropod taxa in five rice paddy sites on Luzon island of the Philippines, including sites where rice has been cultivated for several millennia (e.g. Banaue). Taxonomic composition, food web structure, and arthropod phenology were broadly similar across different sites. In some of these ancient production systems intricate webs of checks and balances have developed that are vulnerable to abrupt external interventions such as the application of insecticides. Schoenly *et al.* (1996b) found in field trials a fourfold pest increase in insecticide-sprayed plots over the unsprayed control plots. In comparison, paddy fields in Bangladesh, Singapore or southern Malaysia have less rich flora and fauna, partly because they have been cultivated for little more than a century (Shajaat Ali, 1987).

In recent years, biodiversity has declined greatly as modern intensive farming systems have been introduced (Edwards *et al.*, 1999). However, some species have been able to persist and even thrive, often because of their ability to adapt to their changing habitats. Although the plants and animals of the 'unplanned diversity' have been recruited from other habitats, evolutionary processes have always been important in matching their life histories more precisely to the conditions in which they find themselves (Baker, 1974). There is plenty of evidence for this evolutionary 'fine tuning' amongst weed species. For example, *Camelina sativa* is a crop mimic which occurs in flax fields, while *Echinochloa crus-galli* is a mimic of rice; in both cases centuries of selection have produced a plant which matches the phenology of the crop very precisely (Gould, 1990). Similarly, hybridization between maize and teosinte, a close relative common in maize fields, produces teosinte weeds that mimic the crop. One form of teosinte has become a mimetic weed by hybridizing with a particular cultivar of maize in the valley of Mexico. Teosinte has developed features that are typical of the maize cultivar used in the region. As a result the teosinte weed is difficult to distinguish from maize and, thus, escapes weeding (Mellon and Rissler, 1998 and references therein).

An entomological example of adaptation to cultural practices is provided by the northern corn root worm, *Diabrotica longicornis barberi*, a major pest of maize in the Midwest USA. It was shown how populations of this species evolved an extended diapause when maize was grown in a 2-year rotation with another crop as a cultural method for alleviating the root worm problem. For many years, farmers successfully also controlled the western corn root worm, *Diabrotica virgifera virgifera*, a relative of the one mentioned above, when rotating maize with soybeans. However, in 1993, farmers using this technique began to suffer serious crop losses. Subsequent studies ruled out a prolonged diapause. Further feeding and behavioural studies revealed that the western corn root worm larvae could actually feed on soybean foliage and even lay eggs in soybean fields. Thus, their host plant range had extended to include soybean, so that the rotation with soybeans no longer disrupted their life cycle (Levine and Oloumi-Sadeghi, 1992; Spencer *et al.*, 1997).

Further evidence of the evolutionary potential of the unplanned biota is provided by the hundreds of crop-pest species which have become resistant to pesticides since synthetic chemicals were first used on a large scale in the 1940s (Gould, 1991). For this reason, many are now worried about the possible development of pest resistance against genetically engineered crop plants expressing the Bt-gene. This gene codes for an insecticidal toxin from *Bacillus thuringiensis*, which is produced throughout the plant in high concentrations and over the whole growing season. Since these plants are grown on millions of hectares (James, 2000) in the USA, they represent an enormous selection pressure on pests to become resistant. Resistance management programmes are now being developed and mandated by the Environmental Protection Agency (EPA) which are intended to delay the development of such resistance (Mellon and Rissler, 1998).

Genetic resistance to herbicides began to appear among weeds in the 1960s and has been reported in at least 84 plant species. Once again, concern exists that outcrossing of the herbicide resistance genes in transgenic crops to weedy relatives or other related cultivars may cause resistant weeds or volunteer crops in subsequent cultures. In 1999, after only 3–4 years of large-scale production of transgenic herbicide-resistant plants, the first triple herbicide-resistant canola plants were confirmed in Canada. These plants had acquired two herbicide resistance transgenes and one conventionally bred herbicide resistance gene (MacArthur, 2000). Potentially, such multiple resistant crops can pose a problem as volunteers in a subsequent crop as the chemical control options are greatly reduced. Reports of resistance by plant pathogens to systemic fungicides are also increasing (Green *et al.*, 1990). The lesson we learn from such studies is that the species which are successful in the agricultural landscape are those whose genetic constitution or breeding system allow them to evolve under the new conditions.

Sustaining Biodiversity in Agriculture: a Challenge for the Future

Figure 13.1 illustrates our hypotheses concerning the main trends in biodiversity in the agricultural landscape. The figure represents the development of biodiversity from the beginnings of agriculture until the 20th century, and the changes which have occurred since then as a result of intensification. A decline in species from natural ecosystems is probably inevitable as soon as agriculture is introduced, though under extensive traditional agriculture a

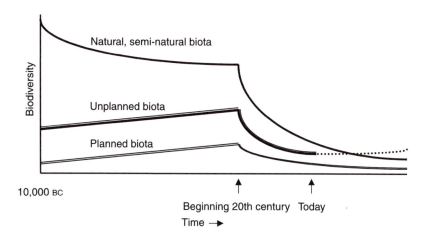

Fig. 13.1. Conceptual diagram illustrating the trends of the three components of biodiversity in agroecosystems.

remarkably high level of diversity can be maintained. In contrast, both planned and unplanned diversity probably tended to increase under some forms of traditional agriculture as a result of new introductions, both deliberate and unintentional. For example, the highly diverse agricultural landscapes of central Europe developed over millennia (Ellenberg, 1988). Over this period, remarkably little of the biological diversity associated with natural ecosystems was lost, and several species of weeds and insects were introduced which thrived in the new agricultural habitats. In addition, the mixed arable–livestock systems practised in many areas provided a great diversity of crop plants, and a wide genetic base of both crops and livestock. Thus, there was probably a net gain in biodiversity, even though a few species – notably the top predators such as wolves and bears – became extinct.

The introduction during the 20th century of modern farming systems dramatically changed this picture. In many areas there has been a drastic reduction in the abundance of species from natural habitats, through changes in land use and the associated fragmentation of natural and semi-natural habitats. The genetic diversity of many crop species has also been drastically reduced, with the giving up of many traditional varieties and their replacement with genetically more uniform 'pure-line' cultivars, often planted in continuous monocultures. The genetic diversity of livestock has suffered a similar fate (see above).

Many species of the unplanned biota have also declined greatly. In Europe, the populations of many farmland bird species have crashed in the last 10–20 years, partly because of a greatly reduced supply of food available to them in the form of insects and seeds of wild plants (Tucker, 1997).

Indeed, birds such as the skylark and corn bunting are perhaps our best monitors of the state of biodiversity on agricultural land. However, we should not assume that the decline of at least some components of the unplanned biota will continue indefinitely. The history of pest control has been one of a long struggle to stay ahead of pest adaptation. Although modern biocides may be effective in suppressing most pest species at present, the capacity of evolution to respond to the challenges posed by the new technologies should not be underestimated. Despite the drastic increase of synthetic insecticides, crop losses due to insect damage have also increased and there are already many species which have evolved resistance to modern biocides. We cautiously suggest that the unplanned biodiversity, at least of the destructive biota also including alien invasive species, will gradually increase again (Fig. 13.1).

The rapidly declining biodiversity of some agroecosystems leads us to pose two questions: firstly, does it matter, and secondly, what can be done about it? Recently, ecologists have begun to address the question of whether species diversity is important for ecosystem functioning and sustainability (Lawton, 1994; Mooney, 1996; Altieri, 1999). There are good reasons to believe that diversity is important for sustainable agriculture (Swift *et al.*, 1996), and that we have gone too far in eliminating biological diversity in the pursuit of more efficient production (Matson *et al.*, 1997; Altieri, 1999). Particular functions of biodiversity in agricultural systems include the reduction of pests and diseases, and the maintenance of soil fertility. For example, there is abundant evidence that genetic uniformity in crops increases the likelihood that pathogens will adapt to the host genotype and cause serious losses (Buddenhagen, 1977). There is also a tendency for densities of certain herbivorous insect species to be higher in monocultures than in mixed cultures and hence crop losses may also be higher. This is especially true for some specialist insects with restricted host ranges which may become major pests. There are several reasons why crop losses may be lower in polycultures. These are related to the greater difficulty an insect may have in finding its host plant and also potentially higher parasitism and predation rates due to the more diverse structure of the invertebrate community (Altieri, 1994).

There is also growing evidence that the unplanned diversity has a role to play in sustainable agriculture; for example, a high diversity of natural enemies can contribute significantly to strategies of integrated pest control, and as a result greatly reduce our dependence upon pesticides (Pickett and Bugg, 1998). Crop heterogeneity is also a possible solution to the vulner-ability of monocultured crops to diseases. In a field trial, Zhu *et al.* (2000) achieved 89% greater yields and 94% less severe blast (*Magnaporthe grisea*) infection when planting disease-susceptible rice varieties in mixtures with resistant varieties than when they were each grown in monoculture. The otherwise regularly applied fungicide treatments were no longer necessary. This is a spectacular example of creating functional diversity by finding an optimal variety mixture. As suggested by Altieri (1999), these effects of diversification can only be determined experimentally across a whole range of agroecosystems. However, Zhu *et al.* (2000) demonstrated that a simple, functional diversity approach to disease control is indeed possible.

The answer to the question of how we can restore biodiversity is different for each of the three components we have described. As a first priority, we should do everything we can to protect the viable areas of remaining natural habitat, especially in areas of intensive agriculture. Of particular importance, we need to create conditions in which species are not caught in a spiral of demographic decline, and in which they can adapt through genetic change to changing environmental conditions (Kark and Blackburn, 2000). However, the recreation of natural habitats is difficult and expensive and, as we have seen, populations of many species are already doomed because their habitats have become too fragmented (Matthies, 2000). In such cases, it may be wiser to focus our efforts on enhancing the biodiversity of the agricultural land itself and such types of functional biodiversity that provide us with key ecological services (Altieri, 1999). In recent years, there have been a number of studies which show that much can be done to restore biodiversity to areas where it has declined through recent intensification. A good example is provided by the Klettgau region of northern Switzerland. Since 1991, farmers in this region have been encouraged to create various types of ecological compensation areas (ECAs) on their land, under a scheme promoted by the Swiss Ornithological Union and the Canton of Schaffhausen. The primary objective of this programme was to improve the intensive arable landscape as a habitat for birds, and particularly the partridge which has declined drastically in the last two decades. However, in intensive arable areas ECAs can enhance species richness more generally even though, as in the Klettgau, they account for only 3% of the total land area. Thus, a total of 234 plant species were recorded in the wildflower strips, of which the great majority, including 35 red list species, appeared spontaneously. The ECAs also provide a habitat for a large diversity of invertebrate species, including 139 heteropteran bug species and about 100 spiders. A comparison of the bug fauna in 1993 and 1999 reveals that the number of bug species increased significantly in the Klettgau area over this period. A very detailed investigation of the population biology and diet of the skylark (*Alauda arvensis*) also showed that wildflower strips are extremely important as sources of invertebrate food for nestlings, and that the growth rates of chicks in territories containing wildflower strips tends to be higher than in territories without them.

Challenges

The great challenge we face is to develop agricultural systems which can sustain a high level of biological diversity without the need for high investment in special conservation measures such as those described for the Klettgau region (the farmers of the Klettgau are paid 3000–5000 SFr ha^{-1} year^{-1} for the ECAs they maintain on their land!). In other words, just as in many traditional agroecosystems, biological diversity should be a property of the system rather than something which is achieved as a result of special policy initiatives. 'Conservation' is not the best word to describe this process of developing biologically diverse agricultural systems, because it has a strong connotation of attempting to preserve something which already exists. This is an important difference from the conservation of natural habitats, when we usually have a clear view of the type of ecosystem which we are trying to protect (e.g. a particular kind of beech forest with a well-defined associated flora and characteristic of acidic soils). In contrast, it is not usually a realistic long-term goal to conserve particular plant or animal communities on agricultural land. As our brief historical account has shown, the structure of the agricultural landscape, and thus also of the associated biota, is highly dynamic and changes continuously in response to changes in agricultural practice. It is futile to attempt to conserve the plant and animal communities which assembled as a response to former land use systems.

There is still a great deal to learn about how we can produce bio-diverse landscapes which are compatible with efficient food production. It is clear that non-crop elements such as wildflower strips, hedgerows, woodland belts and grassy tracks contribute greatly to species diversity at the landscape scale. There is now an urgent priority to find out how much and what form of this 'green-veining' is needed to achieve an acceptable level of diversity and to prevent the excessive fragmentation of populations. We should also develop our understanding of the importance of wildlife resources in economic terms. It is now possible to estimate the financial value of the ecosystem services provided by the unplanned biota – for example, control of pests and diseases or maintenance of soil fertility (Edwards and Abivardi, 1998). In the USA, some biological control experts have begun to place a price tag on the ecosystem services of natural enemies. For example, sugar beet farmers in California have to treat their fields with synthetic insecticides against the beet worm (*Spodoptera exigua*) at costs of about US$1500 per 100 acres. Researchers found that a certain density of two natural enemies in combination will provide a comparable level of control and, thus, must be considered worth the same price (Ehler, 2000). However imperfect and incomplete such a valuation may be, it helps to demonstrate, in terms that decision makers may understand, our essential dependence upon biological diversity.

References

Aebischer, N.J. (1991) Twenty years of monitoring invertebrates and weeds in cereal fields in Sussex. In: Firbank, L.G., Carter, N., Darbyshire, J.F. and Potts, G.R. (eds) *The Ecology of Temperate Cereal Fields*. Blackwell Scientific Publications, London, pp. 305–331.

Altieri, M.A. (1994) *Biodiversity and Pest Management in Agroecosystems*. Food Products Press, New York.

Altieri, M.A. (1999) The ecological role of biodiversity in agroecosystems. *Agriculture, Ecosystems and Environment* 74, 19–31.

Baker, H.G. (1974) The evolution of weeds. *Annual Review of Ecology and Systematics* 5, 1–24.

Buddenhagen, I.W. (1977) Resistance and vulnerability of tropical crops in relation to their evolution and breeding. *Annals of the New York Academy of Sciences* 287, 309–326.

Chrispeels, M.J. and Sadava, D.E. (1994) *Plants, Genes and Agriculture*. Jones and Bartlett Publishers, Boston, Masschusetts, pp. 298–327.

Conedera, M., Müller-Starck, G. and Fineschi, S. (1994) Genetic characterization of cultivated varieties of European Chestnut (*Castanea sativa* Mill.) in Southern Switzerland. (I) Inventory of chestnut varieties: history and perspectives. In: Comunità Montana Monti Martani e Serano of Spoleto; Istituto di Coltivazioni Arboree University of Perugia (eds) *Proceedings of the International Congress on Chestnut, Spoleto, Italy, October 20–23, 1993*, pp. 299–302.

Conklin, H.C. (1954) An ethnological approach to shifting agriculture. *Transactions of the New York Academy of Science* 17, 133–142.

Cox, T.S. and Wood, D. (1999) The nature and role of crop biodiversity. In: Wood, D. and Lenné, J.M. (eds) *Agrobiodiversity: Characterization, Utilization and Management*. CAB International, Wallingford, pp. 35–57.

de Bruijn, O. (1979) Feeding ecology of the barn owl, *Tyto alba*, in the Netherlands. *Limosa* 52, 91–154.

Dennis, P. and Fry, G.L.A. (1992) Field margins: can they enhance natural enemy population densities and general arthropod diversity of farmland? *Agriculture, Ecosystems and Environment* 40, 95–115.

Edwards, P.J. and Abivardi, C. (1998) The value of biodiversity – where ecology and economy blend. *Biological Conservation* 83, 239–246.

Edwards, P.J., Kollmann, J. and Wood, D. (1999) Determinants of agrobiodiversity in the agricultural landscape. In: Wood, D. and Lenné, J.M. (eds) *Agrobiodiversity: Characterization, Utilization and Management*. CAB International, Wallingford, pp. 183–210.

Ehler, L. (2000) Natural enemies provide an ecosystem service. *International Organization for Biological Control (IOBC) Newsletter* 71, 1–2.

Ellenberg, H. (1988) *Vegetation Ecology of Central Europe*. Cambridge University Press, Cambridge.

Gould, F. (1991) The evolutionary potential of crop pests. *American Scientist*, 79, 496–507.

Gould, G. (1990) Ecological genetics and integrated pest management. In: Carroll, C.R., Vandermeer, J.H. and Rosset, P.M. (eds) *Agroecology*. McGraw-Hill, New York, pp. 441–458.

Green, N.G., LeBaron, M. and Moberg, W.K. (1990) *Managing Resistance to Agrochemicals: From Fundamental Research to Practical Strategies*. American Chemical Society, Washington, DC.

Hammer, K. (1998) Agrarbiodiversität und pflanzengenetische Ressourcen. *Schriften zu genetischen Ressourcen* 10, ZADI, Bonn.

Harlan, J.R. (1975) Our vanishing genetic resources. *Science* 188, 618–621.

Harper, J.L. (1977) *Population Biology of Plants*. Academic Press, London.

Harper. J.L. (1982) After description. In: Newman, E.I. (ed.) *The Plant Community as a Working Mechanism.* Blackwell Scientific Publications, Oxford, pp. 11–25.

Heckman, C.W. (1979) *Rice Field Ecology in Northeastern Thailand. The Effect of Wet and Dry Seasons on a Cultivated Aquatic Ecosystem.* Dr Junk, The Hague.

Hillman, G.C. (1990) Domestication rates of wild-type wheats and barley under primitive cultivation. *Biological Journal of the Linnean Society* 39, 39–78.

James, C. (2000) Global status of commercialized transgenic crops: 1999. *ISAAA Briefs No. 17.* ISAAA, Ithaca, New York.

Kark, S. and Blackburn, T.M. (2000) The future of evolution. *Tree* 15, 307–308.

Kéry, M., Diethart, M. and Spillmann, H.-H. (2000) Reduced fecundity and offspring performances in small populations of the declining grassland plants *Primula veris* and *Gentiana lutea. Journal of Ecology* 88, 17–30.

Lawton, J.H. (1994) What do species do in ecosystems? *Oikos* 71, 367–374.

Levine, E., Oloumi-Sadeghi, H. and Fisher, J.R. (1992) Discovery of multiyear diapause in Illinois and South Dakota northern corn rootworm (Coleoptera: Chrysomelidae) eggs and incidence of the prolonged diapause trait in Illinois. *Journal of Economic Entomology* 85, 262–267.

MacArthur, M. (2000) Triple-resistant canola weeds found in Alta. *The Western Producer* February 10, 2000.

Matson, P.A., Parton, W.J., Power, A.G. and Swift, M.J. (1997) Agricultural intensification and ecosystem properties. *Science* 277, 504–509.

Matthies, D. (2000) The genetic and demographic consequences of habitat fragmentation for plants: examples from declining grassland species. *Schriftenreihe für Vegetationskunde* 32, 129–140.

Mellon, M. and Rissler, J. (1998) *Now or Never: Serious New Plans to Save a Natural Pest Control.* Union of Concerned Scientists, Cambridge, Masschusetts.

Mermod, O. (2000) Die endneolithische Seeufersiedlung Saint-Blaise/Bains des Dames NE. Dissertation, Geobotanisches Institut ETH, Zürich.

Mooney, H.A. (ed.) (1996) *Biodiversity and Ecosystem Function.* Springer-Verlag, Berlin, pp. 15–41.

Pickersgill, B. (1989) Cytological and genetic evidence on the domestication and diffusion of crops within the Americas. In: Harris, D.R. and Hillman, G.C. (eds) *Foraging and Farming: the Evolution of Plant Exploitation.* Unwin Hyman, London, pp. 426–439.

Pickett, C. and Bugg, R.L. (eds) (1998) *Enhancing Biological Control.* University of California Press, Berkeley.

Primack, R.B. (1998) *Essentials of Conservation Biology,* 2nd edn. Sinauer Associates, Massachusetts.

Raloff, J. (1997) Dying breeds: livestock are developing a largely unrecognized biodiversity crisis. *Science News Online,* 4 October 1997 (http://www.sciencenews.org/sn_arc97/10_4_97/bobl.htm).

Remmert, H. (1991) The mosaic-cycle concept of ecosystems – an overview. In: Remmert, H. (ed.) *The Mosaic-cycle Concept of Ecosystems. Ecological Studies 85.* Springer-Verlag, Berlin, pp. 1–21.

Rosenzweig, M.L. (1995) *Species Diversity in Space and Time.* Cambridge University Press.

Schoenly, K., Cohen, J.E., Heong, K.L., Litsinger, J.A., Aquino, G.B., Barrion, A.T. and Arida, G. (1996a) Food web dynamics of irrigated rice fields at five elevations in Luzon, Philippines. *Bulletin of Entomological Research* 86, 451–466.

Schoenly, K., Cohen, J.E., Heong, K.L., Arida, G.S., Barrion, A.T. and Litsinger, J.A. (1996b) Quantifying the impact of insecticides on food web structure of rice-arthropod populations in a Philippine farmer's irrigated field: a case study. In: Polis, G.A., and Wisemiller, K. (eds) *Food Webs: Integration of Patterns and Dynamics*. Chapman and Hall, London, pp. 343–351.

Shajaat Ali, A.M. (1987) Intensive paddy agriculture in Shyampur, Bangladesh. In: Turner, B.L. II and Brush, S.B. (eds) *Comparative Farming Systems*. The Guilford Press, New York, pp. 276–305.

Spencer, J., Levine, E. and Isard, S. (1997) Corn rootworm injury to first-year corn: new research findings. In: *1997 Illinois Agricultural Pesticides Conference*. Cooperative Extension Service, University of Illinois at Urbana-Champaign, pp. 73–81.

Swift, M.J. and Anderson, J.M. (1996) Biodiversity and agroecosystem function in agricultural systems. In: Schulze, E.-D. and Mooney, H.A. (eds) *Biodiversity and Ecosystem Function*. Springer-Verlag, Berlin, pp. 15–41.

Swift, M.J., Vandermeer, J., Ramakrishnan, P.S., Anderson, J.M., Ong, C.K. and Hawkins, B.A. (1996) Biodiversity and agroecosystem function. In: Mooney, H.A., Cushman, J.H., Medina, E., Sala, O.E. and Schulze, E.-D. (eds) *Functional Roles of Biodiversity*. John Wiley & Sons, Chichester, pp. 261–298.

Thurston, H.D., Salick, J., Smith, M.E., Trutmann, P., Pham, J.L. and McDowell, R. (1999) Traditional management of agrobiodiversity. In: Wood, D. and Lenné, J.M. (ed.) *Agrobiodiversity: Characterization, Utilization and Management*. CAB International, Wallingford, pp. 211–243.

Tucker, G. (1997) Priorities for bird conservation in Europe: the importance of the farmed landscape. In: Pain, D.J. and Pienkowski, M.W. (eds) *Farming and Birds in Europe: the Common Agricultural Policy and its Implications for Bird Conservation*. Academic Press, San Diego, pp. 79–149.

United Nations Environment Program (UNEP) (1993) *Global Diversity*, Nairobi.

Wang, R.L., Stec, A., Hey, J., Lukens, L. and Doebley, J. (1999) The limits of selection during maize domestication. *Nature* 398, 236–239.

Williams, J.T. and Harper, J.L. (1965) Seed polymorphism and germination. 1. The influence of nitrate and low temperatures on the germination of *Chenopodium album*. *Weed Research* 5, 141–150.

Witcombe, J.R. (1999) Does plant breeding lead to a loss of genetic diversity? In: Wood, D. and Lenné, J.M. (eds) *Agrobiodiversity: Characterization, Utilization and Management*. CAB International, Wallingford.

Wood, D. and Lenné, J.M. (1999) The origins of agrobiodiversity in agriculture. In: Wood, D. and Lenné, J.M. (eds) *Agrobiodiversity: Characterization, Utilization and Management*. CAB International, Wallingford, pp. 15–34.

Wratten, S.D. and Thomas, C.F.G. (1990) Farm-scale spatial dynamics of predators and parasitoids in agricultural landscapes. In: Bunce, R.G.H. and Howard, D.C. (eds) *Species Dispersal in Agricultural Habitats*. Belhaven Press, London, pp. 219–237.

Zhu, Y., Chen, H., Fan, J., Wang, Y., Li, Y., Chen, J., Fan, J.X., Yang, S., Hu, L., Leung, H., Mew, T.W., Teng, P.S., Wang, Z. and Mundt, C.C. (2000) Genetic diversity and disease control in rice. *Nature* 406, 718–722.

Conservation and Utilization of Biodiversity in the Andean Ecoregion

14

W.W. Collins

International Potato Center, PO Box 1558, Lima 12, Peru

Introduction

Biodiversity in this chapter refers to the totality of genetic diversity that resides in and between plant species in the Andes. The topic of utilization of biodiversity, specifically addressing increasing species diversity in landscape and production systems, and the integration of genetic resources (within-species diversity) into high performing breeding materials is a broad one, with literally thousands of researchers worldwide working in direct applications. In general, the high level of activity reflects the importance of not only conserving genetic resources for their intrinsic and current value, but of studying and realizing their potential for the future through utilization in new varieties to meet food, feed and fibre needs. This chapter will present a brief general introduction to plant biodiversity in agroecosystems, and then address the specific work being undertaken by the International Potato Center (CIP) in one very important region of the world – the Andean eco-region. Examples provided rely on information generated and provided by CIP scientists in the course of their research, and which may not yet be published. Further information is available from the CIP website (www.cipotato.org).

CIP is one of the Future Harvest Centers supported by the Consultative Group on International Agricultural Research (CGIAR). Part of the work of CIP deals with genetic resources of potato (*Solanum* spp.), sweetpotato (*Ipomoea batatas*) and a variety of underutilized Andean root and tuber crops. Other activities of CIP focus on the utilization of this genetic diversity to improve potato and sweetpotato varieties and overcome production constraints for poor farmers in developing countries. CIP's objective with respect to Andean root and tuber crops is to maximize the potential of the existing

© 2001 CAB International. *Crop Science*
(eds J. Nösberger, H.H. Geiger and P.C. Struik)

diversity in native landraces to increase market incentives. Along with prod-uctivity and quality increases, CIP's work is aimed at reducing the environ-mental impact of agricultural practices.

Biodiversity in Agroecosystems

The value of biodiversity in all ecosystems, and especially in agroecosystems, has been more widely and dramatically realized through its disappearance rather than through its presence. The loss of biodiversity in cultivated crops and livestock has proceeded at an alarming pace over the past few decades for many reasons. Agroecosystems vary in crop biodiversity components from very simple (for example, the large corporate farms in industrialized countries) to extremely heterogeneous and complex (the small-scale farmer in the rural highlands of the Andes). However, even in those systems described as simple by their crop biodiversity components, a complex system of related biodiversity components exists. These, too, are often in danger, and the losses are less easily recognizable than the crop components by virtue of the lack of understanding of the processes that might be affected. Crop-related biodiversity, both flora and fauna, provides ecological services which are only now beginning to be understood and valued. The loss of these services is magnified at the farming systems level because they are intricately related to both yield and quality outcomes. To add even more complexity to the simplest system, biodiversity is both temporally and spati-ally related throughout a producing agroecosystem. This is especially true in agricultural systems such as those in the Andean highlands as small farmers rely on a diverse pattern of temporal and spatial use of varieties. They are, in effect, utilizing a sophisticated gene pool approach to minimize risks and buffer against environmental disasters.

The characterization of agroecosystems runs from lightly managed rangelands and forage crops all the way to intensively managed, heavily cropped food production systems. It covers the spectrum, as mentioned above, from monocultures to systems which produce multiple crops year-round; and from aggregated small plots of less than 1 hectare to thousands of hectares in more industrialized systems. Agriculture itself can either enhance biodiversity or endanger it through the selected application of agricultural technologies (Collins and Qualset, 1999). Integrating genetic resources into high performing cultivars is one of those technologies that can have positive or negative effects on biodiversity. Consideration is often given to the nega-tive effect of high performing cultivars taking the place of native cultivars that might be more stable over time, but which often have lower production levels. The positive effects of enhancing genetic variability at the farm level can be significant. Introduction of new genes can protect farmers from losses due to rapidly changing diseases or environmental shifts. Increased pro-duction per unit area implies that no new area has to be cleared and used to meet increasing food needs, thus protecting biodiversity habitat. These effects are often seen to be incompatible with the notion of managing gene pools temporally and spatially, when in fact they provide diversity to management options.

An additional impact of reduced diversity in agroecosystems is the corresponding reduction in traditional knowledge of handling biodiversity. As intra-species diversity diminishes, the unique knowledge associated with using and managing that diversity is also lost. The same is true as species diminish, but the impact is greater in agricultural crop systems with the reduction in intra-species diversity. Just as biological diversity lends resilience to production systems, the associated knowledge and traditions lend resilience to the cultural and social systems that underpin agricultural production and the health of agroecosystems.

One way to visualize this interdependence with traditional knowledge is to imagine one crop and its wild relatives. A continuum exists between the crop and its relatives that encompasses the entire spectrum of the ecosystem. The elements of the ecosystem are connected, including agricultural areas, protected areas and non-farm land areas where related species might be found growing wild. Local knowledge cuts across the entire spectrum of managing, handling and using the biodiversity between crop plants and wild relatives (CIP, 2000a).

The Work of the International Potato Center and the Andean Ecoregion

Collectively, the Future Harvest Centers of the CGIAR hold the largest aggregation of genetic resources in the world, the majority of it held in trust for use by anyone. More than 600,000 accessions cover the major food, feed and fibre crops and provide a valuable reserve for addressing the serious problem of assuring food security while preserving the natural resource base on which food security depends.

CIP holds the world collections of potato and related species (*Solanum*), and sweetpotato and related species (*Ipomoea*), and has significant collections of nine other Andean root and tuber crops (Table 14.1). In addition to holding and using extensive *ex situ* (in gene bank) collections of germplasm of these crops and their related species, CIP scientists have worked at the community level to ensure the *in situ* (in-field, on-farm) conservation and maintenance of intra-species diversity. This reflects the Center's philosophy that on-farm conservation of agrobiodiversity by farmers themselves is essential to an integrated and comprehensive strategy for protecting any species complex of crop plants and wild relatives.

Although CIP carries out a programme of global activities, working in more than 40 countries worldwide, this chapter will concentrate on only one of the areas in which it works: the Andean ecoregion. CIP's extensive

Table 14.1. CIP holdings of potato, sweetpotato and ARTC genetic resources.

Class of holdings	Cultivated species (number of accessions)	Wild species (number of accessions)
Potato	5,305	1,755
Sweetpotato	5,870	1,706
Andean root and tuber crop:		
Achira or canna (*Canna edulis*)	60	11
Ahipa (*Pachyrhizus ahipa*)	10	0
Arracacha (*Arracacia xanthorriza*)	50	0
Maca (*Lepidium meyenii*)	35	7
Mashua (*Tropaeolum tuberosum*)	86	3
Mauka (*Mirabilis expansa*)	4	1
Oca (*Oxalis tuberosa*)	553	140
Ulluco (*Ullucus tuberosus*)	462	22
Yacon (*Smallanthus sonchifolius*)	45	2
Total accessions	12,480	3,567

work in Asia and Africa will not be discussed, although many of the research results obtained through work in the Andean ecoregion are non-site specific and do have global application. Instead, the chapter will talk about those crops native to the Andean region with which CIP works. These crops will be used as examples of the importance not only of using biodiversity, but also of assuring that the range of variability is maintained as a dynamic component of the economic and social development process of the region. Only through economic and social development will CIP's role in utilizing biodiversity to help reach the goals of increasing food security and eradicating poverty be realized.

The Andean Ecoregion

The Andean ecoregion is characterized by high levels of poverty, food insecurity, extreme environmental conditions, high-altitude growing areas, a wide range of production systems and small agricultural plots for food production. However, it is also characterized by great diversity in the number of crop species which originated and are cultivated there.

There are fertile areas in the rich inter-Andean valleys – some at high altitudes – and harsher growing conditions on the slopes and in higher valleys. Land degradation, specifically loss of soil and water, is a major problem as agriculture is practised on steep slopes highly vulnerable to soil erosion. Much of the production is at the subsistence level. Tapia and De la Torre (1997) estimated that more than 10,000 peasant communities exist in Peru,

Ecuador and Bolivia. Traditional knowledge and practices are of particular importance in food production systems when dealing with such unpredictable and extreme environmental conditions. Crop production practices rely heavily on women farmers, especially as the selectors and conservationists of genetic resources (Tapia and De La Torre, 1997) used for household food security. In addition, attempts to understand the conservation and use of biodiversity of crop plants in this region must take into consideration the specific uses associated with various agroecologies, including traditional rotational schemes and intercropping, and with various crops. Some of the most common rotational/intercropping schemes include:

- Maize and quinoa in the valley areas.
- Vegetables and potatoes in the warm valleys.
- Potatoes and other tubers on the hillsides.
- Quinoa and haba (*Vicia faba*) on the hillsides.
- Different species of high altitude potatoes (*S. juzepzuki*) in the Puna (> 4000 m).

Production systems in the Andes are highly dependent on potatoes as a central crop at almost levels. At altitudes of 3200 m or above, only potatoes, some cereals and other native root and tuber crops are grown. At approximately 4000 m, where mostly only pastures and sheep or camelids can flourish, farmers produce 'bitter' potatoes of *S. juzepzuki*, which are highly tolerant to the frost conditions that might occur at any time of the year. Other component crops might include barley or quinoa at those altitudes. Moving down in altitude, other types of potatoes are found, mainly accompanied by other root and tuber crops and perhaps maize or other cereals. At lower levels, crops such as maize and some fruits increase in importance, but potatoes always remain a central component and are sometimes very important commercial crops, as in the rich coastal valleys in Peru. Any mechanism to increase food security and reduce poverty in this region must consider the singular importance of potato throughout the production areas.

Potatoes

Conservation and utilization of potato diversity is of critical importance in the Andean ecoregion because of the role of potatoes in cropping systems at all altitudes. At least eight species of *Solanum* are cultivated on the slopes and in the high valleys of the Andean ecoregion. One of CIP's approaches to promoting economic growth and ensuring food security in the region is through exploiting the rich existing diversity in these potatoes and related species in a sustainable manner.

The importance of *Solanum* biodiversity in this centre of diversity has historically been recognized by inhabitants, and is reflected through religious and festive practices which centre on returning the potato to the earth, a practice called tuber seeding, and recapturing the bounty of the earth at harvest (CIP, 2000b). The contribution of this intertwining of crop biodiversity and culture is to protect a wide range of different *Solanum* germplasm and to assure that it is cared for and maintained at the community level. While modern varieties can contribute, even in remote highland villages, respect for the role of native varieties is never underestimated. It is based not only in mysticism, but also in the understanding that they are critical for the life-saving resilience of the food production systems. Management, at the local level, is based on using a gene pool over time and space for different needs and for different situations.

However, there are threats to the continued existence of this gene pool that make it necessary to assure their conservation and the exploitation of their potential. Environmental changes associated with global climate change (prolonged El Niño, for example), years of terrorist activities in the Andean Highlands, and population increases are among the contemporary threats. So also is the extensive migration of young people from the higher, poorer areas to bigger cities where more employment opportunities are perceived. With their migration, the chain is broken in the continuous passing of traditional knowledge from one generation to the next.

The world collection of *Solanum* germplasm is especially significant in providing a mechanism for exploiting the versatility and usefulness of potato. As shown in Table 14.1, CIP holds a very large *ex situ* collection of potato and its related species, which includes many of the cultivated diploids so important to food security in the Andean highlands of Bolivia, Peru and Ecuador. It also includes an extensive collection of wild related species, many of which flourish in the harsh, niche environments of the Andes. These species are often sexually isolated from cultivated species, but tools of modern biotechnology have the power to unleash the potential locked in their genetic codes in terms of adaptation to extreme environmental conditions, as well as for other traits not yet recognized.

CIP's potato genetic resources and breeding work contribute to both on-farm conservation of Andean potato cultivars, through the establishment of Communal Seed Banks (CSB), and to the discovery and use of valuable genes which can be used to address current production constraints.

On-farm conservation: communal seed banks and seed fairs

Biodiversity surveys were conducted in 71 families of 15 communities in the Peruvian Andes. Besides native potatoes, farmers in the area grow three Andean tuber and three root crops. The survey showed a certain degree of genetic erosion in native potato cultivars. More than 40% of farmers who lost native potato cultivars indicated that they do very little to recuperate them. About 26% of farmers indicated that they recuperate lost cultivars by exchange with another farmer or they get a mixture of cultivars as payment for their work; another 20% recaptured diversity through purchasing or exchanging in the markets, local fairs and more recently, in seed fairs; 7% get them from another family member or farmer of the same conservationist group; and 6% recuperated them from another town.

Conservationist farmers participated in the identification of duplicates among samples contributed by farmers for the CSBs. Thus, in the small pueblo of Aymará, Peru, 107 unique cultivars could be differentiated by farmers out of 204 samples. In Collpatambo, Peru, 160 different cultivars were selected out of 380 samples. A preliminary evaluation indicated that nearly 70% of potato cultivars are common to each of these communities. Many of the different cultivars in both communities could be introductions from other districts, provinces or neighbouring departments according to information received from farmers.

Seed fairs at localities in central Peru are being reported as an important source for the recovery of lost genetic diversity. CIP promotes the organization of seed fairs as an alternative to safeguard genetic diversity. The communities where CSBs have been established decided to participate in the seed fairs to show the genetic diversity that they are now conserving. They not only won the first-place awards, but also demonstrated very positively to other farmers the cultivars that had been considered to be lost. This has created great interest by other communities to organize their own CSBs to secure the conservation of existing genetic diversity from their department.

Another approach to encouraging the conservation of these native genetic resources is to encourage a broader scope of utilization at the same time. The culinary quality of native potatoes is one of the characters that could easily appeal to urban markets. The variety of shapes and tastes and appearances has not been exploited for special niche markets and could provide a demand for native growers. Such market incentives can provide reasons for small communities to maintain their native germplasm and utilize it in new markets. CIP is investigating this possibility, as well as helping communities with the processing techniques and machinery to implement it.

These two approaches to linking conservation and utilization of genetic resources have worked extremely well in the highlands of Peru where diversity is high. CIP also uses a similar approach in Bolivia.

Ex situ *conservation: using genetic diversity in breeding*

CIP's use of potatoes and related wild species for breeding underpins all of its research programmes. Two stories that illustrate CIP's utilization of the diversity available in *Solanum* spp. relate to late blight disease and resistance to the potato virus Y.

The most extensive use of the genetic resources of potato by CIP is probably in the search for late blight resistance genes. CIP's strategy for late blight resistance is to build a highly durable and consistent base of quantitatively inherited resistance genes. This resistance, which is the most valuable type for farmers on small parcels who may use many different varieties, can withstand all races of the late blight organism (*Phytophthora infestans*) and provides a level of protection that assures that the entire crop is not lost to the disease. Once these stable types of resistance are incorporated, race-specific resistance (R genes) can be added for additional protection. Part of CIP's breeding work is aimed at locating new sources of resistance genes, outside the normally considered potato relatives that provide R genes for breeding.

The value of incorporating resistance genes can be inferred from the extensive damage caused worldwide by the late blight disease organism (Fig. 14.1). CIP's socio-economic studies indicate that in the developing world alone, farmers lose more than $3.25 billion annually. Part of this is loss of production (US$2.5 billion), with cost of fungicides adding an additional US$750 million (Fig. 14.2). These heavy losses do not even attempt to account for the environmental and related biodiversity damage caused by the heavy, and often impractical, use of toxic pesticides. The increasing ineffectiveness of fungicides makes it even more imperative to locate durable sources of resistance that can withstand the rapid changes of the pathogen.

Starting with a diverse population of *Solanum andigena*, the most important commercial type of potato grown in the Andean ecoregion, CIP scientists first developed a segregating population (population A) which contained both the race-specific genes (those that are fairly short-lived) and genes which are not race-specific (those conferring more durable resistance). As the population continued to be improved for the late blight resistance trait,

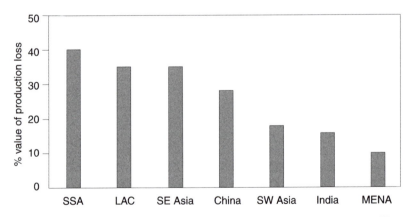

SSA, Sub-saharan Africa; LAC, Latin America and the Caribbean; MENA, Middle East and North Africa.

Fig. 14.1. Relative economic importance of late blight (% value of production loss).

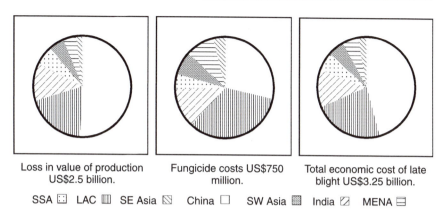

Loss in value of production US$2.5 billion.

Fungicide costs US$750 million.

Total economic cost of late blight US$3.25 billion.

SSA ⊡   LAC ⊞   SE Asia ◨   China ☐   SW Asia ▨   India ▧   MENA ⊟

Fig. 14.2. Late blight damage estimates in developing countries, 1997.

efforts were also made to stress selection for non-race-specific resistance. Since then populations B1 and B2 have been developed. These contain both the *andigena* and *tuberosum* genetic backgrounds. The B2 population is considered free of race-specific genes with high levels of non-race-specific resistance which will be durable over time. The fact that population breeding was used throughout this process has meant that farmers had access to many different genotypes and could choose for themselves the types which best met their needs. In Peru and Ecuador, a programme of participatory plant selection in the early stages of these advanced populations provided planting material to a range of small farmers and communities, with the result that several varieties were released through the farmers' efforts. In addition, an extensive programme of farmer field schools is being carried out in the region, and in other parts of the world, to help farmers understand the basics of late blight disease and to better select their own varieties for their own specific needs.

The utilization of the diversity available to build up the background levels of resistance to late blight has now resulted in approximately 60 elite clones selected for high quality and suitable for deploying in many areas of the Andean ecoregion, as well as in similar agroecological zones in Africa and Asia. The deployment and use of these clones will substantially reduce fungicide applications to control this devastating disease. The farmer field schools mentioned previously also show farmers how to balance resistance with fungicide applications in order to more efficiently let the resistance control the disease.

As a parallel effort to developing advanced clones from the *andigena*- and *tuberosum*-based materials, related diploid species are also being screened for resistance and used in a pre-breeding programme to develop resistance. Already, new sources of race-specific genes have been located in species where they have not been previously found. The genes could provide a new and unique resistance mechanism to add to the genes already being used.

A very different utilization of *Solanum* biodiversity has been deployed in CIP's search for resistance to potato virus Y (PVY) and potato virus X (PVX) which illustrates the value of new technologies to the effective utilization of species diversity. Potato viruses were among the early models used in transformation with viral coat protein genes for virus resistance. CIP took a different approach and searched various species for extreme resistance to the viruses. Single genes conferring extreme resistance to either PVY or PVX have been located, mapped using precision DNA-mapping tools, and now cloned. They are now being used to transform popular varieties which lack resistance to one or both viruses. This illustrates the value of characterization and evaluation of genetic resources, and reinforces the reasons why conservation is needed, even when potential use is not readily apparent at first glance.

Andean Root and Tuber Crops

CIP employs yet another approach in the utilization of Andean root and tuber crop germplasm. CIP conserves and maintains germplasm of nine little known and underutilized Andean root and tuber crops (Table 14.1). As with potatoes and sweetpotatoes, CIP also concentrates on *in situ* conservation. *In situ* conservation is actually more important at the moment for these so-called lost crops, because they have traditionally played little or no role in regional and national markets and thus there is little economic advantage for an institutional conservation role. CIP's work combines support for *in situ* conservation with identification and growth of local and regional markets so that farmers can realize benefits associated with conserving a wide range of biodiversity. Most Andean farmers who grow these crops do not commercialize them extensively. Diversity is preserved by the dynamic interrelationships between communities and farmers in exchanging their crops, and also by the extreme need in the Andes to protect against the consequences of a very harsh environment, one that includes drought, frost, floods and earthquakes. One of the most important considerations in CIP's work is to balance the search for income-producing markets with the effects of these markets on this local, community protection of biodiversity in this harsh agroecosystem.

Three very brief examples of the successful move to markets for these crops are described below.

The maca story

Maca (*Lepidium meyenii*) is cultivated above 4000 m in the Andes. It has long been used by local communities for human and animal fertility.

Working closely with CIP, a Peruvian pharmaceutical company that has invested more than \$1 million in research and product development has concluded that maca not only improves fertility but also increases energy and relieves stress. Demand for the crop has risen dramatically with its export to other countries in the world. The demand for other products, including alcohol, sweets, flours and medicinal capsules has skyrocketed and underscores the importance of the diversity of genotypes available.

The ulluco story

Ulluco (*Ullucus tuberosus*) is one of the most brilliantly attractive tubers produced by plants. Once it was found only occasionally in a few city markets. However, after studies indicated that it is highly desirable for its taste and its attractiveness, it can be found year-round in the most fashionable supermarkets in Lima, Peru. The possibilities opened by this introduction to new markets provide small-scale farmers with a highly valuable commodity, while encouraging the maintenance of the diversity that is so desirable to the consumer.

The yacon story

Yacon (*Smallanthus sonchifolius*) is regularly eaten uncooked. It is rich in fructans, which can be particularly helpful to dieters and diabetics. Yacon is now being researched as a non-glucose sweetener and its properties are already being appreciated internationally. CIP is involved in pilot studies to produce products such as syrup from yacon, including the start-up of a local pilot plant to work with small farmers in the Andes.

Each of these success stories underscores the importance of biodiversity in the agroecosystems of the Andes. It is unlikely that any one of these crops could sustain farmers in the Andes, and drastically reduce the levels of poverty that seem to be inherent in this agroecosystem. However, skillful management of the biodiversity – both intra-species and inter-species – could provide this agroecosystem with the broader potential for a thriving economic situation. These crops are highly specialized for these environments; they require virtually no additional fertilization or pesticide use. The negative impact of their culture is extremely low. They also demonstrate the potential importance of biodiversity as a key to development through their economic use.

Conclusions

According to the recent publication by the Crucible II Group (2000): 'Under any scenario, genetic resources for food and agriculture, the role of farming communities who nurture and develop diversity, and the vital contributions of formal sector plant breeders, all assume critical importance (to achieve food security)'. This is never truer than in the Andean ecoregion, where conditions are most often not amenable to monocultures or even to significant acreages of any one genotype. The evolution of so many crop species in this area was no doubt due to necessity, to buffer against extreme environmental effects or to exploit small niches of opportunity, for example. The difference now is that utilization of biodiversity can be directed much more precisely to exactly those needs that lead to food security. The understanding of the biological basis for diversity is greater and for the exploitation of that base to provide the desirable outcomes. The roles of conservation and utilization are intertwined, interdependent and equally important if the full potential of the diversity is to be realized.

The challenges for the future are nevertheless equally as great as the opportunities. The level of poverty in the Andes, combined with increasing population and environmental pressure on the land, does not present farmers with optimal incentives to protect biodiversity on their own. Lack of funding for agricultural research and development is also a threat, especially in Latin America which is seen sometimes to be less poor and less needy than other regions of the world such as sub-Saharan Africa. Local NGOs are often filling the gaps, but struggle themselves to find the funding necessary. International research organizations, such as CIP, are seeing declines in funding and increased demand from the global donor community for immediate impact in poverty alleviation. Conservation of biodiversity, and fulfilling its potential to alleviate poverty and increase food security, deserves a long-term approach which should not be at the mercy of short-term needs.

References

CIP (International Potato Center) (2000a) *Contribution of Local Knowledge and* in situ *Conservation Strategies on the Management and Use of Biodiversity. Report of the International Potato Center to the United Nations Environment Program of a Workshop Held in Copacabana, Bolivia, January 8–13, 2000.* CIP, Lima.

CIP (International Potato Center) (2000b) *La Papa: Tresor de los Andes.* CIP, Lima.

Collins, W.W. and Qualset, C.O. (1999) *Biodiversity in Agroecosystems.* CRC Press, New York.

Crucible II Group (2000) *Seeding Solutions*, Vol. 1, *Policy for Genetic Resources (People, Plants and Patents Revisited).* IDRC, Ottawa.

Tapia, M.E. (1999) *Agrobiodiversidad en los Andes.* Friedrich Ebert Stiftung, Lima.

Tapia, M.E. and De la Torre, A. (1997) *Women Farmers and Andean Seed.* IPGRI and the Food and Agriculture Organization of the United Nations, Rome.

The Role of Landscape Heterogeneity in the Sustainability of Cropping Systems

15

J. Baudry[1] and F. Papy[2]

[1]*Institut National de la Recherche Agronomique, SAD-Armorique, 65 rue de Saint-Brieuc, F-35042 Rennes Cédex, France;* [2]*Institut National de la Recherche Agronomique, SAD APT, BP 01, F-78850 Thiverval-Grignon, France*

Introduction

In rural areas, landscapes are made of different fields, successions of crops, grassland, woods along the road, some lines of trees, some strips of grass. These mosaics and networks form the basis of our everyday experience of landscape heterogeneity. In some areas, crops seem set in very specific places in the landscape, while in other regions randomness seems the main driving factor. In urban areas the diversity is also striking, but out of the scope of this chapter. Though this spatial heterogeneity is a major component of our environment, it is rarely a keyword in crop science research. As few variables drive their behaviour, homogeneous systems are easier to deal with, they can be manipulated and are subject to experiment. The field homogeneity paradigm has permitted increases in crop production, mastering of inputs and so forth. In the meantime, pollution of water, air and soil was attributed to the intensification (increase of fertilizers and pesticides) of farming systems, as well as losses in biodiversity and landscape amenities.

After World War II, Western Europe was far from being self-sufficient in terms of food production. In the early 1960s the setting up of the Common Agricultural Policy singled out production as the main and only goal for agriculture. Overproduction, awareness of the loss of biodiversity, and water pollution came together to stress other needs of people. International agreements (e.g. the Rio Convention, the Common Agricultural Policy), national laws and regional regulations stress the need to integrate farming activities with other functions of the landscape, such as providing habitats for flora and fauna, cleaning water, etc. Nature conservation is no more confined to reserves or environmentally sensitive areas, but spread all over the land with such schemes as pan-European corridors and the greening of agricultural policies.

Low-input farming, precision agriculture and organic agriculture may be routes to solve some of the problems. Nevertheless, these approaches utilize paradigms based upon a view of homogeneous, independent management units. This is not sufficient to obtain sustainable farming systems because interactions among fields and landscape heterogeneity play a major role in the control of numerous fluxes of nutrients and energy as well as on the dynamics of plant and animal populations. The development of landscape ecology since the early 1980s emphasizes this role of spatial patterns (Forman, 1995; Burel and Baudry, 1999). Those are key processes to sustain or disrupt ecological systems, hence for a sustainable development.

The purpose of this chapter is to give an overview of the role of heterogeneity in landscapes characterized by cropping systems, its measurement and its management. Agricultural landscapes are heterogeneous for a series of reasons including the physical and chemical heterogeneity of soil, the diversity of production systems, farming methods and healthy practices of crop succession. Thence, knowledge of the functioning of farming systems is a requisite to understand and design landscapes to enhance their ecological sustainability.

Agricultural Modernization and Changing Landscape Patterns

One of the most impressive features of the evolution of agriculture during the last few years is the increase in work productivity. From 1971 to 1995, in France, work productivity increased by 260%, due to more and more efficient machines and larger and larger areas per worker (APCA, 1998). This led to an important increase in farm sizes; they doubled during the period considered, along with many disappearances, an increase in the size of the parcels, a specialization of production systems and spatial redistribution of grasslands. All of these changes have important consequences for the landscape.

Increases in farm size do not occur in a concentric way around the central buildings, but by addition of several distant parcels of land. Since powerful machines allow time to be gained on the parcels and are hard to move from one area to another, farmers have kept increasing the size of their holdings. The great capacity of current sprayers also induces this increase, since the size of the area is calculated so that a whole tank of pesticides can be used. Thence, hedges and field boundaries tend to disappear; their network becomes looser and looser. Removal is especially intense during re-allotment programmes. The spatial organization of farm activities is rearranged to fit the greater distances between the parcels and the buildings. However, these increased distances make it harder to watch over the most remote parcels, and lead to a decreasing knowledge of one's neighbour, yet are useful to understand matters of proximity, such as control of water flow.

The parcels are bigger, thus more heterogeneous, which influences the crops, because of the greater or lesser disposability of nutrients, and the spatial variation of water retention capacity, weed seeds and pathogenic germs. But precision agriculture techniques can automatically modulate the doses of input and thus correct land heterogeneity. This then leads to the standardization of large patches.

In the case of intensive stock farms, the grazing parcels increase in size because the herds become bigger. But, in this kind of production, the share of grazing decreases while mechanically harvested areas increase for the same reason as cash crops. As for extensive animal production, it is developed on large grasslands to allow labour saving. Thus, whatever the production system, the elementary unit has kept on extending during the last decades. This is one of the principal features of the evolution of agricultural landscape.

It is true, at least in France, that the regions have not really specialized, except Brittany where intensive indoor production has been developed, but, within each region, the farms tend to be more and more specialized. Animal and crop production differentiate as they increase in size. Within the latter, the number of cultivated species decreases, since the use of various pesticides allows the succession rules to be simplified. Grasslands are redistributed, since land drainage allows the ploughing of lands formerly dedicated to grass because of their hydromorphy. Thus, the combination of mixed cropping and animal production decreases on the farm level, but remains within the wider regions: the scale of heterogeneity has changed.

Characterization of Landscape Composition and Structure

To describe and manage landscapes for functional purposes, such as species conservation, one needs to have methods for describing those landscapes and their heterogeneity.

An empirical definition of heterogeneity is 'non-homogeneous', a set of dissimilar objects. A landscape appears heterogeneous because we perceive different types of crops, non-crop patches, as well as field boundaries, be they grassy strips or hedgerows. Scientific inquiries on the effects of heterogeneity and landscape patterns, more generally, require formal definitions and assessments. The development of geographic information systems (GIS) permits numerous measurements to be made on maps to compare landscapes in space and time.

Heterogeneity was a crucial initial concept; for most authors, it refers to both the diversity of land-cover types and the complexity of their spatial arrangement. A highly heterogeneous landscape, measured with systems derived from information theory, such as the Shannon–Weaver index, is a landscape with numerous types of land cover displayed in numerous fragments randomly distributed (Fig. 15.1). *Per se*, this does not provide very meaningful information from an ecological standpoint, because the value of the measure does not change if one type of land cover is replaced by another.

Fragmentation and connectivity are more useful concepts, which are components of heterogeneity. They usually refer to a single type of land cover. Fragmentation is the process by which a large patch of a given land cover is split into smaller patches. Deforestation is the most studied process of fragmentation. On the contrary, increase in field size has led to less fragmented crop cover. Connectivity is the key concept in landscape ecology that is related to flow from one element to a different one. Connectivity is the measure of exchange of individuals or propagules and matter between two landscape elements, in a given landscape. For example, two maize fields are connected for a pest species if an individual of that species can move from

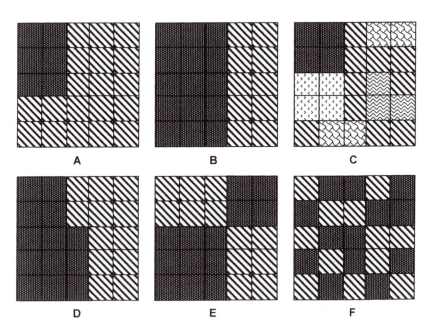

Fig. 15.1. The components of heterogeneity. From A to C, heterogeneity increases by change in the number and proportion of components; from D to F, it increases by changes in the spatial distribution of components. Adapted from Burel and Baudry (1999).

one field to the next. Obviously, wise landscape management aims at diminishing connectivity for pests. Connectivity refers to landscape features such as corridors, which are linear elements (e.g. hedgerows, road verges, etc.) linking two patches (forest, grassland, etc.) of a similar type. Within or along a corridor, individuals can move. Ecotone is another important concept describing the fact that when two different land covers are adjacent, specific processes occur at the interface. If the boundary between the two land covers is permeable, individuals can move from one cover to the next. Buffers are special ecotone zones that are not permeable to a flow or movement and therefore limit or prevent flows from one element to another element. Riparian zones, the stretch of land along a stream, usually flooded during the rainy season, are crucial buffers between streams and upland, as described below.

The concept of scale is central in studying heterogeneity. Measurement of the different parameters depends on the resolution and the extent of the maps. This renders comparisons across landscapes difficult. This phenomenon of scale dependence is a major characteristic of heterogeneous systems; it is also apparent in the analysis of the factors driving biodiversity components. At different scales, different factors are dominant, which leads to the recognition of hierarchies of levels of organization (field, set of fields, landscapes).

Among the main variables affecting agricultural landscape patterns are the diversity of crops, the size and spatial arrangement of fields and the importance of non-crop cover, including field boundaries. Those field boundaries, though of a limited area, play a major role as habitats and in controlling fluxes; road verges are part of field boundaries. Hedgerows and shelterbelts around fields or farmsteads are common features in many landscapes of the world, including the most open areas. A recent symposium on field margins (Boatman, 1994) provides numerous examples.

Influence of Patterns on Biodiversity, Nutrient Retention and Erosion

Empirical studies, as well as models, demonstrate that landscape patterns affect energy flows as well as the fate of nutrients or the dynamics of biodiversity. It is worth considering the different types of flows together to enhance our understanding of landscape-scale processes and to develop design principles. In the course of the history of agriculture, farmers have developed means to protect their crops against wind and water, implementing features such as hedgerows or ditches that nowadays are praised for their ecological values. The management of one type of flux cannot be easily dissociated from the management of another.

Biodiversity dynamics

Landscape patterns control species composition, species abundance and species interactions, especially trophic interactions. Insects and invertebrates in general deserve a special interest as they constitute the most abundant group of species, both pest and beneficial. Two recent books (Paoletti, 1999; Ekbom et al., 2000) provide an overview of the question.

Species composition and abundance depend upon the type of land covers and their spatial relationships. Individual crops/land covers are potential resources for a limited set of species. The development of cropping systems in the temperate and tropical zones represents a major shift in habitats as forest is cleared and replaced by herbaceous, mostly annual species. For this reason, land use changes are recognized as one aspect of the global changes affecting the Earth (Lubchenco et al., 1991).

Two theories form the background of most research. The first is the island biogeography theory developed by MacArthur and Wilson (1967) to account for the differences in the number of species among oceanic islands. The central statement of the theory is that the larger and the closer to the continent an island is, the more species can thrive on it. The theory has been applied to terrestrial islands (patches of a given land cover in a matrix of contrasting land cover). It was first tested on birds in woods by Forman et al. (1976). They show that more species inhabit large woods. Kruess and Tscharntke (in Ekbom et al., 2000) have tested the theory for endophagous insects. They found a significant effect of meadow area and isolation in a cropland area on species richness on Vicia sepium and Trifolium pratense for herbivores and parasitoids.

The metapopulation theory was put forward to explain the fact that a species is absent in apparently suitable habitats. Populations became locally extinct because of demographic or stochastic events (flood, frost, fire, etc.). A metapopulation is, then, a set of local populations that exchange individuals so that vacant habitats are recolonized. This involves a medium- to long-range movement, far less frequent than the movement involved in daily activities such as mating or foraging.

In both theories, the dispersal ability of species is central to explain their distribution. This applies to both animals and plants, the latter being frequently dispersed by animals. The means (walking, flying, being transported by wind, etc.) as well as the range of dispersal vary greatly among species. This explains some of the differences in their distribution. Landscape pattern interferes with dispersal capacity in several ways. If the habitat of a species is continuous, it is easier for that species to move across a landscape than if the habitat is fragmented. Surrogates to continuous habitats may exist, such as corridors, which are linear features. The most studied corridors are hedgerows linking forest habitat. Even if the habitat is fragmented, individuals may move from patch to patch if they are not too far apart and if the land cover in between is permeable. A hedgerow that is a corridor for forest species is often a barrier for open field species.

Species dispersal and use of landscape elements

Baudry and Burel (1998), in their review of the concept of dispersal, noted that there is some evidence for adaptation between landscape structure and dynamics and dispersal behaviour of species. In isolated patches of suitable habitat, dispersal is limited and dispersing individuals are unlikely to be replaced by immigrants. One may then expect species to have lost flight ability in such a situation. On the other hand, patchy and ephemeral habitats will select for high dispersal ability. An extreme case is aphid species, which rapidly affect the quality of their host plants. They have evolved a level of long-distance dispersiveness proportional to their impact on host plants; when a resource is depleted, individuals move to other patches.

Carabids using wooded corridors exemplify the other extreme. For example, the flightless forest carabid, *Abax parallelepipedus* (Pitt), is organized in metapopulations within the agricultural landscape. Small local populations occur in woods and large nodes of the network. Hedgerows act as corridors for individual dispersal. Their efficiency as dispersal corridors depends on their vegetation structure, and thus on the way farmers manage them.

Most of the crops of the steppe are species that harbour other steppe plants, often labelled weeds, as well as insects, mammals and birds. In steppe landscapes, the introduction of agriculture may not be a major shift in physiognomy, but a change in species composition and a species impoverishment, including the disappearance of perennial species.

Wild plants associated with crops are generally annual species more sensitive to field-scale practices, such as the use of phytocides, than to landscape structure. Field boundaries may be a refuge for such species, but only if the boundaries are disturbed to facilitate the reproduction of annual plants. In this case they are regarded as a source of weeds, difficult to manage, by farmers. A better conservation strategy for annual plants is conservation headlands, i.e. a strip of crop not sprayed with herbicide around the field (Sotherton, 1991). Conservation headlands also provide seeds for game birds and habitats for insects, such as butterflies (Dover, in Boatman, 1994).

Field boundaries are more suitable as a habitat for perennial plants. To protect plants in boundaries, it is necessary to avoid spraying of either herbicides or fertilizers. Herbicides kill plants, while nutrient enrichment selects for eutrophic species (*Urtica* spp., *Gallium aparine*, etc.). There are no means to maximize richness at a single boundary scale. The diversity of boundary structures is another important element at the landscape scale. In her unpublished thesis, C. Moonen exemplifies this point. Plotting the cumulative number of species recorded over a set of field margins (a boundary has

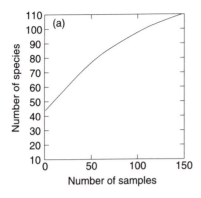
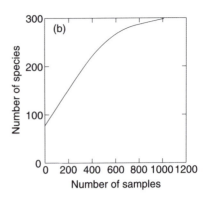

Fig. 15.2. Cumulated number of herbaceous plant species in field boundaries of two landscapes (unpublished data of Moonen and Le Cœur).

two margins), in an area of 200 ha, does not show any threshold (Fig. 15.2a). When the vegetation records of a nearby landscape characterized by hedgerows are added, no limit shows either (Fig. 15.2b). Similar patterns are found for insects or birds.

The case of vegetation in field margins exemplifies the hierarchical organization of biodiversity in landscapes. A field margin is included in two kinds of hierarchical systems. The first is the farming system; a margin is part of a boundary managed according to the adjacent fields used, it is also part of the more encompassing system of the whole farm. The second is the landscape ecological hierarchy; again, it is part of a boundary and of the boundary/field system, then it is included in networks of boundaries along which individuals and seeds move. A detailed analysis of field margin species composition demonstrates that factors at boundary (structure, management), field/boundary (crop succession), farm (type of production system) and landscape (type of landscape structure) level explain, independently, the diversity of vegetation. Farmers are distinct in terms of their overall management of margins, but manage each different margin in a specific way, often depending upon the crop cultivated in a given year. Therefore, the management techniques may change from year to year and though being similar for a given year in fields of different farmers, their cumulative effect over time is different due to differences in farm-level strategies.

From the ecological standpoint, species composition is driven by: (i) the structure of the boundary (e.g. presence or absence of trees); (ii) crop successions in adjacent fields (food resource or not); and (iii) possibility for individual to use routes like corridors.

Hedgerows are special types of field boundaries that provide habitats to species that do not thrive in cropland, and also act as corridors allowing movement across farmland. Arrangement of hedgerows into a network is necessary to insure a sustainable functioning at the landscape scale (Burel,

1996). Grassy strips between crops are also used as corridors, especially if they are rich in flowers and are used by nectar-feeding insects such as butterflies. They may also be corridors for pests, such as the small mammals, *Microtus*.

Many insects (coleoptera) overwinter in field boundaries and move into fields during the summer. Dennis and Fry (1992) found that in June, aphid-feeding arthropods move more frequently from boundaries into fields than from field to boundary.

Ouin *et al*. (2000) demonstrated that the small mammals, *Apodemus sylvaticus*, leave the boundaries during the spring and become more abundant in crops. They are not equally distributed among crops; they prefer wheat and peas to maize and carrots.

The barrier effect has been detected for butterflies and carabids. Maure-mooto *et al*. (1994) found that typical narrow English hedgerows slow down carabid movement, but do not prevent crossing. This was done in an experiment to test the ability for carabids (*Harpalus rifipes*, *Pterostichus melanarius* and *Pterostichus madidus*) to move from field to field across hedgerows. They also found that starvation, before release, increases speed of movement. Jepson (in Boatman, 1994) reviewed the potential barrier effect of field boundaries and concluded that they may be important, even to create local extinction. The barrier effect of field boundaries is more dramatic for butterflies (Fry and Robson, in Boatman, 1994). Dense and high hedgerows are not crossed by butterflies; for some species even lines of shrubs 1 m high reduce movements between fields. Observations of actual movements of butterflies reveal that they tend to concentrate and move along boundaries, crossing when meeting a gap.

Trophic interactions and pest–predator relationships

Wilson *et al*. (1999) analysed correlations between potential food-source decline and bird decline. They found that declining groups of invertebrates such as *Coleoptera*, *Diptera*, *Lepitodptera*, *Hymenoptera*, *Hemiptera* and *Arachnida* account for a significant part of the diet of declining birds. Birds also suffer from the disappearance of seeds of *Polygonum*, *Stellaria* and *Chenopodium*. These weeds vanish from arable fields receiving herbicides. They conclude that uncultivated grasslands, even strips of grass along field boundaries, are important as habitats for arthropods part of birds' diet.

Kruess and Tscharntke (in Ekbom *et al*., 2000) have demonstrated that isolation and decreasing size of grassland negatively affect species diversity, as well as interactions among endophagous herbivores and their parasitoids. The percentage parasitism rate of stem borers of *Trifolium pratense* and *Vicia sepium* decreased from 85% in the largest meadow (90 ha) to 40% in the smallest (0.03 ha). In a manipulated experiment, they showed that beyond 100 m from potential sources, the rate of parasitism is close to zero. Similarly, Elliott *et al*. (1999) showed that in Dakota, USA, the (relatively) fine-

grain landscapes harbour more aphid predators than coarse-grain landscapes. Marino and Landis (in Ekbom *et al.*, 2000) reported an evolutionary
perspective on the interactions between crops, pests and parasitoids in north-
central USA. There, 93% of the lepidopteran pests attacking crops were
native to the region and 67% are polyphagous. The authors also found that
generalist parasitoids were the main parasitoids. They reported the example
of the true armyworm (*Pseudaletia unipuncta* Haworth); it has 35 potential
parasitoid species, but a single generalist (*Meterus communis* Cresson)
accounts for most differences in parasitism between heterogeneous and
homogeneous landscapes. One explanation is that the parasitoids find
alternative hosts in native trees and shrubs. So when crops were
implemented in the region, local herbivore species adapted to them and
eventually became pests; these pests are controlled by local parasitoids.

Along the same lines, Wratten and Van Emden (1995) demonstrated
that wheat had fewer aphids (*Sitobion avenae*) in fields bordered by flowering
margins than in fields not bordered by any margin. They also reported several studies showing the role of stable vegetation as an overwintering habitat
for predators or parasitoids invading fields and limiting herbivorous populations in crops. We lack evidence to decide if this pattern is widespread or
not, but it points out the importance of heterogeneity in maintaining a complete food web within an agricultural landscape.

Biogeochemical and physical barriers

The two major fluxes of concern are soil erosion and nitrogen that threaten
both soil and water quality. Their control requires two different types of
barriers; for erosion, it must be physical barriers across slopes to diminish
water flow and arrest soil particle movement; for nitrogen, the barriers are
biogeochemical. In this latter case, biodiversity has a clear functional meaning. Very often these barriers, the buffer zones, can be both physical and
biological (Haycock *et al.*, 1997). Biogeochemical barriers are places where
nitrates are transformed into nitrogen, which returns to the atmosphere by
the action of microorganisms. They utilize the oxygen of nitrates for respiration, therefore transforming NO_3^- into N_2. If the denitrification process is
not completed, it is a greenhouse gas (N_2O) that goes into the atmosphere.
Nitrate flow is a striking case of scale dependence due to spatial heterogeneity. Lowrance *et al.* (1984) studied the nutrient budget of four watersheds,
between 1600 and 2200 ha, where farmland occupies 40–50% of the land.
They calculated a balance from input by precipitation and fertilization minus
output by harvest and stream flow. They found, for nitrogen, that this equation yielded a yearly net balance deficit between 34 and 45 kg ha^{-1} year^{-1};
they concluded the existence of sinks where nitrogen accumulates or leaves
the watershed in a gaseous form. Peterjohn and Correll (1984) found that

there was a loss of nitrogen in groundwater in forested riparian zones. Their hypotheses included uptake by vegetation and denitrification.

These findings promoted a wealth of research on riparian zones and their role as a buffer, especially through denitrification (Haycock *et al.*, 1997). Corell (in Haycock *et al.*, 1997) reported that in the coastal plain of the eastern USA, mass balances indicate that nitrogen retention in riparian zones can be as much as 74 kg ha^{-1} year^{-1}, an amount equal to 90% of the inputs. Thus, inputs of mineral nitrogen to stream water can be considerably less than nitrogen leached from fields. Denitrification–riparian processes are likely to be more efficient in agricultural landscapes because the required nitrogen is more abundant than in forested landscapes.

Erosive phenomena have got worse during the last decades: catastrophic floods occur more often (Papy and Souchere, 1993), hydrological systems fill in and surface water becomes eutrophic; underground systems degrade too. The waterproofing of built-up areas and road systems induces an increase in the volume and speed of the streaming water, and thus of erosion risks. However, it is true that the recent evolution of agricultural landscape has emphasized erosion risks and those linked to it, and has therefore influenced the water quality (turbidity, pathogens, pesticides, etc.). In erosive systems as different as those of Lauragais (south-west France) or Pays de Caux (Normandy, France), two principal causes can be highlighted: the disappearance of grasslands which protect the soil from raindrops or erosion by the streaming water; and the parcels increasing in size. In the first case, associated mechanisms are diffuse erosion and gully erosion running parallel down slopes within a single parcel; the quantity of eroded soil brought to the surface increases with the length of the slope. In the second case are combined diffuse erosion and concentration of water in thalwegs (lines of steepest descent from any point on the land surface, where water accumulates to form zero order streams). Erosion risks increase as the area of the catchment of the thalwegs, where surface runoff occurs, increases. This is the case when the whole catchment is as a single parcel, as in ploughing periods, or when soil surface is degraded in the different fields, such as bare soil in winter.

It is worth stressing that the farming processes leading to increased risks of erosion are the same as those leading to risks of decreasing biodiversity: field enlargement, removal of perennial vegetation, and bare soil in winter. Erosion is also a means by which farming activities threaten aquatic biodiversity (Frissel *et al.*, 1986).

Management of Landscape Patterns

Sound management of landscape is necessary to control fluxes of species and matter. How does that fit into farming activities?

Agricultural activities and the construction of landscape heterogeneity

Farms are the smallest entities of agricultural production organization; within farms, various activities are developed that generate heterogeneity. The first factor is that the environmental features and the cropping abilities they induce are not distributed homogeneously on the farm territory. Other factors can explain the production diversity as well: the necessity to divide the climatic risks, since agricultural productions are sensitive to climatic hazards, and to share out the annual amount of work, since agricultural activities are seasonal (Allen and Lueck, 1998). The desire to divide the commercial risks should also be considered, but this is true of any kind of enterprise, and not only of farm production.

The heterogeneity of the physical environment can be taken advantage of to favour the desirable diversity of activities. Land cover patterns (annual crops, grasslands, trees, etc.) on the farm territory depend upon the physical characteristics. Some areas are only used for grass because the steepness or poor accessibility prevents the use of machines; where machines can be used, the most demanding crops are grown on the most suitable land.

While making the most out of landscape heterogeneity, agricultural activities also take part in its construction:

1. Land division is an important factor of this heterogeneity. The size of the parcels is established so that the cropping operations can be accomplished in a day, thus guaranteeing more or less constant meteorological conditions. Therefore, the size depends mostly upon the speed of action allowed by the equipment.

2. Given a certain field pattern, several principles of organization of production explain how agricultural activities take part in the construction of landscape heterogeneity. Land cover does not depend only upon physical environment differences, but also on the accessibility, the distance from the farm, etc., of the parcels. Thus, since milking cows have to be milked twice a day, they are placed on the grasslands closest to the farm; since the transportation of beet crops requires cumbersome machines, the parcels of land have to be easily accessible. Therefore, the mutual emplacements of the parcels, their distance from the buildings, their size and their dispersion induce a spatial organization of the agricultural activities, which produces landscape heterogeneity.

3. After the harvest of an annual crop, the environment is not left in its primitive state. Thus, the succession of crops is not haphazard; it forms a series of cycles (or rotations) taking into account minimum delays between two similar crops and the way each crop can improve the environmental state left by the previous one.

4. Taking into account crop hierarchies (which ones are the more important for the farm economy), principles of land cover linked to the crop abilities of the land, and the rules that have just been presented, the farmer applies different land cover patterns and, on the land dedicated to annual crops, different rotations. From one year to the other, all the crops of the same rotation are moved to another spot. Thus, patterns change over time.

Once the rules proceeding from farm organization are settled, it is possible to use them to understand landscape patterns on agricultural land occupied by many farms. It can then be understood how the spatial distribution of land cover and practices is linked to that of land, parcels, habitat and communication network.

Collective maintenance of landscape pattern for ecological sustainability: a requisite for sustainable agriculture

Landscape patterns, as seen before, result from choices made at individual farm levels. Farmers act, most of the time, independently of one another. Thence, the global landscape does not result from any particular collective goal. If proper measures to induce the farmer to reorganize his farm might be useful, they are not sufficient, because ecological processes ignore field or farm boundaries; thus, environmental issues linked to biotope maintenance or management of stream quality require a cooperation of the actors within physical units defined by ecological and hydrological processes (Papy, 1999). Landscape management must result from concerted, collective decisions.

The two erosive situations described before can illustrate this point of view. It has been shown that land division can be an efficient way of controlling erosion risks. In Lauragais, where a slope is cultivated by a single farmer it has to be divided into two fields; a stipulation inscribed in a farm-territory development plan can be sufficient. But in the Pays de Caux, the land division has to be planned in a more collective way on a larger area including several contiguous zero or first order watersheds.

In a more general way, it is necessary to coordinate crop practices, arrangements (such as land division, hedge planting, grassy strips implementation, etc.) and maintenance of these arrangements, over continuous stretches of land. It is not easy to apply collective organizations, because, contrary to what happens when farmers cooperate to sell products, the collective control of ecological processes does not allow the choice of who to cooperate with. Thus, to maintain or rebuild a corridor, or to protect a thalweg from erosion, all the adjacent farmers and landowners must cooperate. Experience shows that the collective thinking of farmers on these issues often has to be initiated by specialists recognized for their competence and impartiality.

Spatial modelling

Proposals for landscape planning must rely on an increased knowledge based on landscape-scale processes. Most often, it is difficult to conduct experiments at the landscape scale to analyse the effects of spatial or temporal patterns. Comparisons of species distribution along gradients of landscape is one way to collect information (Elliott *et al.*, 1999), as is comparison of different situations for erosion. Another way to conduct research is to build models. Spatially explicit models (Turner *et al.*, 1995) are now being widely developed to create landscape patterns and analyse their effects on species movements and population dynamics. Though the basic biological knowledge of species behaviour is often scarce, this permits the definition of a threshold for species survival.

Modelling of spatial dynamics as driven by farming activities is still in its infancy. Its development depends on the possibility of defining rules for land-use allocation over a given landscape. The merging of both modelling and GIS holds the promise of simulation models as a support for decision making. Such models will incorporate crop successions, pointing to the shifting nature of the crop mosaic. The movement of a crop from field to field is a means to connect fields over a short period of time; this may facilitate the dispersal of pests (Baudry and Burel, 1998).

Conclusion

At regular intervals, scientists emphasize the importance of blending ecology and agronomy to promote a sustainable development. This brief overview of ecological processes as controlled by cropping patterns and management goes along these lines. Ecology proposes concepts and methods to analyse and assess the mosaic of habitats resulting from the management of cropping systems. In turn, crop science and agronomy in general provide means to understand how to manage landscape in a sustainable manner. What is, at first sight, perceived as heterogeneity by an ecologist is, in fact, an ordered pattern produced by farming activities. A proper management of landscape and ecosystems does not consist of imposing a design from the sole ecological perspective, but to implement an ecological standpoint in the process of organizing crops in space and managing them in time. A multiple-scale approach (field/landscape; field/farm/group of farms) is a requisite for both understanding and managing processes.

Heterogeneity is just a starting point; in itself, it is neither good nor bad. It is the process of deciphering heterogeneous patterns that leads to the discovery of organized and disrupted systems. There is a long way to go to merge different lines of thinking, not necessarily to blend them in a single approach, but to produce a dialogue enhancing our understanding of the fate of a mastered nature.

Several reasons permit one to be optimistic. The first is that sustaining biodiversity or water quality does necessitate landscape features and patterns that are also adequate to protect soil – the basic farming resource. The same is true for the positive interactions between farm sustainability, biodiversity and the diversity of crops grown on a farm (Vandermeer *et al.*, 1998). The second is that services provided by fauna and flora such as recycling of organic matter, denitrification, pollination, pest control, etc. are more and more widely recognized (Altieri, 1998). The third one is that the paradigm shift in ecology from 'balance of nature' to 'dynamic landscapes' should facilitate the linking of spatial ecological models and spatial cropping models.

Challenges

The incorporation of landscape-scale management into agricultural practices faces two major challenges: the first is to combine scientific approaches relying on experiment (crop growth, protection, etc.) and approaches that are at too coarse a scale to permit experiments. Landscape comparisons provide insights, but require sampling combining independent variables at several scales in space (field, field boundary, mosaic structure) and time (crop successions and past mosaic structures). Modelling is a complement that can help to decide where to sample for biodiversity or water quality.

The second challenge is the building of interdisciplinary approaches. Agronomy or crop sciences cannot solve the problems, and not even give enough information to foresee and understand the problems. Landscape ecology and population biology provide means to evaluate the fate of biodiversity, as hydrology and biogeochemistry provide the means of evaluating water quality. Farming system approaches give an understanding of how the farmer makes decisions on his different fields to share resources (labour, machines, etc.), and social sciences analyse the decision-making process as well as interactions among farmers and between farmers and the society as a whole. The challenge is to educate people to understand each other's discipline, not to do someone else's job, but to know the concepts, how data can be articulated to make a model. Sometimes, environmental scientists think that social sciences must seek ways to 'make people understand' the right technical solution. This is not the point; the point is to understand why people act as they do.

This challenge is a key to success for young agronomists and land managers, and interactions with a wide range of disciplines is a requisite for their training.

Acknowledgements

This chapter is paper no. 1 of the FORTE programme of the SAD Department. Financial support for research was provided by INRA, the ministry in charge of the environment, and the environment programme of the CNRS. We thank Guillaume Pain and two anonymous reviewers for comments and Sandrine Baudry for editing.

References

Allen, D.W. and Lueck, D. (1998) The nature of the farm. *Journal of Law and Economics* 41, 343–386.

Altieri, M.A. (1998) The ecological role of biodiversity in agroecosystems. *Agriculture, Ecosystems and Environment* 74, 19–31.

APCA (1998) La localisation de l'agriculture française. *Chambres d'agriculture* 872, 13–36.

Baudry, J. and Burel, F. (1998) Dispersal, movement, connectivity and land use processes. In: Dover, J. and Bunce, R.G.H. (eds) *Key Concepts in Landscape Ecology*. International Association of Landscape Ecology, pp. 232–339.

Boatman, N. (ed.) (1994) *Field Margins: Integrating Agriculture and Conservation*. British Crop Protection Council, Farnham.

Burel, F. (1996) Hedgerows and their role in agricultural landscapes. *Critical Review in Plant Sciences* 15, 169–190.

Burel, F. and Baudry, J. (1999) *Ecologie du Paysage: Concepts, Méthodes et Applications*. Lavoisier, Paris.

Dennis, P. and Fry, G.L. (1992) Field margins: can they enhance natural enemy population densities and general arthropod diversity on farmland? *Agriculture, Ecosystems and Environment* 40, 95–116.

Ekbom, B., Irwin, M.E. and Robert, Y. (eds) (2000) *Interchanges of Insects Between Agricultural and Surrouding Landscapes*. Kluwer Academic Publishers, Dordrecht.

Elliott, N.C., Kieckhefer, R.W., Lee, J.-H. and French, B.W. (1999) Influence of within-field and landscape factors on aphid predator populations in wheat. *Landscape Ecology* 14, 239–252.

Forman, R.T.T. (1995) *Land Mosaic. The Ecology of Landscapes and Regions*. Cambridge University Press, Cambridge.

Forman, R.T.T., Galli, A.E. and Leck, C.F. (1976) Forest size and avian diversity in New Jersey woodlots with some land use implication. *Oecologia (Berlin)* 26, 1–8.

Frissel, C.A., Liss, W.J., Warren, C.E. and Hurley, M.D. (1986) A hierarchical framework for stream habitat classification: viewing streams in a watershed context. *Environmental Management* 10, 199–214.

Haycock, N.E., Burt, T., Goulding, K.W.T. and Pinay, G. (eds) (1997) *Buffer Zones: Their Processes and Potential in Water Protection*. Quest Environment Publisher, Harpenden.

Lowrance, R., Todd, R., Fail, J. Jr, Hendrickson, O., Leonard, R. and Asmussen, L. (1984) Riparian forests as nutrient filters in agricultural watersheds. *BioScience* 34, 374–377.

Lubchenco, J., Olson, A., Brubaker, L.B., Carpenter, S.R., Holland, M.M., Hubell, S.P., Levin, S.A., MacMahon, J.A., Matson, P.A., Melillo, J.M., Mooney, H.A., Peterson, C.H., Pulliam, H.R., Real, L.A., Regal, P.J. and Risser, P.G. (1991) The sustainable biosphere initiative: an ecological research agenda. *Ecology* 72, 371–412.

MacArthur, R.H. and Wilson, E.O. (1967) *The Theory of Island Biogeography*. Princeton University Press, Princeton, New Jersey.

Mauremooto, J.R., Wratten, S.D., Worner, S.P. and Fry, G.L.A. (1994) Permeability of hedgerows to predatory beetles. *Agriculture, Ecosystems and Environment* 52, 141–148.

Ouin, A., Paillat, G., Butet, A. and Burel, F. (2000) Spatial dynamics of wood mouse

(*Apodemus sylvaticus*) in an agricultural landscape under intensive use in the Mont Saint Michel Bay (France). *Agriculture, Ecosystems and Environment* 78, 159–165.

Paoletti, M.G. (ed.) (1999) *Invertebrate Biodiversity as Bioindicators of Sustainable Landscapes*. Agriculture, Ecosystems and Environment, Elsevier Science.

Papy, F. (1999) Agriculture et organisation du territoire par les exploitations agricoles: enjeux, concepts, questions de recherche. *Comptes Rendus de l'Académie d'Agriculture de France* 85, 233–244.

Papy, F. and Souchere, V. (1993) Control of overland runoff and thalweg erosion. A land management approach. In: Brossier, J., De Bonneval, L. and Landais, E. (eds) *Systems Studies in Agriculture and Rural Development*. INRA, Paris, pp. 87–98.

Peterjohn, W.T. and Correll, D.L. (1984) Nutrient dynamics in an agricultural watershed: observations on the role of a riparian forest. *Ecology* 65, 1466–1475.

Sotherton, N.W. (1991) Conservation headlands: a practical combination of intensive cereal farming and conservation. *The Ecology of Temperate Cereal Fields*. Blackwell Scientific Publications, Oxford, pp. 193–197.

Turner, M.G., Arthaud, G.J., Engstrom, R.T., Hejl, S.J., Liu, J., Loeb, S. and McKelvey, K. (1995) Usefulness of spatially explicit population models in land management. *Ecological Applications* 5, 12–16.

Vandermeer, J., van Noordwijk, M., Anderson, J., Ong, C. and Perfecto, Y. (1998) Global change and multi-species agroecosystems: concepts and issues. *Agriculture, Ecosystems and Environment* 67, 1–22.

Wilson, J.D., Morris, A.J., Arroyo, B.E., Clark, S.C. and Bradbury, R.B. (1999) A review of the abundance and diversity of invertebrate and plant foods of granivorous birds in northern Europe in relation to agricultural change. *Agriculture, Ecosystems and Environment* 75, 13–30.

Wratten, S.D. and Van Emden, H.F. (1995) Habitat management for enhanced activity of natural enemies of insect pests. In: Glen, D.M., Greaves, M.P. and Anderson, H.M. (eds) *Ecology and Integrated Farming Systems*. John Wiley & Sons, Chichester, pp. 117–145.

Cropping Systems for the Future

16

J. Boiffin[1], E. Malezieux[2] and D. Picard[3]

[1]*Institut National de la Recherche Agronomique (INRA), 147 rue de l'Université, F-75338 Paris Cédex 07, France;* [2]*Centre de Coopération Internationale en Recherche Agronomique pour le Développement (CIRAD), BP 5035, 34032 Montpellier Cédex, France;* [3]*Institut National de la Recherche Agronomique (INRA), Centre de Recherche de Versailles, Route de Saint Cyr, 78026 Versailles Cédex, France*

Introduction

Farmers manage their crops and fields through sequences of technical acts which are interdependent since each of these acts has multiple and prolonged effects on the agroecosystem dynamic. Any given crop is selected and managed in relation to the preceding and subsequent crops, and a given technical operation is decided and implemented in relation to other techniques involved in the crop management. This gave rise to the concept of cropping system, a term that needs to be carefully defined since it encompasses meanings and corresponding scales that differ according to authors. Here we consider a cropping system as a set of management procedures applied to a given, uniformly treated agricultural area: this can be a field, part of a field or a group of fields (Sebillotte, 1990). Following this definition, cropping systems – an agronomic notion – are components of farming systems, which refer to agroeconomic entities. They are identified by a sequence and/or a spatial combination of crops and the corresponding technical operations, involving not only the crops themselves, but also between-crop periods with bare soil or a plant cover. Cropping systems defined in this way vary widely throughout the world, from simple patterns of monoculture and monocropping, to complex multiple cropping systems in which several crops are grown simultaneously on the same field.

World trends liable to induce and/or restrict changes in cropping systems have been extensively discussed during the Third International Crop Science Congress. They include quantitative and qualitative changes in demand for food, evolution of world trade in agricultural products, restrictions on cropland extension, competition for water, increase in atmospheric CO_2 concentration and associated climatic changes, the nature and rhythm of agrotechnical innovation, as well as changing public policies concerning agriculture, the environment and biotechnological innovations. Most of these trends involve contrasting scenarios and regional patterns. Overall, they should result in more unstable cropping systems, at least for a transitional period of change in economic regimes starting from the next international negotiations on world trade. An 'ideal' world farming system that would adjust directly to the world food demand without expanding on to fragile lands or forest areas should involve a general intensification of cropping systems with an increase in invested capital per unit area of land. A further adjustment would be the partial substitution of cereals such as rice and of grasslands with cereals that can be incorporated into animal feeds (maize and wheat). In fact, there will probably be no further intensification, and possibly even a decrease in the use of industrial inputs per unit area in Western Europe. At the same time, great intensification could occur in areas where potential yields and economic resources make it possible and profitable, such as new irrigated areas, or areas close to large markets for food crops. Between these two extreme situations, a wide range of types and degrees of intensification will occur according to local market conditions and available resources. The extreme variability of cropping systems combined with the diversity of their dynamics means that any attempt to describe existing cropping systems across the world is extremely difficult, and that forecasting future cropping systems is almost impossible. Thus, the following chapter does not pretend to be predictive, but seeks to be prospective. We must therefore first identify some of the basic features of cropping systems evolution, then look at critical issues for mastering their change, and finally, examine some related challenges for crop science.

Evolution of Cropping Systems: Some Basic Features

Cropping system changes on the farm

Whatever the intensity of external forces, cropping systems cannot be adjusted to a changing context freely and in real time, because their evolution takes place in a farm framework and interacts with agroecological factors, as shown by the geographical differentiation of cropping systems. Table 16.1 shows a case study in western Africa illustrating how differences or similarities in millet-seeding strategies correspond to a common constraint imposed by the farm context – a restricted availability of labour – interacting with features of the climatic and ecological context.

At the field scale, any change in crop sequences or crop management techniques creates new and unexplored situations. A slight change in the

Table 16.1. Millet-seeding strategies in the Sudan–Sahele region of Africa (Milleville, 1998).

Agroclimatic situation (average yearly rainfall, mm)	Critical resources and problems	Strategy
North Burkina (400)	Human labour Scarce rainfall and low soil moisture	S1: Seed early and fast (a)
South Senegal (1200)	Human labour Weeds	S2: Lengthen the seeding period (b)
Gambia (600–800)	Human labour Climatic variability	S1 when rainy season is late, S2 when rainy season is early

(a) S1: seeding as early as possible in the rainy season, to make the best of the short period during which the soil is easily workable and allow fastest seeding, and to decrease the risk of drought at the end of the crop cycle. (b) S2: a prolonged seeding period allows the great subsequent demand for labour for weeding to be spread over a longer time.

sowing period of a crop may have significant effects on the biological cycles of pests and diseases, and on the dynamics of abiotic factors. *A fortiori*, modifying a crop sequence implies revisions of all aspects of the management of each crop. Then a change in cropping systems is not a gradual process, but a discontinuity.

At the farm level, a given combination of cropping systems corresponds to a given allocation of land, water, labour and equipment resources, and to a given pattern of work organization, with critical periods when labour and equipment resources are close to being saturated by the demands made on them (Fig. 16.1). Then changes in cropping systems are always related, either as causes or as results, to more general changes in the farm functioning. This remark can also apply to situations of more or less collective farming systems, where resource allocation and crop management are decided at the level of a rural community.

Another important aspect conditioning changes in cropping systems is their link with specific decision processes (Sebillotte, 1990; Papy, 1994; Hardaker *et al.*, 1997). Any given cropping system results from the implementation of a combination of decisions taken by an individual farmer or by a community. This more or less implicit decision process incorporates external factors that depend on the economic, technological or ecological context. It also integrates the targets and needs of the farmers' families, labour and the machinery resources available on the farm, and the farmers' knowledge of and information about their crops and soils. Figure 16.2 shows that cropping systems and decision processes are really coupled, and this coupling works

(a) Main features of cropping systems

Cropping system	Corresponding fields	Crop sequence	Crop management		
			Soil tillage	Weeding	Other features
A	1, 2	WW / M	Conventional (deep ploughing)	Selective herbicides a, b, c	No irrigation, same patterns for fertilization, crop protection etc.
B	3, 4	WW / M	Reduced (direct drilling)	Selective herbicides a, b, and glyphosate	
C	5, 6	Continuous M	Conventional (deep ploughing)	Selective herbicides b, c	

(b) Farmland allocation to crops   (c) Labour requirements and critical periods

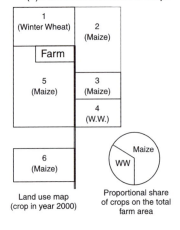

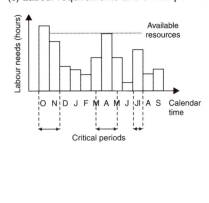

Fig. 16.1. Cropping systems on a virtual farm growing winter wheat (WW) and maize (M). Three cropping systems (A,B and C) are identified (a). Systems A and B have the same crop sequence but different tillage and weeding patterns, and they differ from C by the crop sequence. The corresponding farmland allocation is described in (b): each year, respectively about two-thirds and one-third of the total area is grown to maize and wheat. This induces a time distribution of labour needs for field work (c), with critical periods in autumn (maize harvest, wheat sowing) and spring (fertilizer, growth regulator and pesticide applications on wheat and maize sowing).

at different time intervals. Decisions may be divided into strategic choices – planned crop sequences, the main guidelines for crop management – and tactical or operational choices that adapt strategies to the climatic, agricultural and economic events encountered by the farmers. These two kinds of choices determine the actual sequences of crops and technical operations, and these act upon the functioning of the agroecosystem. In turn, this functioning produces information – for instance the presence of a pest – that may be captured by the farmers as input data to influence their next

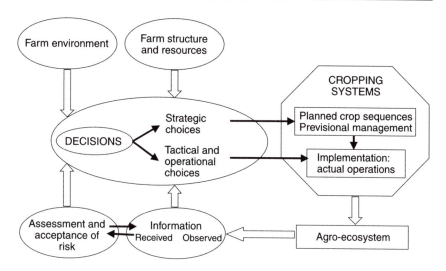

Fig. 16.2. The coupling of decision making and cropping systems.

decisions – for instance when deciding on pesticide application – or stored in their memory, then contributing to risk assessment.

The basic components of decision processes for cropping systems are decision rules, which differ according to the scale they address (Aubry *et al.*, 1998; Le Gal and Papy, 1998). Decision rules regulating field scale operations are often of the 'if–then' type, and are based on indicators derived from field observations. Of course the efficiency of crop management depends on the relevance of these indicators and on the quality of observation. Decision rules addressing farm-scale problems involve a wider range of cases, including coordination or priorities for resource allocation. Describing the decision process not only implies identifying decision rules, but also understanding how they are assembled, and activated in a given order. It is now possible to produce effective models of these aggregates, as a result of advances in information science and technology, particularly artificial intelligence.

New crops, or crop sequences, or crop management techniques mean new decisions to be taken, and new procedures for risk assessment. Farmers tend to make few changes to their decision procedures, because any decision-maker facing an uncertain context cannot react adequately to hazards without using pre-programmed pieces of software. Decision processes that would be permanently upset might be poorly adapted to unusual climatic, agronomic or economic events. A faster change in cropping systems should then be accompanied by more attention to the decision processes that control them.

Innovation for cropping systems

As a consequence of its discontinuous and systemic character, innovation for cropping systems needs a long, complex and collective trial-and-error approach. This starts with the production of new plant material and technologies by research, industry or the farmers themselves, leading to new varieties, chemicals, equipment, and more comprehensive packages – as was involved in the 'Green Revolution'. They cannot be adopted without an evaluation stage that is designed to test, select and adapt these new 'raw materials' to a given agroecological context. Part of this 'pre-innovative' evaluation process is carried out by extension services, through multi-site and multi-year field experiments. But many problems and/or evaluation criteria cannot be taken into account by such experiments, simply because they are not submitted to the same constraints as real farm fields. The role of farmers themselves in testing, adapting and generating agrotechnical innovation is always decisive, even in developed countries with powerful agricultural research and extension services.

The generation of new cropping systems results from the integration of these pre-selected new plant materials and technologies into existing cropping systems, but also from farmers trying new combinations of crops and technologies (Buckles and Triomphe, 1999). This process is generally accompanied by exchanges of a great deal of information within the farming community, which allows the sharing of errors and successes obtained by individual testers. Neglecting the farmers' role and the on-farm stages of innovation in cropping systems can lead to a misunderstanding of farmer response to new techniques or new technical packages. Sociological and educational factors are too frequently invoked to explain 'farmers' reluctance' to a given innovation, without any serious analysis of the impacts of this innovation on field and farm functioning, including work, organization and decision procedures. This impact can partly explain the differences in adoption that may occur for different types of innovations, or different groups of farmers towards a given type of innovation. Innovations having a relatively low impact on other components of the cropping systems (e.g. new cultivars with higher performances but the same development pattern as previous ones) may be rapidly adopted. Others having a greater impact may expand more slowly or more irregularly, even when they clearly have an economic advantage over traditional practices. These considerations typically apply to the adoption of reduced or no-tillage techniques (Nowak, 1992). The reluctance of farmers to adopt these techniques in areas of intensive arable cropping (e.g. northern France) is mainly due to the increased difficulty in controlling some pests and diseases and the greater risk of cumulative soil structure degradation, depending on the crop sequence and on the soil type. In dry areas where few financial resources are available (e.g. Mexico or Madagascar), farm-scale problems such as competition for crop residue between cattle feeding and mulching, or the absence of financial resources to buy herbicides, have proved to be bottlenecks that must be analysed and solved before changing soil tillage (Affholder *et al.*, 1998).

Up to now, research on cropping systems has contributed to the generation of new cropping systems through studying and evaluating existing or emerging systems, rather than by designing new ones from theoretical principles, except for conservation tillage. This companion rather than proactive attitude may have a number of causes, including the complexity of interactive processes involved in a cropping system and the related difficulty of testing prototypes of cropping systems in a rigorous experimental way. The unlimited number of combinations to be checked means that thousands of farmers may conduct the exploration of cropping systems more efficiently than a few agricultural scientists, and that those scientists target their efforts on selected cropping systems.

Mastering Change in Cropping Systems: Critical Issues

The above considerations would suggest that it is more relevant to examine how to improve the mechanisms by which cropping systems are generated and modified, rather than to design 'good' cropping systems that will never occur as steady states. As seen before, these mechanisms involve agrotechnological innovation, decision processes and socio-economic regulations.

Agrotechnological innovation: how to promote an integrated pattern?

The intensification of cropping systems that has occurred to date has followed a 'segmented' pattern of innovation with crop management evolving towards addition of specialized actions, each one targeted to a specific limiting factor. A restricted use of external inputs, for either financial or environmental reasons, means that a given stress or limiting factor must be controlled not just by a single technique but by the whole cropping system, as illustrated by integrated crop protection (Jordan *et al.*, 1997). We can expect plant biotechnology to contribute to an 'integrated pattern' of innovation by providing plants that are more resistant to pests and diseases. Manipulating or introducing other plant properties may also extend the range of technical options that can be exploited, in order to design integrated cropping systems. For instance, the tolerance of some crops to non-specific herbicides may allow some evolutions that were prohibited by the pre-existing weed flora, e.g. the reduction of soil tillage, with associated environmental or organizational benefits. But full advantage cannot be drawn from this potential if the efficiency and relevance of evaluation processes are not considerably improved. Basically, new material – especially crop varieties and improved seeds – are and will be delivered at an

increased rate, which means an increase in labour and costs of testing. Moreover, evaluation should consist not only in comparative tests of plant performances, but in a more global evaluation of the new cropping systems that can
be developed owing to the new or improved plant properties. This implies less
standard, and probably heavier, experimental approaches. The danger is that
technical institutes and extension services risk being submerged by agrotechnological innovation and transfer only a very small part of it. This difficulty
is obviously even greater in countries where such institutions are weaker or
non-existent.

A second bottleneck results from the fact that new targets and criteria
disqualify a trial and error approach, and this may hamper the generation
of new sustainable cropping systems. Farmers are able to assume the leader-
ship in generating new cropping systems as long as they can follow simple
guidelines (for instance increasing crop yields) and check their initiatives
from the appearance of the crop. But the criteria used to evaluate new cropping systems have become much less obvious, since they take into account
environmental objectives or harvest quality and corresponding processes.
Farmers cannot easily check nitrate leaching, gaseous losses or the protein
contents of grains. Hence, a trial-and-error approach no longer works.

This leads to a new responsibility for agronomic research, i.e. the prototyping of cropping systems. For us, prototyping means deriving crop sequences
and crop management methods to satisfy specific targets and constraints,
which is the reverse of evaluation. Recent studies in this domain indicate that
the approach might consist of the following steps (Vereijken, 1992):

- Identifying constraints and purposes, and grading them from the most
  global (availability of resources, access to market, etc.) to the most detailed
  (decision rules and indicators for specific techniques).
- Specifying the targets corresponding to each level, starting from the most
  general and building logical chains between them. These chains correspond to different types of cropping systems.
- Attributing adequate values to all variables and parameters involved in
  the selected chains, for instance total amounts of available water, yield
  targets, threshold values of indicators for triggering pesticide applications,
  etc.

A case study illustrating this approach is summarized in Table 16.2. Of
course, the design phase must be followed by adequate testing, combining
simulation, field experiments, and iterations with farmers and technicians.

A number of models can now be used, at least in research institutions,
to simulate and analyse the various effects of cropping systems on the functioning of the agroecosystem (Boote *et al.*, 1996; Rossing *et al.*, 1997; Gary
et al., 1998). They are useful for evaluating innovative cropping systems,
but this does not solve the problem of designing prototypes. One way of
designing prototypes is to use the models in an inversion mode, and so identify values of input data that lead to target values of output data, for instance

Table 16.2. Prototyping cropping systems: a case study in South-Western France (Debaeke *et al.*, 2000).

Level of choice (most global → most detailed)		Targets and constraints imposed on the cropping system		
		System I	System II	System III
Global farm resources	Access to water for irrigation	Unlimited	Restricted	No
	Financial resources	High	Medium	Low
	Labour	High	High	Low (double activity)
Global strategy for cropping system		High productivity preserving the environment	Extensification	Rusticity and simplicity
Possible crops		Maize, soybean, etc.	Pea, soybean, durum wheat, etc.	Winter wheat, sunflower, sorghum, faba bean, etc.
Global strategy for crop management		Reaching potential yields	Rationing	Stress avoidance
Strategy for specific techniques	e.g. criteria for choice of cultivars	Productivity	Productivity and tolerance	Resistance and earliness
	e.g. *a priori* limitations on chemical crop protection	No	Partial (e.g. only one fungicide application per crop)	Strong (e.g. no fungicide at all)
Reasoning for specific operations	e.g. fungicide application	f(alerts, advice and field observations)	f(alerts, advice and field observations)	—

yields, residual nitrogen, and so on. But again, this is still far from a full design of cropping systems, i.e. formal combinations of technical acts if we accept the definition given above.

Support for decision making

A faster change may lead to sustainable agriculture only if the decision processes that generate cropping systems incorporate more information, take into account a wider range of criteria, and allow for faster adaptation. This makes support for decision making a crucial issue, which involves improving decision rules, improving their implementation, and building decision support systems.

Improving decision rules means introducing, replacing or modifying decision rules to better satisfy the constraints and targets that are assigned to the cropping systems. This can be illustrated by the example of nitrogen applications on annual crops in northern France (Table 16.3). The recommended rates of nitrogen fertilizer in force until the late 1970s were empirical norms derived from field response curves. The present dominant system (several authors in Lemaire and Nicolardot, 1997) is derived from a balance equation whose terms represent different types of ecological processes. This new set of rules has led to more efficient nitrogen inputs that have resulted in financial savings and/or improved yields. It has also led to greater generality and flexibility, better control of harvest quality, better control of water and air pollution, greater transparency, and finally better evolutivity, since the balance equation can incorporate new processes and submodels. These advantages result from an explicit description of the nitrogen dynamics in the soil–plant system. In turn, these dynamics must be documented through an appropriate mix of field indicators, plant and soil analyses, experimental references and incorporated sub-models. Then the agronomic references needed for implementing this set of rules, and the methods for obtaining them, have become totally different from those for the establishment of N response curves. Other examples of improved decision rules, for instance those involved in integrated crop protection, confirm a general trend towards flexible algorithms based on scientific knowledge and field observations, instead of rigid sets of norms corresponding to delimited areas of validity.

Very good decision rules may fail to be adopted if conditions are not suitable for their implementation, and this is why the 'nitrogen-balance method' became a success story only about 10 years after its delivery by research. The basic condition is an access to input data, which can be boosted by recent advances in remote sensing, weather forecasting or epidemiological alerts. Other conditions involve software for data processing, and appropriate off-farm services such as laboratories, databases or consultants. They can even include the automated control of equipment. Precision

Table 16.3. An example of knowledge-based decision rules: the nitrogen fertilization of annual crops in France.

General algorithm (all terms in kg mineral nitrogen ha^{-1})

$$X = PN - [(IS - RS) + NMSOM + NMEOM]$$ (1)

Plant needs   Soil supply

Specific algorithms (case of winter wheat)

$$X = X1 + X2 + X3$$ (2)

With $X1 = 0$ to 80 *according to IS* (3)

$X2 = X - X1 - 40$ (4)

$X3 = 0$ or 40 *according to N index* (5)

Meaning and origin of terms

Term	Meaning	Determination
X	Total advised rate of fertilization	Calculated from (1), then modified according to (2) and (5)
PN	Plant needs: total absorption needed for reaching yield and quality targets with $P \geq$ given value	Crop models and experiments
IS	Initial stock of mineral nitrogen in the soil, measured before the beginning of intense N absorption, cumulated from the soil surface to the maximum rooting depth	Field sampling, soil analysis, references on rooting depths
RS	Residual stock, estimated as the minimum soil N content at harvest (unextractable mineral N) on a same soil volume as *IS*	Models and experiments on N absorption, references on rooting depths
NMSOM	Net mineralization from soil organic matter	Soil science models and experiments
NMEOM	Net mineralization from exogeneous organic matter (crop residues and organic wastes and fertilizers)	Soil science models and experiments
X1, X2, X3	Fertilization rates at tillering, beginning of stem elongation, ear emergence	Calculated from (3), (4), (5)
N index	Nutrition index measured during stem elongation	Field sampling and plant analysis

agriculture corresponds to this extreme case and illustrates the interdependence between agronomy and technology. Formalized rules are needed to make the best of technological progresses regarding sensors, global positioning, on-board computers and equipment control. At present, the bottleneck in precision agriculture is going to be the lack of decision rules needed to convert spatial data into operating instructions, rather than anything else.

The next steps in decision support no longer concern the elementary decision rules and corresponding operations, but involve aggregates of rules designed to manage or choose cropping systems. They lead to the production of decision support systems (DSSs), based on advances in information technology (Attonaty et al., 1999). These DSSs simulate the performances of given configurations of farming and/or cropping systems (Rossing et al., 1997). They ideally comprise four components (Fig. 16.3). The first is a decisional module that gives instructions to the second, an operating module. This produces a dynamic description of how these decisions are implemented with the means available on the farm. The third component is an agroecological model that describes the response of the soil–plant system to these operations, and predicts yields and other biophysical outputs. The last component is an evaluation module that converts these outputs into agrotechnical, economic and environmental performance. It may also process these data using sorting or optimizing procedures. This chain must be reiterated for different climatic scenarios, allowing evaluation of the climate-induced risks.

Such DSSs have been extensively developed and used for consulting in domains other than agriculture, especially for logistics. DSSs for cropping systems were developed as long ago as the 1950s, mainly based on linear

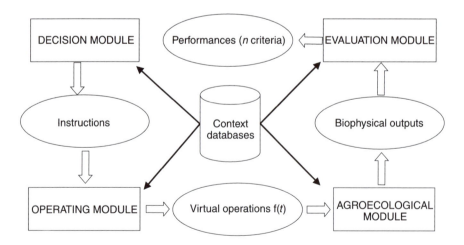

Fig. 16.3. General structure of decision support systems for farming and cropping systems.

programming, in order to optimize farm land allocation to different crops; they also addressed fertilizer planning at the farm scale. More recently, DSSs have been developed for a much wider range of domains (several authors in ten Berge and Stein, 1997), such as irrigation strategies, glasshouse management, choice of equipment and fieldwork organization, pasture management, orchard or arable crop management. Most DSSs are designed for consultants rather than for the farmers themselves. They should be considered to be tools for a richer and easier dialogue between researchers, consultants and farmers, rather than for discovering the 'best' solution (Cox, 1996).

Socio-economic mechanisms: how to make sustainability profitable?

Access to market and relative prices of agricultural outputs and inputs determine a large part of a farmer's choices concerning crops, crop sequences and crop management. Then regional patterns of changes in cropping systems depend heavily on how local market prices of the various agricultural products and inputs evolve.

Many contradictory agricultural policies have been applied in different parts of the world over the past 30 years. On the whole they have resulted in a very fast increase in gross yields, but also in a large number of unsustainable patterns of agricultural development, especially in the management of natural resources. One example is the intensification that occurred in north-western Europe as a result of coupling public subsidies to the amount of product through artificial prices. The changes in land use that took place in Argentina, on the fringes of the Amazon forest, in Eastern Europe and in many other areas all show that totally different policies may lead to changes in cropping systems where long-term management is more or less ignored. This gives rise to the general question of how to design socio-economic mechanisms and still make sustainable cropping systems profitable for farmers.

The European experience indicates that restricting intensification and/ or alleviating its detrimental effects is not sufficient to ensure sustainability. Even with lower agricultural prices (about one-half of those c. 1980) and more severe regulations on the agricultural use of chemicals, an optimized and rational use of fertilizers and pesticides can still cause pollution. Then agronomic measures specifically designed to maintain soil, water and air quality are necessary. For instance, keeping the nitrate content of drainage water to less than 50 mg l^{-1} requires not only an optimized and reduced application of fertilizers, but also the planting of catch crops during the winter. Parts of the hydrological basins in many areas should be withdrawn from arable cropping and turned into grasslands or forests (several authors in Lemaire and Nicolardot, 1997). Preventing runoff erosion and the associated pollution of surface water (especially by pesticides) needs grassland strips, ditches or other structures placed in suitable strategic locations in a catchment. Again, similar conclusions could be drawn about many other environmental targets, such as biodiversity, or landscape quality and accessibility.

This points out the need to promote in a positive way, and even to remunerate, some unproductive functions of cropping systems, or even unproductive cropping systems. Many of the corresponding measures require the collective management of rural space. This means that there is a need for economic or legal mechanisms that support this collective management. Of course, such mechanisms must be based on a sound knowledge of the spatial processes to be controlled, and of the way they are related to crop and space management. Spatial models allowing for the testing of land use scenarios are a very useful contribution of agroecology.

Basically, crop scientists must alert economists and policy makers to the need to preserve a local diversity of crops, then to avoid extreme regional specialization. Increasing world trade, together with the breaking up of national protection systems, could lead to more intense worldwide competition between producers, with a decrease in production costs and regional specialization. This liberal trend could thus restrict the number of profitable crops in a given area to those for which local producers have ecological, logistic or some other type of comparative advantage. It is difficult to predict where and to what extent such trends might be counterbalanced by successful strategies of diversification, based on market segmentation and the differentiation of plant products according to quality. On the whole this might result in a loss of diversity in cropping systems, with subsequent agroecological drawbacks, because this diversity is closely linked to sustainability in many ways (Karlen et al., 1994; Jordan et al., 1997). One of these drawbacks is a risk of pest development, which in turn is linked to the potential for pesticide use. Integrated crop protection is widely based on long intervals between growing the same crop on a given field, and avoiding covering large areas of the landscape with the same plant. Other links between crop diversity and sustainability concern work organization, runoff and erosion control, gene-flow control, landscape appearance and other environmental and socio-economic topics.

Challenges for Crop Science

Providing degrees of freedom in designing cropping systems

The contribution of crop science to future cropping systems is not only to improve crop performance, but also to widen the spectrum of improved plant properties, in order to increase the flexibility of the whole cropping system, and the potential diversity in cropping systems. The development of integrated crop protection is conditioned by successes that will be obtained in improving plant resistance to pests and diseases, and making it more durable. Improving plant tolerance to abiotic stresses such as mineral or water deficiencies, salt toxicity, waterlogging and hypoxia, mechanical obstacles to seedling emergence or root development, may also create new options in crop management and crop sequences. More generally, increasing the range of profitable crops in a given area is a direct way to promote the diversity of crop sequences, then of cropping systems. In this prospect, one valuable contribution of crop science is in improving or introducing properties that can make plants competitive for a wider range of functions. These functions may relate to environmental purposes (phytoremediation), industry (biofuels, biolubricants, plastics, etc.), medicine (vaccines, antibodies, etc.) or to specific aspects of animal or human nutrition (Dunwell, 1999).

Another source of possible diversification is the relevant application of crop physiology to crop management, especially if it is targeted towards minor crops that have been given less attention than the main crop species. Even in contexts of intensive agriculture with high rates of agrotechnical innovation, some crops are still being managed in very unsophisticated ways. As a result, actual yields remain far below potential yields, leading to low competitivity with other crops and consequently to decreased experimental and breeding efforts. In order to break this vicious circle, crop physiology can help to diagnose and correct the weaknesses of traditional crop management patterns. A typical example is the management of oil flax in northern France. The most usual practices are in fact directly transposed from the techniques developed for fibre flax, i.e. for obtaining thin, regular and straight stems, instead of a maximum seed harvest. An innovative management pattern, based on much lower plant density, has been shown to increase the crop response to nitrogen, and finally the oil yield (F. Flenet, France, 2000, personal communication). Rapeseed is another instance of a crop for which in-depth changes have occurred in management patterns – and associated plant morphology – and contributed to making the crop competitive in new areas.

A contribution to the evaluation and design of cropping systems

Increasing the flexibility and diversity of cropping systems also concerns decision and pre-innovative processes that generate cropping systems and their changes. Crop science should then try to provide material and methods to make these processes more efficient, more explicit and more transferable from one context to another.

Crop science can make a significant contribution by developing indicators for use in decision rules for crop management or in *a posteriori* diagnosis. Such indicators concern a wide range of processes and related crop characteristics, such as present or past stresses, pests and diseases, the presence of contaminants, predictable yields and harvest qualities. Making the diagnoses more and more early and specific is an example of what modern biology can contribute to decision support in agriculture. In this prospect, parallel investigations of gene expression and whole plant behaviour are a promising approach, as illustrated by Muller *et al.* (2000).

Another type of contribution is the development of integrated crop models as tools for evaluation, exploration and learning. The aim is to combine virtual and field experiments in order to renovate the trial and error approach that precedes the adoption of new cropping systems (Rossing *et al.*, 1997). Agroecological models designed for this type of use must fulfil several challenging criteria, such as:

- A capacity for connection upstream to decision rules, and downstream to economic evaluation.
- A dynamic description of stresses arising from the restricted use of inputs, with a particular emphasis on biotic stresses due to constraints on pesticide applications, and water stresses due to limited access to irrigation.
- A capacity to predict environmental outputs as well as crop production. This concerns a range of ecological processes that are closely linked to crop physiology, such as pollen emission and reception, carbon deposition and exudation into the soil, interactions between plants and microorganisms in the rhizosphere, and many other processes that are not yet incorporated into crop models.

A wide operational use obviously implies specific constraints, in terms of ergonomics and data availability, and this demands that we simplify the description of agroecological processes down to their essential features (Passioura, 1996; Sinclair and Seligman, 1996).

Such specifications imply a great improvement in the description of biological and ecological processes rather than a so-called 'degradation' of research models.

Conclusion

A number of factors will lead to massive, widespread changes in cropping systems worldwide. This change will occur in a less predictable, more unstable and more widely open context, in which cropping systems will be required to produce more, but also to meet targets other than food security, crop production and related benefits. They also should be designed so as to ensure the long-term management of natural resources – soil conservation, water quality and quantity, carbon sequestration, preservation of biodiversity – and to satisfy other social needs such as waste recycling or landscape quality and accessibility.

This changing framework leads to predicting more or less significant disruptions in the generative and adaptive processes that enable cropping systems to evolve in response to socio-economic, environmental or techno-logical changes. A crucial issue must then be to renovate the trial-and-error approach which characterizes this evolution, by incorporating more explicit evaluation or simulation procedures.

This corresponds to the development of a 'decision agriculture' (Miflin, 1997) that is increasingly knowledge-based, and increasingly rooted in the information and communication sciences and technologies. This does not, however, mean a technology-driven process of innovation, but on the con-trary increased feedback of action and decision into the design of innovation.

Such a 'decision agriculture' runs the risk of being restricted to devel-oped countries with powerful research and extension services in the absence of wider international educational programmes. It can also, however, prove an opportunity for wider and more equitable sharing of agronomic knowl-edge, as the proximity of farmers, technical advisors, agronomists and crop scientists is becoming less and less a matter of geographical distance. On the other hand, such a decision agriculture could also make farmers the captives of off-farm institutions or companies that will master information flows, pro-cess the data and convert them into instructions for farm management. Again, in the face of such risks, scientists, i.e. those working in crop science research as well as those working in cropping system research, need to adopt an ethical position.

Acknowledgements

We thank the following for their help, comments and information: Affholder, F., Attonaty, J.M., Beaudoin, N., Benoit, M., Bergez, J.E., Bonny, S., Debaeke, P., Deybe, D., Dufumier, M., Dugué, P., Dürr, C., Duru, M., Faloya, V., Fayet, G., Gabrielle, B., Garcia, F., Gary, C., Gosse, G., Griffon, M., Guiard, J., Guy-omard, H., Habib, R., Happillon, J., Le Gal, P.Y., Lemaire, G., Lescourret, F., Levert, F., Lucas, P., Ludwig, B., Martin Clouaire, R., Mary, B., Meynard, J.M., Millet, G., Munier-Jolain, N., Nolot, J.M., Papy, F., Parkes, O., Pichot, J., Rellier, J.P., Renoir, S., Richard, G., Robert, M., Scopel, E., Sebillotte, M., Sourie, J.C., Tardieu, F., Tchamitchian, M., Trebuil, G., Voltz, M., Wallach, D. and Wery, J.

References

Affholder, F., Bonnal, P., Jourdain, D. and Scopel, E. (1998) Small scale farming diversity and bioeconomic variability: a modelling approach. *Proceedings of the 15th International Symposium of the Association for Farming Systems Research-Extension, Pretoria*, pp. 952–959.

Attonaty, J.M., Chatelin, M.H. and Garcia, F. (1999) Interactive simulation modelling in farm decision-making. *Computers and Electronics in Agriculture* 22, 157–170.

Aubry, C., Papy, F. and Capillon, A. (1998) Modelling decision-making processes for annual crop management. *Agricultural Systems* 56, pp. 45–65.

Boote, K.J., Jones, J.W. and Pickering, N.B. (1996) Potential uses and limitations of crop models. *Agronomy Journal* 88, 704–716.

Buckles, D. and Triomphe, B. (1999) Adoption of Mucuna in the farming systems in northern Honduras. *Agroforestry Systems* 47, 67–91.

Cox, P.G. (1996) Some issues in the design of agricultural decision support systems. *Agricultural Systems* 52, 355–381.

Debaeke, P., Nolot, J.M. and Wallach, D. (2000) Adapting the management of spring-sown crops to water availability. I Development and testing of crop management systems for grain sorghum in south-west France. *Agricultural Systems* (in press).

Dunwell, J.M. (1999) Transgenic crops: the next generation or an example of 2020 vision. *Annals of Botany* 84, 269–277.

Gary, C., Jones, J.W. and Tchamitchian, M. (1998) Crop modelling in horticulture: state of the art. *Scientia Horticulturae* 74, 3–20.

Hardaker, J.B., Huirne, R.B.M. and Anderson, J.R. (1997) *Coping with Risk in Agriculture*. CAB International, Wallingford.

Jordan, V.W.L., Hutcheon, J.A. and Donaldson, G.V. (1997) The role of integrated arable production systems in reducing synthetic inputs. *Aspects of Applied Biology* 50, 419–429.

Karlen, D.L., Varvel, G.E., Bullock, D.G. and Cruise, R.M. (1994) Crop rotations for the 21st century. *Advances in Agronomy* 53, 1–45.

Le Gal, P.Y. and Papy, F. (1998) Coordination processes in a collectively managed cropping system: double cropping of irrigated rice in Senegal. *Agricultural Systems* 57, 135–159.

Lemaire, G. and Nicolardot, B. (eds) (1997) *Maîtrise de l'Azote dans les Agrosystèmes*. INRA Editions, Paris.

Miflin, B.J. (1997) Sugar beet production: strategies for the future. In: *Proceedings of the 60th IIRB Congress*, IIRB, Brussels, pp. 253–262.

Milleville, P. (1998) Conduite des cultures pluviales et organisation du travail en Afrique soudano-sahélienne: des déterminants climatiques aux rapports sociaux de production. In: Biarnès, A. (ed.) *La Conduite du Champ Cultivé. Points de Vue d'Agronomes*. Editions de l'Orstom, Paris, pp. 165–180.

Muller, B., Tardieu, F., Ghazi Y., Doumas, P. and Rossignol, M. (2000) Response of the root system to phosphorus deprivation in *Arabidopsis thaliana*. Parallel analyses of changes in architectural program and gene expression at a large scale. *Journal of Experimental Botany* 51 (suppl.), 8.

Nowak, P. (1992) Why farmers adopt production technology. *Journal of Soil and Water Conservation* 47, 14–16.

Papy, F. (1994) Working knowledge concerning technical systems and decision support. In: Dent, J.B. and McGregor, M.J. (eds) *Rural and Farming Systems Analysis. European Perspectives*. CAB International, Wallingford, pp. 222–235.

Passioura, J.B. (1996) Simulation models: science, snake oil, education or engineering? *Agronomy Journal* 88, 690–694.

Rossing, W.A.H., Meynard, J.M. and Van Ittersum, M.K. (1997) Model-based explorations to support development of sustainable farming systems: case studies from France and the Netherlands. *European Journal of Agronomy* 7, 271–283.

Sebillotte, M. (1990) Some concepts for analysing farming and cropping systems and for understanding their different effects. In: Scaife, A. (ed.) *Proceedings of the First Congress of European Society of Agronomy*, Vol. 5, European Society of Agronomy, Colmar, pp. 1–16.

Sinclair, T.R. and Seligman, N.G. (1996) Crop modeling: from infancy to maturity. *Agronomy Journal* 88, 698–704.

ten Berge, H.F.M. and Stein, A. (eds) (1997) Model-based decision support in agriculture. *Quantitative Approaches in Systems Analysis 15*, DLO-WAU, Wageningen.

Vereijken, P. (1992) A methodic way to more sustainable farming systems. *Netherlands Journal of Agricultural Science* 40, 209–223.

Will Yield Barriers Limit Future Rice Production?

17

J.E. Sheehy

International Rice Research Institute (IRRI), MCPO Box 3127, Makati City 1271, Philippines

Introduction

The study of maximum yields and how to attain them is a fascinating intellectual pursuit in its own right; what gives it importance now is its link to food production. In that context the statistics of world population are of great concern: 1300 million when the oldest continuous agricultural experiments were set up in 1843 in the UK at Rothamsted Experimental Station; 6000 million reached in 1999; and a predicted 8909 million in 50 years' time (UNFPA, 1999).

Dyson (1999) suggested that in order to meet the future demand for food the continuation of existing trends in cereal yield would be sufficient. Several important questions then arise. Can the rate of increase in average yield be sustained? Will a biophysical yield barrier prevent it from continuing? What is the likely role of genetic engineering in the pursuit of higher yields and improvements in dietary quality? Are higher yields sustainable from an environmental viewpoint? How will global climate change affect yields? Given the importance of rice for humans, I intend to address briefly the above questions for rice. Considerations of limitations caused by pests and diseases are left to others.

Rice Area and Ecosystems

Rice is the most important food for 60% of the world's population and 90% of the crop is grown in Asia in probably the most varied of crop ecosystems, ranging from flooded to droughted. Irrigated rice systems in Asia date back several thousand years. Triple cropping probably started 600 years ago in the Yangtze river region (Greenland, 1997). Currently, the irrigated ecosys-

Table 17.1. Global rice ecosystems, their area and share of world production (93% of rice area is in Asia). Total world production in 1998–1999 was approximately 561 million tons.

Ecosystem	Area (ha × 10^6)	Rice production (% of total)
Irrigated	79	75
Rainfed	36	18
Upland	19	4
Deepwater and tidal wetland	12	3

tem produces 75% of total rice production and much of the required improvement in yields must come from this ecosystem (Table 17.1). There are land losses to agriculture associated with developments of new housing, infrastructure, recreation and increasing industrialization as well as degradation caused by erosion and salinization (Cardenas *et al.*, 1996; Evans, 1998). This is a serious problem given that in Asia no new significant areas are available for cultivation. It is difficult to estimate how much land will be lost during the next 50 years. Evans (1998) for example used a figure of 1% per annum. Currently, each hectare of rice land in Asia provides rice for 27 people. Fifty years from now, food support from that land will be demanded by at least 38 people ha^{-1} and 43 people ha^{-1} if there is a 10% land loss. The impact of these changes in increased demand for rice on its yield and quality will be examined below.

Yield Definitions

It is worth noting the definitions of yield, yield potential and potential yield given by Evans and Fischer (1999). Here I modify them slightly for the purpose of continuity with other information published for rice. Briefly, yield is the mass of rough rice at harvest quoted at 14% moisture content. Yield potential is the maximum yield achievable experimentally in a given environment with a given cultivar for a growing season of defined duration. Potential yield is the maximum yield predicted by a computer model for that cultivar growing without stress. To that list I add 'ceiling yield', here defined as the maximum achievable yield for an ideal cultivar growing in a defined environment for a season of fixed duration; in other words, the predicted upper limit to yield based on biophysical considerations. Ceiling yield is the ultimate yield barrier.

Rice Quality

Rice is most commonly eaten after milling and polishing, processes that lead to losses of some micronutrients and fibre, but improve storage characteristics and digestibility. Many quantitative characteristics can be used to define quality. However, many are subjective, often reflecting strong cultural preferences such as colour, translucency, chalkiness, grain shape and size, cooking quality, stickiness and aroma. Others are concerned with industrial processing such as milling and brewing quality. Rice has the lowest protein and fibre content of the cereals and contains little or no vitamin A, C or D. An average adult requires daily an energy intake of about 10.5 MJ and about 52 g of protein, equivalent to approximately 700 g of rice.

Potrykus *et al.* (1995) suggested that the introduction, by genetic engineering, to the endosperm of rice of the genes coding for the synthesis of β-carotene would prevent xerophthalmia (dry eyes), a disease from which 500,000 children lose their sight annually. The β-carotene would turn the rice somewhat yellow. Early indications suggest that transformed rice synthesizing β-carotene at the appropriate level, 2 µg β-carotene g^{-1} in milled rice (about 25% that of maize), will become available in the not too distant future (Ye *et al.*, 2000). Given the low levels required in the grain, it is unlikely that the introduction of the extra biosynthetic pathway would have any significant impact on yield.

Yield Trends

First, I want to address two questions: (i) is the rate of increase in cereal yield slower in developing countries than in developed countries? and (ii) can the rate of increase in yield match the increase in population? From a study of long-term yield databases it is clear that from the middle of the 19th century until about 1940–1950 yields changed slightly and then started to increase progressively. Weather fluctuations cause changes in annual yields and are largely responsible for the variation that occurs around that systematic upward trend in yield (Geng, 1999). Linear yield trends are characteristic of rice, wheat and maize growing in both developing and developed countries (Fig. 17.1). It is remarkable that the systematic trend is linear across decades, countries and crops. None the less, what governs or forces the actual linear rate of increase is obscure. The factors influencing the rate of increase (genetic 29%, N fertilizer 48% and others 24%) have been described by Bell *et al.* (1995) for wheat in the Yaqui valley in Mexico. Increasing concentrations of atmospheric CO_2 may have also contributed to the yield increases. However, simulation models suggest that the fertilizing effect of increasing concentrations of CO_2 over the next 50 years could be more than offset when accompanied by the likely 2–6°C increase in air temperature (Matthews *et al.*, 1995; Ziska *et al.*, 1997).

The annual rates of increase in rice yields in Arkansas (USA) and Central Luzon (Philippines, dry season) are the same, although the absolute yields in Arkansas are higher than in Central Luzon in any given year (Fig. 17.1d). The rate of increase in wheat yields in the Yaqui valley of Mexico is similar

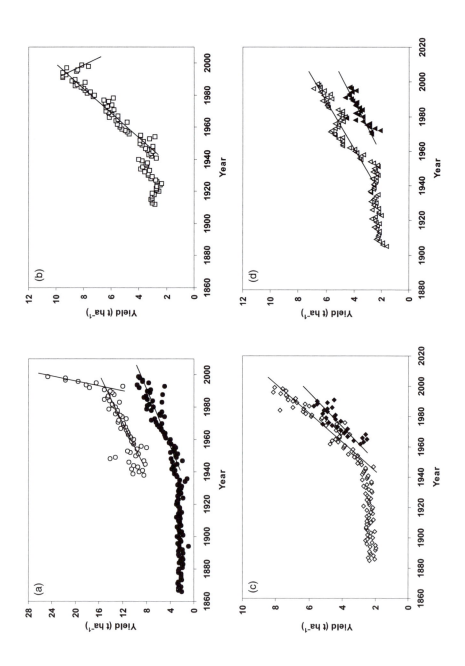

to that in the UK and is greater than in the USA (USDA, 2000). It is clear that rates of increase in yield in developing countries can be comparable with those in developed countries. The rate of increase in California rice yield (Fig. 17.1b) was higher than that of Arkansas, but it started to decline linearly from about 1994, probably as a result of state legislation concerning environmental issues in relation to burning straw and use of chemicals. It is salutary to note that legislation regulating environmental impacts of farming could reverse yield increases.

A comparison of the yields obtained by the winner of the Iowa Master's Corn Grower's Contest and the state average shows that both increased linearly, with the winner's line having a slightly greater rate of increase. Abruptly, in 1992, the rate of increase in the yield of the winner of the competition increased by a factor of about ten (Fig. 17.1a), a rate change attributable to the string of successes of a single farmer. The data show that it is possible to change markedly the rate at which the yield gap between potential and actual yields is closed. The change in rate is probably brought about by incentives, use of the most productive hybrids and extremely high standards of crop and resource management (Duvick and Cassman, 1999). Yield increases in developing countries began 15–20 years later than in developed countries (Fig. 17.1c, d); probably as a result of the establishing of the International Centers and their initiation of the Green Revolution.

Currently, in Asia, approximately 551 million tons of rice support 3682 million people. If the average annual rate of increase in Asian rice yield continues for the next 50 years at 0.06 t ha^{-1} (FAO, 2000), an extra 396 million tons will be produced. This means that if the rice were evenly distributed, all Asians would currently receive 410 g day^{-1}, and in 2050, their allowance would actually increase to 493 g day^{-1}. As now, the bulk of future

---

Fig. 17.1. (Opposite) Historical yield (t ha^{-1}) trends for maize, wheat and rice for various countries and regions.
(a) State average (●) (1950–1991, $y = 0.12 \times -222.3$, $r^2 = 0.86$) and Master's competition winners (○) for rainfed maize in Iowa, USA (1950–1991, $y = 0.13 \times -251.8$, $r^2 = 0.86$; 1992–1999, $y = 1.3 \times -2642$, $r^2 = 0.71$).
(b) California state average (□) for rice (1950–1994, $y = 0.13 \times -251.8$; 1994–1999, $y = 502.2 \times -0.25$, $r^2 = 0.44$).
(c) UK country average (◇) for wheat (1950–1999, $y = 0.10 \times -220.4$, $r^2 = 0.90$) and Yaqui Valley, Mexico (◆) for irrigated wheat (1962–1991, $y = 0.09 \times -171.2$, $r^2 = 0.73$).
(d) Arkansas state average (△) for rice (1950–1999, $y = 0.07 \times -139.6$, $r^2 = 0.82$) and dry-season rice yield in Central Luzon, Philippines (▲) (1970–1997, $y = 0.07 \times 124.5$, $r^2 = 0.82$).
Sources: (a) USDA (2000) and Meis (2000); (b) USDA (2000); (c) MAFF (2000) (UK) and Fischer (1993) (Mexico); and (d) USDA (2000) (Arkansas) and PhilRice (1994) (Central Luzon).

yield increases will have to largely come from the irrigated system (Table 17.1).

Extrapolating the data for irrigated rice (Hossain and Pingali, 1998; Fig. 17.2) suggests that an average yield of 8 t ha⁻¹ will be reached in 2019 and 10.8 t ha⁻¹ in 2050. To determine whether such a rate of increase in average yield can be sustained, we can examine the yield trend in a little more detail. If it is assumed that the upper quartile of farmers obtain yields 25% greater than the average in 2019 their yield must equal 10 t ha⁻¹. In 2050, when the average yield would have to be 10.8 t ha⁻¹, the yield for the upper quartile would have to be 13.5 t ha⁻¹ to maintain the trend. These projected yields are greater than the yield potential of indica rice in the tropics (Kropff *et al.*, 1994). A land loss of 10% could mean that the yields of the best farmers would have to increase to 15 t ha⁻¹ in 2050 to maintain the current rate of increase in irrigated rice production. Figure 17.2 shows the yield trends for irrigated rice extrapolated to 2050 assuming there is no limit to yield. Also shown in Fig. 17.2 are the limits to achievable yields and these

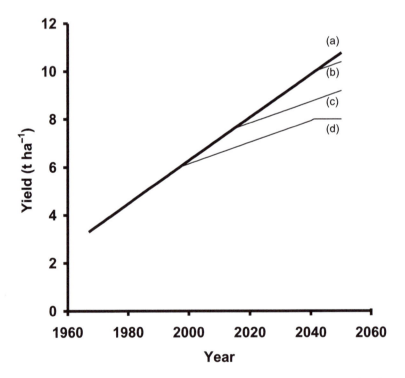

Fig. 17.2. (a) The yield trend for irrigated rice ($y = 0.09 \times -174$) extrapolated to 2050 (Hossain and Pingali, 1998). The trend lines assume farmers can achieve 80% of potential yield for potential yields of (b) 15 t ha⁻¹; (c) 12 t ha⁻¹; and (d) 10 t ha⁻¹.

have been calculated as 80% of potential yield; the potential yields used were $10\ t\ ha^{-1}$, $12\ t\ ha^{-1}$ and $15\ t\ ha^{-1}$. To determine which of the above scenarios is likely to be correct, we must examine the issue of upper biophysical limits to yield.

Limits to Yield

It is almost a matter of intuition that yield has limits. At these limits, the demands by the crop for the dilute resources (carbon dioxide, soil nutrients) that become fixed and concentrated in dry matter cannot exceed the maximum physical rate at which they can be supplied, either by mass transport or diffusion. Within the crop, there are limits to the rates of processes that produce dry matter and assemble it further into cells, tissues and organs. Because our understanding of the processes is based on existing species and cultivars, the limits to yield must be calculated using models, guided by general laws of physics and chemistry. More difficult than calculating rates of mass accumulation is estimating the partitioning of captured resources into the structural, architectural and mechanistic properties required to sustain maximum rates of resource capture and to physically support high yields.

Despite scepticism concerning modelling, it is generally believed that current maximum yields can be increased further. The plausibility of modelling calculations depends not only on the accuracy with which the parameters were measured, but also on whether the environmental circumstances in which the measurements were made reflect the real-world growth environment. Measurements made in environmental conditions dissimilar to those normally experienced provide information about abnormal plant behaviour that can mislead rather than assist the unwary. Mechanistic modelling of the root mass, number of plants, tillers, leaves and spikelets per unit ground area in relation to crop function and lodging resistance is surprisingly uncommon, but almost essential to support progress in the genetic improvement of yield. The great value of mechanistic models is that they aid understanding and point out knowledge gaps. Indeed, their greatest value often lies in the gap between observed and predicted values of yield. To predict yield, simple empirical or scientific rule-of-thumb models are usually good enough and it is in that spirit that I move to the next section.

A simple yield equation

It is not my purpose here to model in detail the growth of a rice crop; rather, I want to derive some simple rules of thumb and provide an overview of the important features of a high yielding rice system. A systems analysis approach to the complex issues surrounding yield limits can help in deciding what factors determine yield and their relative importance (Sheehy, 2000). Using daily time steps, growth rate can be written in units of CH_2O as

$$dW/dt = P_g(t) - R_s(t) - R_m(t) - D(t) \tag{17.1}$$

where P_g is gross photosynthesis (shoot net photosynthesis + shoot respiration for the daylight hours), R_s is the crop synthetic respiration (shoots + roots), R_m is the crop maintenance respiration and D is the daily loss of matter through detachment, assuming that each variable has been measured for a day and t is time.

The common agronomic equation for grain yield, Y, links above-ground biomass to grain yield through the harvest index, H. A simple yield equation can be obtained as shown by Sheehy (2000) by integrating growth rate over the growing season from transplanting (*tr*) to maturity (*tm*) and multiplying it by H, giving Y (17.2) assuming the weight at transplanting is negligible.

$$Y = H \int_{tr}^{tm} (0.64P_g(t) - m_T W_s(t) - D_s(t)) \, dt \tag{17.2}$$

where m_T is the respiration coefficient at temperature T, W_s is the shoot weight and D_s is the rate of detachment of shoot weight. It is sometimes suggested that the maximum theoretical value of H may be close to the value calculated for wheat by Austin *et al.* (1980) of 0.6. High yielding rice is characterized by values of H close to 0.5, this is a function of the mass balance of the crop in terms of structural and physiological requirements for high yield. It is clear, that to maximize yield, P_g must be maintained at its highest value throughout the growing season, and m_T and D_s must be minimized over that period. To maintain crop photosynthesis at its highest value requires the appropriate partitioning of assimilates between leaves and other organs, maintenance of high individual leaf rates of photosynthesis and an erect leaf canopy as discussed below.

Limits and radiation use efficiency

Early theoretical work at the International Rice Research Institute (IRRI) suggested that grain yields of 15 t ha^{-1} were possible. This suggestion rested on erroneous values for the efficiency of radiation conversion of rice used by Yoshida (1981). The instantaneous value of the radiation conversion factor (or radiation use efficiency), ε, expressed as g DW MJ^{-1}, can be written as:

$$\varepsilon = \frac{dW_s(t) / dt}{I_{int}(t)}. \tag{17.3}$$

Following Sheehy (2000) ε can be written as:

$$\varepsilon = \frac{0.64P_g(t) - m_T W_s(t) - D_s(t)}{I_{int}}, \tag{17.4}$$

where I_{int} is the daily total of intercepted PAR (photosynthetically active radiation; MJ m^{-2} day^{-1}). Calculations of ε at a range of irradiances (Sheehy *et al.*, 1998b) for a rice crop (shoots + roots) intercepting 60% of incident PAR showed ε is negative when crop respiration exceeds photosynthesis, it reaches a maximum at about 4 MJ m^{-2} day^{-1} and then gradually declines because canopy photosynthesis increases less rapidly than irradiance. The average daily irradiance (PAR) in the dry season (January–May) at IRRI is 10.3 MJ m^{-2}; at that irradiance and a daily average temperature of 30°C, the estimated value of ε is 2.2 g DW MJ^{-1}.

Making the common assumption that ε is approximately constant, the general equation of crop yield can be developed by integrating over the growing season from transplanting (*tr*) to maturity (*tm*) so that:

$$Y = H\varepsilon \int_{tr}^{tm} (Q(t)\ f(t))\ dt \qquad (17.5)$$

where *tr* is date of transplanting, *tm* is time of maturity, $Q(t)$ is the photosynthetically active radiation incident on the crop at time t (MJ m^{-2}); and $f(t)$ is the fraction of radiation intercepted at time t. It is this equation that is often used as the basis of simple crop models of yield. Mitchell *et al.* (1998) concluded from a literature survey that the typical radiation conversion factor (RCF) for rice was 2.2 g DW MJ^{-1} and that was consistent with a rice yield in the tropics of approximately 10 t ha^{-1}. In the same survey the RCF of wheat, grown with adequate water and nutrients, was 2.7 g DW MJ^{-1} and the mean value for maize was reported as 3.3 g DW MJ^{-1}. Loomis and Amthor (1999) reported a range of values for maize, 5.5–4.0 g DW MJ^{-1} based on theoretical calculations. It should be noted that for the same RCF value, yield in a temperate environment should be greater than in a tropical environment because of the longer crop duration. At an RCF of 2.2 g DW MJ^{-1} a tropical yield might typically be 10 t ha^{-1} and a temperate yield 15 t ha^{-1}.

Experimentation in the dry season at IRRI using high rates of nitrogen fertilizer (>450 kg N ha^{-1}) and nets to support the crop increased yield to 11.6 t ha^{-1} (Sheehy *et al.*, 2000) and ε to 2.6 g DW MJ^{-1}. However, unsupported indica cultivars lodged at a yield of approximately 10 t ha^{-1}. It is hoped that new plant types derived from tropical japonicas will have sufficient lodging resistance to enable them to express their full yield potential.

Hybrid rice has a yield advantage of about 5–30% at farm levels of production in China and very high yields have been reported for small plots (Yuan, 1994, 1998). It should be noted that the growing season in China is longer than in the tropics and, with the same RCF, extra yield is simply a function of longer crop duration rather than improved physiology. As yield ceilings are approached, both inbreds and hybrids must become limited by the same biophysical factors, in other words, available variation for yield will have been fully exploited in both plant types (Bingham and Austin, 1993). Khush *et al.* (1998) stressed that a major aim was to develop indica/japonica hybrids that would have greater heterosis because of diversity between the parents; the yield limits of such hybrids remain to be determined.

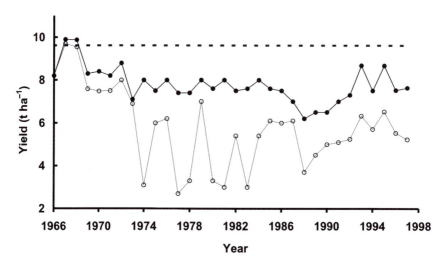

Fig. 17.3. Yields of the best entry in breeders' trials at IRRI (•) and of IR8 (○) for the past 30 years in the dry season based on data of Kropff *et al.* (1994). Also shown is the predicted potential yield (- - -) based on the model of Mitchell *et al.* (1998).

The concept of a yield barrier (Cassman, 1994) for rice in the tropics largely rests on the data for the dry season showing the yield of the best entry in breeders' trials at IRRI for the past 30 years (Fig. 17.3). It can be seen that the maximum yield never exceeded 10 t ha^{-1} and that the long-term average was 7.8 t ha^{-1}. The average potential yields for the dry and wet seasons (9.6 and 7.7 t ha^{-1}) were calculated over the same period using an RCF of 2.2 g DW MJ^{-1} in the model of Mitchell *et al.* (1998). Yields are lower in the wet season because there is 15% less solar irradiance owing to cloud cover; the level of solar irradiance imposes a limit on yield. Yield potential appears to be approximately 80% of potential yield. The breakdown in the resistance to diseases and pests of the early Green Revolution cultivar IR8 (released in 1966) is reflected in the decline and the large fluctuations in yield across the 30-year period. To obtain yields of 15 t ha^{-1}, the RCF would have to rise to 3.3 g DW MJ^{-1} and this is unlikely to be achieved without significant improvements in canopy photosynthesis.

Limits to canopy photosynthesis

As shown above, canopy photosynthesis is an important determinant of yield and RCF. To clarify some of the issues addressed in this chapter concerning yield and canopy photosynthesis, it is convenient to divide leaf area into three components. Monteith (1965) considered two categories of photosynthetically active leaf area: a sunlit part, A_0, and a once-shaded part, A_1. Leaves shaded more than once compose a third component, L_N, that is unimportant photosynthetically, but important as a nitrogen store (Sinclair and Sheehy, 1999). The difference between the live leaf area index (LAI) (L) and A_0 and A_1 is the area L_N used for storing nitrogen (Sheehy, 2000), given by:

$$L_N = L - (A_0 + A_1). \tag{17.6}$$

The interception and distribution of irradiance in a Monteithian canopy is defined by a parameter s, where s defines the fraction of irradiance that passes through unit LAI without interception. The relationship between the extinction coefficient k (Bouguer-Lambert law) and s was defined by Sheehy and Johnson (1988). The relationship between various properties of a Monteithian canopy is shown in Table 17.2.

It must be remembered that, all other things being equal at low LAI (<4), extremely erect leaf canopies do not grow as rapidly as more prostrate ones because of poor light interception. The early growth advantage persists for nearly 60 days in temperate grasses (Sheehy *et al.*, 1980). In contrast, beyond LAIs of approximately 4.0, the greater the level of irradiance, the greater the photosynthetic advantage of more erectophile canopies (Monteith, 1965). Canopies with moderately erect leaves perhaps offer the best compromise for light use when leaf angle is a fixed characteristic of a cultivar.

The relationship between the photosynthetically active LAI and total LAI for a somewhat idealized canopy with $s = 0.7$ is shown in Fig. 17.4. It can be seen that the asymptotic value of photosynthetically active LAI is approximately 6.6. In the absence of rice plants with leaves that successively

Table 17.2. The properties of a Monteithian canopy: s is the light transmission parameter, k is the extinction coefficient, β is the mean leaf elevation in degrees, LAI is the leaf area index, A_0 is the sunlit leaf are index, A_1 is the once-shaded leaf area index and L_N is the leaf area index used for storing nitrogen. On a typical dry-season day with an average irradiance of 1096 μmol PAR m^{-2} s^{-1}, the irradiance below an LAI of 6.3 and above 11 ranges from 151 to 35 μmol PAR m^{-2} s^{-1}, values close to the light compensation point, thus preventing senescence (Weei-Pirng *et al.*, 1986; Thomas, 1994).

Property	New plant type canopy	Idealized canopy
s	0.6	0.7
k	0.45	0.31
β	66	73
LAI	11.2	11.2
A_0	2.5	3.3
A_1	2.4	3.0
L_N	6.3	4.9

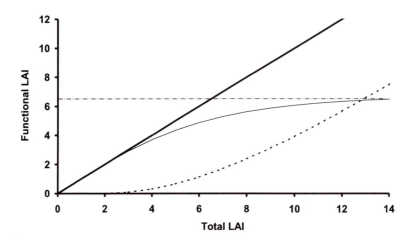

Fig. 17.4. The relationship among the total live leaf area index (LAI, ——), the leaf area index used mainly for nitrogen storage (L_N, - - -), the sunlit and once-shaded leaf area indices ($A_0 + A_1$, ——) for a canopy with s = 0.7 and the photosynthetic leaf area limit (–·–·–·–·–).

change from prostrate to very erect, the highest photosynthetically active LAI achievable in the foreseeable future is going to be approximately 6.6 in practice. This limit on LAI results in a limit to the maximum achievable rate of gross canopy photosynthesis, P_{gmax}, given by:

$$P_{gmax} = A_{0max}\, p_{mo} + A_{1max}\, p_{m1}, \text{ as } I \to \infty \qquad (17.7)$$

where I is the instantaneous value of irradiance (PAR), p_{mo} is the maximum rate of individual leaf photosynthesis for young sunlit leaves, p_{m1} is the maximum rate of individual leaf photosynthesis for older shaded leaves and A_{0max} and A_{1max} are the maximum values of the sunlit and once-shaded leaf area indices (both are 3.3).

The maximum quantum efficiency of a crop intercepting all incident PAR is given by:

$$dP_g/dI_0 = \alpha\, (1 + \tau), \text{ as } I \to 0 \qquad (17.8)$$

where α is the photochemical efficiency (quantum yield) of an individual leaf and τ the transmissivity of a leaf. It can be seen that the quantum yield of individual leaves and the rates of maximum individual leaf photosynthesis are another set of limits to growth and yield to be added to that imposed by available solar irradiance.

Peng et al. (1995) showed that the relationship between leaf photosynthesis and leaf nitrogen content was linear for rice, suggesting the importance of nitrogen and that maximum achievable rates had not been attained. Very high rates of individual leaf photosynthesis, 52 μmol m^{-2} s^{-1}, have been reported for lucerne. The rates were not correlated with whole-plant photosynthesis, but whole-plant photosynthesis was correlated with whole-plant nitrogen content (Sheehy *et al.*, 1980). The lack of correlation is not uncommon because such rates are associated with small, thick and narrow leaves (Bingham and Austin, 1993). Not surprisingly (equation 17.4), high concentrations of nitrogen in leaves are associated with higher values of the RCF (Sinclair and Horie, 1989).

The rates of photosynthesis of the whole canopy and the sunlit and once-shaded leaves were calculated for high irradiance conditions in the dry season using the parameters for leaf photosynthesis described by Sheehy (2000). The total rate of canopy gross photosynthesis was 10.6 g CH_2O m^{-2} h^{-1}, the shaded leaves contributing 22% of that rate. It is interesting to note that the loss of the once-shaded area and the leaf area associated with stored nitrogen ($L_N + A_1$) would reduce canopy photosynthesis by only 22%. This emphasizes the point that nitrogen and carbohydrates can be recycled from approximately 70% of the leaves with canopy photosynthesis declining by only about 25%. I return to this issue below.

Yield and redesigned photosynthesis

On the basis of current evidence, unless the link between leaf size and high rates of leaf photosynthesis in C_3 plants can be broken, it would appear that a 15 t ha^{-1} yield is probably beyond the capacity of tropical short-duration rice with its current photosynthetic pathway. As a consequence, achieving yields beyond 12.5 t ha^{-1} in the next 50 years could require the introduction of the C_4 photosynthetic pathway to rice (Edwards, 1999; Ku *et al.*, 1999; Sheehy, 2000). At the relatively high temperatures experienced by rice, equations 17.7 and 17.8 suggest that a C_4 rice would have a significantly improved rate of canopy photosynthesis. It is often suggested that increases in atmospheric CO_2 (Schimel *et al.*, 1996) could produce a similar yield increase in C_3 rice by about 2050. However, such suggestions usually ignore the possible effects, on canopy photosynthesis, of the temperature increases that will accompany increases in atmospheric CO_2. It was suggested by Long (1999) that at current concentrations of CO_2, increases in temperature above about 24°C depress rates of canopy photosynthesis in C_3 crops, but not in C_4 crops. However, Ziska *et al.* (1997) reported increased biomass in rice growing under elevated CO_2 (+200 µl l^{-1}), and no change in biomass when the elevated CO_2 was accompanied by a 4°C temperature increase. Those results suggest that the increased temperature was unlikely to have depressed canopy photosynthesis in the elevated CO_2 treatment.

Several potential obstacles could prevent a C_4, or high-CO_2 C_3, rice crop from using the extra photosynthate to increase grain yield without reducing grain quality. It is possible that the number of spikelets could be smaller in a C_4 plant than in a C_3 plant and this might lead to a limitation of yield potential. Furthermore, the C_4 syndrome might reduce the protein content of the grain. Evidence of such problems can be found because a C_3 crop growing under elevated CO_2 conditions with suppressed photorespiration may resemble a C_4 crop in terms of its N content (Nakano et al., 1997) and yield. Several experiments have been conducted to investigate such effects. Many such experiments have been carried out in pots or enclosures of limited size and so the results are somewhat difficult to extrapolate to normal field conditions, especially for ceiling yield scenarios. Ziska et al. (1997), using open-top field enclosures, observed increases in grain yield with elevated CO_2, but when accompanied by increases in temperature harvest index and yields decreased. They also reported that protein content in the grain declined significantly with increasing CO_2. If a C_4 rice plant were to be developed and atmospheric CO_2 concentrations increased, so that C_3 and C_4 rice photosynthesized at the same rate, the C_4 rice would use significantly less water in transpiration than the C_3 rice. For example, if CO_2 levels increased to 510 ppm the stomatal conductance of C_4 rice could fall by more than 50% whilst maintaining the same rate of photosynthesis per unit leaf area as would occur at current concentrations. The sunlit leaves of a C_4 rice would be unable to shed as much heat under such conditions and it can be estimated that their temperature would rise by about 2°C (Sheehy et al., 1998c). Somewhat surprisingly, in a future world of elevated CO_2, the greatest advantage of a C_4 rice may well flow from its potential drought tolerance in upland and rainfed ecosystems.

Nutrient Capture and Yield

Plants and their functional growth mechanisms are partly made of mineral elements. To meet the nutritional requirements of humans in turn, the nutrient demands of the crop must be met. Increases in yield cannot be achieved without increases in nutrient capture. Naturally occurring sedimentation, nutrient inflow by irrigation, organic residues, biological N_2 fixation and carbon assimilation by flood-water flora and fauna played an important role in securing the sustainability of traditional irrigated rice systems at yield levels of about 1 t ha^{-1} (Greenland, 1997). The large amounts of mineral elements required in high yielding crops (Table 17.3) far exceed soil supply capacity and the prevention of 'mining' requires the use of inorganic fertilizers.

Table 17.3. The mineral element content (kg ha^{-1}) of a rice crop yielding 12 t ha^{-1} (14% moisture content) with a harvest index of 0.5.

N	P	K	Ca	Mg	S	Cl	Si	Fe	Mn	B	Zn	Cu
234	56	377	33.3	42.5	18	119	1086	6	1.6	0.8	0.43	0.04

The equation of mass conservation can be applied to nutrient management. Let E_{grain} and E_{straw} represent the fractional values (g g^{-1} DW) of a particular nutrient, or element, in the grain and straw at harvest, and W_{grain} and W_{straw} represent their respective dry weights. The quantity of the element captured by the crop from intrinsic soil sources is E_{soil} and that derived from fertilizer is $F_e E_{fert}$, where E_{fert} is the total amount of element E applied as fertilizer and F_e is the fraction captured by the crop. Thus, we can write the following equation describing the relationship between supply and demand:

$$W_{grain} E_{grain} + W_{straw} E_{straw} = E_{soil} + F_e E_{fert}. \qquad (17.9)$$

The capture from soil sources represents an integrated result of several different processes, e.g. the daily mineralization of nitrogen and its diffusion to the roots. If there is an extremely large amount of E available in the soil relative to the amount captured by the crop, clearly it does not have to be added as fertilizer and $E_{fert} = 0$. However, this is not the case with the macroelements (N, P, K), and in many rice soils microelements such as zinc have to be added.

Because of their great abundance in the soil relative to their uptake, many of the elements do not need to be added as fertilizer. The fertilizer use efficiency for the other elements is highly variable, depending on soil type and environmental factors, and site-specific nutrient management is advocated for increasing yield. Calculations of the supply rate of other mineral elements is not simple, but a recovery rate of 25% P and 50% K has been observed in rice soils (Dobermann *et al.*, 1998).

Nitrogen is often the key limiting factor and it has a low fertilizer use efficiency in irrigated rice systems. At IRRI, for a 100-day growing period, N_{soil} is typically about 70 kg N and N_{grain} varies from 1.0% to 1.8% depending on cultivar and rate of fertilizer N applied. The N content of the aboveground biomass required for a 12 t ha^{-1} yield is 234 kg N ha^{-1}. If the soil contributes 70 kg N, then 328 kg N must be applied at a fertilizer recovery efficiency of 50% to ensure no N limitation to yield. If this N is not supplied, such a crop would not be able to reach its yield potential. Formulation of precise fertilizer recommendations is complicated because of the large spatial and temporal variations in N_{soil} (Cassman *et al.*, 1996; Olk *et al.*, 1999). Yield declines can also occur because of nitrogen sequestration in intensively cropped aerobic soils. These declines can be reversed by using additional N or other management strategies (Cassman *et al.*, 1995).

Of concern are the negative K balances in about 80% of the intensive rice fields in Asia. Dobermann *et al.* (1998) suggested that it is only a matter of time until the indigenous supply becomes a limiting factor on the most fertile lowland rice soils. The issue of meeting increasing demands for food and attempting to decrease losses of agricultural chemicals into the environment would seem to be in conflict. Asian farmers typically lose 60% of applied nitrogen fertilizer. How to improve nutrient use efficiency is a relatively unglamorous research area worthy of investment.

Nitrogen reservoirs and sink size

The size of the sink in rice is determined before flowering and, unlike wheat or barley, the weight of an individual rice grain for a cultivar is almost constant. Consequently, yield improvements in rice result from an increase in grain number per unit ground area (grains are filled spikelets). The maximum number of juvenile spikelets on a developing panicle is observed at the late differentiation stage, about 10 days before heading. The maximum number of juvenile spikelets observed per square metre in irrigated, field-grown rice, with high nitrogen inputs, in the dry season was approximately 115,000 (unpublished). For high-yielding indicas, actual spikelet number is approximately 50,000, about 75% of which are filled grains. The potential sink size (juvenile spikelets) is much greater than the actual sink size and not all of the actual sink is utilized by the crop. Potential sink size is much greater than required to produce the highest yields discussed in this chapter and cannot be considered a limit to yield. The limitation perhaps has more to do with turning potential sinks into actual sinks at maturity. Resource availability, in particular the nitrogen content of the plant at the late differentiation stage, influences spikelet number (Horie *et al.*, 1997). Yoshida (1973) elevated CO_2 in small open-top field enclosures during different growth stages and showed that it was possible to increase percentage grain filling by increasing the concentration of CO_2 during the grain-filling period. The rate at which the developing panicle acquires nitrogen exceeds the rate at which the crop acquires it through its roots during grain filling (Sheehy *et al.*, 1998a), emphasizing the requirement for a large 'reservoir' in the vegetative tissues. If the crop cannot capture sufficient nitrogen through its roots to meet this demand, nitrogen must be withdrawn from the vegetative portions of the crop, assuming there are no losses from the root tissues. The limit to the nitrogen extracted comes when the concentration in the vegetative portions falls to the concentrations associated with the structural tissues (0.8%) (Makarim *et al.*, 1994).

By using the critical N model of Greenwood *et al.* (1990), crops can be managed to ensure that their N content is optimal and that grain yield is not limited by nitrogen availability. The critical nitrogen concentration is defined as the minimum nitrogen concentration in the crop required for maximum growth rate at any time. Sheehy *et al.* (1998a) used the model to predict that half the total nitrogen accumulated by the crop must be acquired by the time the crop has acquired one quarter of its biomass and that early nitrogen availability was important for achieving high yield. Using data describing the accumulation of above-ground biomass for a high-yielding new plant-type crop (Fig. 17.5a), the daily demand for nitrogen by a crop growing with critical N content was estimated using the Greenwood equation. The estimated daily rate of N acquisition by the crop reached a maximum of 0.38 g N m^{-2} day^{-1} and declined to zero thereafter (Fig. 17.5b). The daily acquisition rate of nitrogen by the panicle reached a maximum of

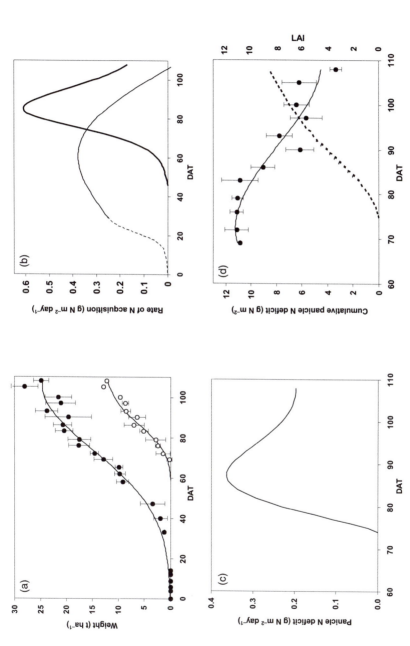

Fig. 17.5. (a) The dry weights of the total above-ground biomass (●) and the panicle (○) of a new plant type (IR65564-44-4-1) throughout the dry season (1998). The lines represent fitted curves (r^2 >0.97) and the data points have corresponding standard errors. (b) The daily rate of nitrogen acquisition of the panicle (——) and the above-ground biomass based on experimental data (- - -) and the Greenwood *et al.* (1990) model (——). (c) The daily nitrogen deficit of the panicle. (d) The accumulated nitrogen deficit of the panicle (- - -) and the decline in LAI (●) after panicle emergence; the line represents the fitted curve (r^2 >0.95) and the data points have corresponding standard errors. DAT = days after transplanting.

0.62 g N m^{-2} day^{-1} on day 80 and declined to 0.17 g N m^{-2} day^{-1} on day 105. The difference between the daily rates of nitrogen acquisition of the above-ground biomass and the panicle shows an increasing deficit with time (Fig. 17.5c). This deficit emphasizes the importance of the reservoir of nitrogen accumulated during the vegetative stage of development. The photosynthetic model above showed that an LAI ($L_N + A_1$) of 7.9 could be lost and canopy photosynthesis could decline by only 22%. It is interesting to note that in the experiment from which some of those data are drawn, the LAI lost during panicle development was 7.8. The total nitrogen deficit accumulated by the panicle was 8.6 g N m^{-2}, and if this were to be met by the nitrogen withdrawn from a senescent LAI of 7.8, the N withdrawn per unit LAI would be 1.1 g N m^{-2}. Sinclair and Sheehy (1999) used a value of 1 g N m^{-2} and, at this N content, an LAI of 8.6 would have to become senescent; and at an irradiance of 2070 μmol m^{-2} s^{-1}, canopy photosynthesis would fall by 40% of its maximum value. The cumulative nitrogen deficit of the panicle and the decline in LAI associated with the supply of nitrogen to compensate for the deficit are shown in Fig. 17.5d. It is crucial to understand that senescence is essential for supplying stored nutrients such as N for high yields. However, to ensure grain filling, somewhere between an LAI of 1.5 and 3.0 must remain photosynthetically active until maturity. In canopies of high-yielding crops, approximately 70% of the leaves must become senescent and transfer nitrogen to the grain, for a loss of about 25% of the maximum rate of canopy gross photosynthesis. The relatively lower intrinsic N content of a C_4 crop may mean that the nitrogen reservoir in its shoots would be inadequate to support the nitrogen demand of the developing grains. A C_4 rice might have a reduced sink size and grains with low protein content (Sheehy, 2000). The combination of nitrogen management and new cultivars possessing superior mechanical strength, high LAI and low specific leaf area is the way to achieve yields close to the limit of rice with conventional and redesigned photosynthetic pathways. In monsoon Asia, crops are often under-fertilized to prevent lodging, which remains a significant problem. As a result, prevention of lodging in post-Green Revolution rice requires research investment.

Global Climate Change

In a world of nearly doubled carbon dioxide concentration, a general warming of 1.5–4.5°C is expected in the latter half of this century (Mitchell *et al.*, 1990). Increased atmospheric CO_2 can increase yields with current cultivars only when there are no significant temperature increases (Baker *et al.*, 1992). In the absence of increased transpiration, plant temperatures will be higher and cross thresholds for damage more often. Rice spikelets show a sigmoid response curve to air temperature from 100% fertility at 33°C to 0% at 40°C (Horie *et al.*, 1997). Since 33°C air temperature does commonly occur in the tropics, rice crops are on the threshold of declining fertility with temperature. Thermal damage to rice spikelets results from particular combinations of temperature, radiation, wind speed and inadequate transpirational cooling. It is unclear how high-temperature sterility can be avoided, or whether there is variation in tolerance for high temperatures or avoidance by earlier time of day of flowering among rice plants from various climates. High temperatures can also reduce canopy photosynthesis, but this reduction can be offset by increased CO_2 (Long, 1999).

Reducing greenhouse gas emissions such as methane and nitrous oxide, and preserving surface water and groundwater quality are becoming increasingly important targets. Legislation may impose restrictions on straw burning, total N input–output budgets at the farm level, crop management practices related to nitrate accumulation and leaching, and pesticide use. As argued earlier, such legislation could lead to a slowdown or halt in the annual increase in rice yields.

Nitrous oxide emissions from rice fields occur as a result of nitrification–denitrification during periods of alternating wetting and drying. Emissions are usually small in irrigated rice systems with good water control and small to moderate inputs of fresh organic material (Bronson *et al.*, 1997a, b). If intensification of rice agriculture involves more crop diversification, N_2O emissions and NO_3^- leaching may become greater problems. Wetland rice has contributed to the increase in atmospheric methane (CH_4) concentration, which has more than doubled during the past 200 years and continues to rise at a rate of 1% year^{-1} (Neue, 1993). Other pollutants such as ozone can significantly reduce yields in rice (Maggs and Ashmore, 1998), but how significant this effect is away from urban sites requires investigation.

Furthermore, it should be remembered that what feeds people is production: area of crop multiplied by yield. During periods of adverse weather such as El Niño droughts (Cole *et al.*, 2000), farmers do not always plant crops and production falls.

Conclusions

To answer the question posed in the title (will yield barriers limit future rice production?) the simple answer is yes. Unless investment continues to support research associated with high-yielding crops and cropping systems, approaching ceiling yields will be impossible. That ceiling yields exist is not widely appreciated, possibly because it is difficult to determine precisely what those limits are for different ecosystems and crops. However, calculating the limits, albeit crudely, is crucial for setting realistic yield targets and designing elite germplasm with progressively higher yield potential to meet increases in demand for food. I worry that many are over-optimistic in their belief that yield will simply continue to rise to meet demand unchecked by any biophysical limits. I do not think it will do so unless substantial scientific advances are made in plant science, breeding and extension work. Many challenges lie ahead in trying to meet the growing demand for food especially in a world in which climate is changing rapidly. We must therefore make use of all scientific tools to ensure that those demands can be met (Lipton, 2000) in a sustainable and environmentally friendly manner.

Challenges

- To ensure that the current rate of increase in rice yield continues into the future. In the tropics, there are three major obstacles to yield increases: the first is the yield potential of indicas (8 t ha^{-1}), the second is the potential yield of C$_3$ rice (12.5 t ha^{-1}) and the third is the ceiling yield of C$_4$ rice (15 t ha^{-1}).
- To breed a higher yielding new rice plant type within the next 5 years. A new plant type will have to have an RCF of 2.7 g MJ^{-1}, 25% higher than what is currently achieved in the tropics and taking rice close to the limit for C$_3$ crops ($\approx$12.5 t ha^{-1}, 110-day duration) under current global climate conditions. This might be achieved by improving C$_3$ photosynthesis. High rates of leaf photosynthesis are often observed in leaves with low specific leaf area (i.e. less area per unit dry matter, a 'thicker' leaf). Plants that can maintain low specific leaf area and high leaf photosynthesis when part of a dense canopy would be desirable, assuming competition from weeds is negligible.
- Given that temperature is likely to increase as a result of global climate change, it is vital to invent rice that avoids high-temperature sterility. Introducing clock genes, making rice flower earlier in the day (thus avoiding higher temperatures) might do this. Another possible approach would be to discover and introduce genes conferring high temperature tolerance on the temperature-sensitive processes of anthesis.
- To invent a rice with a C$_4$ photosynthetic pathway. There is no reason to think that crop plants have evolved or have been selected to the best possible state. For example, the C$_4$ pathway has evolved relatively recently in response to a period of low atmospheric concentration of carbon dioxide (Sage and Monson, 1999). By chance, or for some other reason, the C$_4$ pathway has not evolved in rice (or related genera) even though it would appear to offer productivity advantages to a tropical crop. The C$_4$ pathway in irrigated rice would increase photosynthesis and its temperature optimum. Perhaps most importantly, in a future world of increased temperatures and CO$_2$, such a C$_4$ rice could produce an increased yield of grain using half the transpired water and a third less nitrogen than current elite C$_3$ indicas and japonicas. Such a plant could revolutionize rice farming, particularly in rainfed and more drought-prone ecosystems, by reducing risk, production costs and pollution. None the less, the nitrogen content

of a C_4 rice crop could limit sink size, yield and the protein content of the rice plant.

- Ensuring an adequate supply of water for irrigating rice. Competition for water between agriculture and the expanding urban and industrial complexes of Asia is growing. For high yields, it is vital that the transpirational demands of the crop are fully met and other forms of water loss are minimized or eliminated.
- To invent forms of fertilizer and methods and timings of application that greatly enhance recovery by minimizing losses (volatilization or percolation, etc.).

Acknowledgements

I would like to thank Rudy Rabbinge for setting the challenges in this chapter and A. Ferrer, J. Dionora, A. Elmido, G. Centeno and Bill Hardy for help in preparing this manuscript.

References

Austin, R.B., Bingham, J., Blackwell, R.D., Evans, L.T., Ford, M.A., Morgan, C.L. and Taylor, M. (1980) Genetic improvements in winter wheat yields since 1900 and associated physiological changes. *Journal of Agricultural Science* 94, 675–689.

Baker, J.T., Allen, L.H. Jr and Boote, K.J. (1992) Temperature effects on rice at elevated CO_2 concentration. *Journal of Experimental Botany* 43, 959–964.

Bell, M.A., Fisher, R.A., Byerlee, D. and Sayre, K. (1995) Genetic and agronomic contributions to yield gains: a case study for wheat. *Field Crops Research* 44, 55–65.

Bingham, J. and Austin, R.B. (1993) Achievements and limitations of wheat breeding in the United Kingdom. In: Buxton, D.R., Shibles, R., Forsberg, R.A., Blad, B.L., Say, K.H., Paulsen, G.M. and Wilson, R.F. (eds) *International Crop Science I*. Crop Science Society of America, Madison, Wisconsin, pp. 181–187.

Bronson, K.F., Neue, H.U., Singh, U. and Abao, E.B. Jr (1997a) Automated chamber measurements of methane and nitrous oxide flux in a flooded rice soil. I. Residue, nitrogen, and water management. *Soil Science Society of America Journal* 61, 981–987.

Bronson, K.F., Singh, U., Neue, H.U. and Abao, E.B. Jr (1997b) Automated chamber measurements of methane and nitrous oxide flux in a flooded rice soil. II. Fallow period emissions. *Soil Science Society of America Journal* 61, 988–993.

Cardenas, D.C., Briones, N.D., Francisco, H.A., Rola, A.C. and Pacardo, E.P. (1996) Socio-economic aspects of land conversion in Cavite. *Journal of Agricultural Economics and Development* 24, 1–2.

Cassman, K.G. (1994) *Breaking the Yield Barrier. Proceedings of a Workshop on Rice Yield Potential in Favorable Environments*. International Rice Research Institute, Los Baños.

Cassman, K.G., De Datta, S.K., Olk, D.C., Alcantara, J., Samson, M., Descalsota, J. and Dizon, M. (1995) Yield decline and the nitrogen economy of long-term experiments on continuous, irrigated rice systems in the tropics. In: Lal, R. and Stewart, B.A. (eds) *Soil Management: Experimental Basis for Sustainability and Environmental Quality.* CRC Press, Boca Raton, Florida, pp. 181–218.

Cassman, K.G., Dobermann, A., Sta Cruz, P.C., Gines, G.C., Samson, M.I., Descalsota, J.P., Alcantara, J.M., Dizon, M.A. and Olk, D.C. (1996) Soil organic matter and the indigenous nitrogen supply of intensive irrigated rice systems in the tropics. *Plant and Soil* 182, 267–278.

Cole, J.E., Dunbar, R.B., McClanahan, T.R. and Muthiga, N.A. (2000) Tropical pacific forcing of decadal SST variability in the western Indian ocean over the past two centuries. *Science* 287, 617–619.

Dobermann, A., Cassman, K.G., Mamaril, C.P. and Sheehy, J.E. (1998) Management of phosphorus, potassium, and sulfur in intensive, irrigated lowland rice. *Field Crops Research* 56, 113–138.

Duvick, D.N. and Cassman, K.G. (1999) Post-Green Revolution trends in yield potential of temperate maize in the North-Central United States. *Crop Science* 39, 1622–1630.

Dyson, T. (1999) World food trends and prospects to 2025. *Proceedings of the National Academy of Sciences USA* 96, 5929–5936.

Edwards, G. (1999) Tuning up crop photosynthesis. *Nature Biotechnology* 17, 22–23.

Evans, L.T. (1998) Processes, genes, and yield potential. In: Buxton, D.R., Shibles, R., Forsberg, R.A., Blad, B.L., Say, K.H., Paulsen, G.M. and Wilson, R.F. (eds) *International Crop Science I.* Crop Science Society of America, Madison, Wisconsin, pp. 687–696.

Evans, L.T. and Fischer, R.A. (1999) Yield potential: its definition, measurements and significance. *Crop Science* 39, 1544–1551.

FAO (2000) Food and Agriculture Organization of the United Nations. FAO Statistical Databases, http://apps.fao.org/

Fischer, R.A. (1993) Cereal breeding in developing countries: progress and prospects. In: Buxton, D.R., Shibles, R., Forsberg, R.A., Blad, B.L., Say, K.H., Paulsen, G.M. and Wilson, R.F. (eds) *International Crop Science I.* Crop Science Society of America, Madison, Wisconsin, pp. 201–209.

Geng, S. (1999) Potential impact of global environment change on world food production. In: Horie, T., Geng, S., Amano, T., Inamura, T. and Shiraiwa, T. (eds) *World Food Security and Crop Production Technologies for Tomorrow.* Kyoto, pp. 25–30.

Greenland, D.J. (1997) *The Sustainability of Rice Farming.* CAB International, Wallingford.

Greenwood, D.J., Lemaire, G., Gosse, G., Cruz, P., Draycott, A. and Neeteson, J.J. (1990) Decline in percentage N of C_3 and C_4 crops with increasing plant mass. *Annals of Botany* 66, 425–436.

Horie, T., Ohnishi, M., Angus, J.F., Lewin, L.G., Tsukaguchi, T. and Matano, T. (1997) Physiological characteristics of high yielding rice inferred from cross location experiments. *Field Crops Research* 52, 55–57.

Hossain, M. and Pingali, P.L. (1998) Rice research, technological progress, and impact on productivity and poverty: an overview. In: Pingali, P.L. and Hossain, M. (eds) *Impact of Rice Research. Proceedings of the International Conference on the Impact of Rice Research, Bangkok.*

Khush, G.S., Brar, D.S. and Bennett, J. (1998) Apomixis in rice and prospects for its use in heterosis breeding. In: Virmani, S.S., Siddiq, E.A. and Muralidharan, K. (eds) *Advances in Hybrid Rice Technology. Proceedings of the 3rd International Symposium on Hybrid Rice, IRRI, Manila*, pp. 297–309.

Kropff, M.J., Cassman, K.G., Peng, S., Matthews, R.B. and Setter, T.L. (1994) Quantitative understanding of yield potential. In: Cassman, K.G. (ed.) *Breaking the Yield Barrier. Proceedings of a Workshop on Rice Yield Potential in Favorable Environments*. International Rice Research Institute, Los Baños, pp. 21–38.

Ku, M.S.B., Agarie, S., Nomura, M., Fukayama, H., Tsuchida, H., Ono, K., Toki, S., Miyao, M. and Matsuoka, M. (1999) High-level expression of maize phosphoenolpyruvate carboxylase in transgenic rice plants. *Nature Biotechnology* 17, 76–80.

Lipton, M. (2000) *Reviving Global Poverty Reduction: What Role for Genetically Modified Plants?* Consultative Group on International Agricultural Research; the 1999 Sir John Crawford Memorial Lecture given on 28 October 1999, Washington, DC.

Long, S.P. (1999) Environmental response. In: Sage, R.F. and Monson, R.K. (eds) C_4 *Plant Biology*. Academic Press, London, pp. 215–249.

Loomis, R.S. and Amthor, J.S. (1999) Yield potential, plant assimilatory capacity, and metabolic efficiencies. *Crop Science* 39, 1584–1596.

MAFF (2000) Ministry of Agriculture, Fisheries and Food, UK. http://www.maff. gov.uk/

Maggs, R. and Ashmore, M.R. (1998) Growth and yield response of Pakistan rice (*Oryza sativa* L.) cultivars to O_3 and NO_2. *Environmental Pollution* 103, 159–170.

Makarim, A.K., Roechan, S. and Pw, P. (1994) Nitrogen requirement of irrigated rice at different growth stages. In: ten Berge, H.F.M., Wopereis, M.C.S. and Shin, J.C. (eds) *Nitrogen Economy of Irrigated Rice: Field and Simulation Studies. SARP Research Proceedings*, International Rice Research Institute, Los Baños, pp. 70–75.

Matthews, R.B., Kropff, M.J., Bachelet, D. and van Laar, H.H. (1995) *Modelling the Impact of Climate Change on Rice Production in Asia*. CAB International, Wallingford in association with the IRRI, Manila.

Meis, J. (2000) Iowa Crop Improvement Association, Ames, Iowa. http://www.agron.iastate.edu/icia/

Mitchell, J.F.B., Manabe, S., Meleshko V. and Tokioka, T. (1990) Equilibrium climate change and its implications for the future. In: Houghton, J.T., Jenkins, G.J. and Ephraums, J.J. (eds) *Climate Change: the IPCC Scientific Assessment*. Cambridge University Press, Cambridge, pp. 131–174.

Mitchell, P.L., Sheehy, J.E. and Woodward, F.I. (1998) Potential yields and the efficiency of radiation use in rice. *IRRI Discussion Paper Series No. 32*. Manila.

Monteith, J.L. (1965) Light distribution and photosynthesis in field crops. *Annals of Botany* 29, 17–37.

Nakano, H., Makino, A. and Mae, T. (1997) The effect of elevated partial pressures of CO_2 on the relationship between photosynthetic capacity and N content in rice leaves. *Plant Physiology* 115, 191–198.

Neue, H.U. (1993) Methane emission from rice fields. *BioScience* 43, 466–474.

Olk, D.C., Cassman, K.G., Simbahan, G., Sta Cruz, P.C., Abdulrachman, S., Nagarajan, R., Pham Sy, T. and Satawathananont, S. (1999) Interpreting fertilizer-use efficiency in relation to soil nutrient-supplying capacity, factor productivity, and agronomic efficiency. *Nutrient Cycling in Agroecosystems* 53, 35–41.

Peng, S., Cassman, K.G. and Kropff, M.J. (1995) Relationship between leaf photosynthesis and nitrogen content of field-grown rice in the tropics. *Crop Science* 35, 1627–1630.

PhilRice (1994) *Philippine Rice Statistics 1970–1996*. PhilRice Bureau of Agricultural Statistics.

Potrykus, I., Burkhardt, P.K., Datta, S.K., Futterer, J., Ghosh-Biswas, G.C., Kloti, A., Spangenberg, G. and Wunn, J. (1995) Gene technology for developing countries: genetic engineering of indica rice. In: Terzi, M. (ed.) *Plant Molecular and Cellular Biology*. Kluwer Academic Publishers, Dordrecht, pp. 253–262.

Sage, R.F. and Monson, R.K. (1999) C_4 *Plant Biology*. Academic Press, San Diego, California.

Schimel, D., Alves, D., Enting, I., Heimann, M., Joos, F., Raynaud, D., Wigley, T., Prather, M., Derwent, R., Enhalt, D., Fraser, P., Sanhueza, E., Zhou, X., Jonas, P., Charlson, R., Rodhe, H., Sadasivan, S., Shine, K.P., Fouquart, Y., Ramaswamy, V., Solomon, S., Srinivasan, J., Albritton, D., Isaksen, I., Lal, M. and Wuebbles, D. (1996) Radiative forcing of climate change. In: Houghton, J.T., Meira Filho, L.G., Callander, B.A., Harris, N., Kattenberg, A. and Maskell, K. (eds) *Climate Change 1995: The Science of Climate Change*. Contribution of Working Group I to the Second Assessment Report of the Intergovernmental Panel on Climate Change. Cambridge University Press, Cambridge, pp. 65–131.

Sheehy, J.E. (2000) Limits to yield for C_3 and C_4 rice: an agronomist's view. In: Sheehy, J.E., Mitchell, P.L. and Hardy, B. (2000) *Studies in Plant Science. Redesigning Rice Photosynthesis to Increase Yield*. Elsevier Science, Amsterdam B.V.

Sheehy, J.E. and Johnson, I.R. (1988) Physiological models of grass growth. In: Jones, M.B. and Lazenby, A. (eds) *The Grass Crop*. Chapman and Hall, New York.

Sheehy, J.E., Fishbeck, K.A. and Philips, D.A. (1980) Relationships between apparent nitrogen fixation and carbon exchange rate in alfalfa. *Crop Science* 20, 491–495.

Sheehy, J.E., Dionora, M.J.A., Mitchell, P.L., Peng, S., Cassman, K.G., Lemaire, G. and Williams, R.L. (1998a) Critical nitrogen concentrations: implications for high yielding rice (*Oryza sativa* L.) cultivars in the tropics. *Field Crops Research* 59, 31–41.

Sheehy, J.E., Mitchell, P.L., Tsukaguchi, T., Dionora, J., Ferrer, A. and Torres, R. (1998b) Breaking the yield barrier in rice: problems and solutions – does radiation use efficiency limit rice (*Oryza sativa* L.) yields in the tropics. In: Horie, T., Geng, S., Amano, T., Inamura, T. and Shiraiwa, T. (eds) *World Food Security and Crop Production Technologies for Tomorrow*. Kyoto, pp. 147–151.

Sheehy, J.E., Mitchell, P.L., Beerling, D.J., Tsukaguchi, T. and Woodward, F.I. (1998c) Temperature of rice spikelets: thermal damage and the concept of a thermal burden. *Agronomie* 18, 449–460.

Sheehy, J.E., Mitchell, P.L., Dionora, M.J., Tadashi, T., Peng, S.B. and Khush, G.S. (2000) Unlocking the yield barrier in rice through a nitrogen-led improvement in the radiation conversion factor. *Plant Production Science* 3, 372–374.

Sinclair, T.R. and Horie, T. (1989) Leaf nitrogen, photosynthesis and crop radiation-use efficiency: a review. *Crop Science* 29, 90–98.

Sinclair, T.R. and Sheehy, J.E. (1999) Erect leaves and photosynthesis in rice. *Science* 283, 1456–1457.

Thomas, H. (1994) Resource rejection by higher plants. In: Monteith, J.L., Scott, R.K.

and Unsworth, M.H. (eds) *Resource Capture by Crops*. Nottingham University Press, UK, pp. 375–385.

UNFPA (1999) The state of world population 1999, United Nations Population Fund, New York. http://www.unfpa.org/swp/1999.chapter2fig5.htm

USDA (2000) US Department of Agriculture, National Agricultural Statistics Service. http://www.usda.gov/nass/

Weei-Pirng, H., Ling-Yuan, S. and Ching-Huei, K. (1986) Senescence of rice leaves: XVI. Regulation by light. *Botanical Bulletin of Academia Sinica* 27, 163–174.

Ye, X., Al-Babili, S., Klöti, A., Zhang, J., Lucca, P., Beyer, P. and Potrykus, I. (2000) Engineering the provitamin A (beta-carotene) biosynthetic pathway into (carotenoid-free) rice endosperm. *Science* 287, 303–305.

Yoshida, S. (1973) Effects of CO_2 enrichment at different stages of panicle development on yield components and yield of rice (*Oryza sativa* L.). *Soil Science and Plant Nutrition* 19, 311–316.

Yoshida, S. (1981) *Fundamentals of Rice Crop Science*, International Rice Research Institute, Manila.

Yuan, L.P. (1994) Increasing yield potential in rice by exploitation of heterosis. In: Virmani, S.S. (ed.) *Hybrid Rice Technology*. Selected papers from the International Rice Research Conference, Manila, pp. 1–6.

Yuan, L.P (1998) Hybrid rice breeding in China. In: Virmani, S.S., Siddiq, E.A. and Muralidharan, K. (eds) *Advances in Hybrid Rice Technology. Proceedings of the 3rd International Symposium on Hybrid Rice*. International Rice Research Institute, Manila, pp. 27–33.

Ziska, L.H., Namuco, O., Moya, T. and Quilang, J. (1997) Growth and yield response of field-grown rice to increasing carbon dioxide and air temperature. *Agronomy Journal* 89, 45–53.

New Crops for the 21st Century

18

J. Janick

Center for New Crops and Plant Products, Purdue University, West Lafayette, IN 478907-1165, USA

Crop Resources

Vital to our agricultural systems is the choice of servant species to sustain us. The options are prodigious. Of an estimated 350,000 plant species in the world, about 80,000 are edible by humans. However, at present only about 150 species are actively cultivated, directly for human food or as feed for animals and, of these, 30 produce 95% of human calories and proteins (Table 18.1). About half of our food derives from only four plant species – rice (*Oryza sativa*), maize (*Zea mays*), wheat (*Triticum* spp.), and potato

Table 18.1. The 30 major food crops, 1995 (megatonnes of fresh harvested product). Comparisons between agronomic and horticultural groupings are not directly comparable because of the difference in dry matter content.

Food crop	Fresh harvested product (Mt)
Cereals	Wheat (554), rice (551), maize (515), barley (143), sorghum (54), oat (29), millet (27), rye (23)
Oilseeds and legumes	Soybean (126), cottonseed (58), coconut (47), rapeseed/canola (35), groundnut (29), sunflower (27)
Vegetables	Tomato (84), cabbage (46), watermelon (40), onion (37), bean (18)
Fruits	Banana/plantain (85), orange (57), grape (55), apple (50), mango (19)
Tubers	Potato (285), cassava (164), sweetpotato (136), yam (33)
Sugar crops	Sugarcane (1168), sugarbeet (265)

Table 18.2. Contribution of various commodity groups to world production of edible dry matter and protein, adapted from Allard (1999).

Commodity group	Total edible dry matter (%)	Total protein (%)	Main species
Cereals	69	55	Wheat, maize, rice, barley
Legumes	6	13	Soybean
Vegetable oil seeds	3	2	Rapeseed
Tubers and starch crops	8	5	Potato, cassava, sweetpotato
Sugar crops	4	0	Cane, beet
Vegetables	2	<1	Tomato, cabbage, onion
Fruit	<1	<1	Grape, apple, coconut, orange
Meats and products	6	13	Cattle, swine, poultry
Fish	<1	4	
Other	2	7	

(*Solanum tuberosum*). Similarly, only three species (cattle, swine and chicken) dominate animal food and, although many marine species are still gathered from the sea, this technology is changing with the rise of aquaculture. The contribution of different commodity groups to edible dry matter and protein is listed in Table 18.2, but it is important to note that the value of fruits and vegetables to human nutrition and health via nutrients, vitamins and minerals is ignored. While many plant species provide non-food uses such as fibres, timber, industrial chemicals and medicinal uses, even here only a few cultivated species dominate, such as pine (*Pinus* spp.), cotton (*Gossypium* spp.) and rubber (*Hevea brasiliensis*).

Practically all of the plants and animals that support civilization were selected and domesticated in prehistory by anonymous and unsung farmers and herders, not agricultural scientists. Despite the tremendous advance by the scientific revolution, the discovery or creation of new crops is a rare and unusual event. We are, in fact, dependent upon Stone Age (Neolithic) crops and animals. Have our forebears made, in fact, the best choices of servant species? Are the food species that we now depend upon sufficient and adequate for the future? Are we hostage to the solutions of the past or can we begin anew? One is awed by the conservatism of the human species seemingly held captive by the resource base of the past. One might intuitively expect, in light of increasing population pressure, that we would be expanding the number of species to sustain and nourish us. The opposite is true; diversity is decreasing with fewer and fewer species providing world sustenance. However, it is also true that the yield of the favourite species that support us increased enormously in the 20th century due to a combination of improvements in agricultural technology (irrigation, fertilizers,

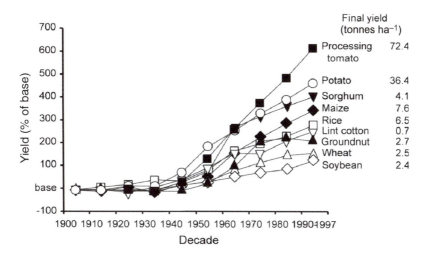

Fig. 18.1. Change in average US crop yield ha^{-1} by decade from 1900 (tomato from 1920, sorghum and soybean from 1930). Last decade until 1997. Source: Warren (1998).

shift to better land, weed, disease and pest control) and genetic gain. This is particularly true in the developed world economies such as the USA (Fig. 18.1). Note that most of this improvement has come about in the last 50 years.

Crop Classifications

Crops may be classified in many ways, many not mutually exclusive. Traditionally food crops are classified by a combination of use, affinity or edible portion such as cereals, legumes, oilseeds, tubers, root, starch crops, vegetables, fruits and sugar crops. Within these groupings crops are often referred to as 'new' or 'old' crops, but the distinction is unclear. The fact is that there are relatively few completely new crops despite the common use of the term. One's 'old' crop may be another's 'new' crop. The term new crop has been applied in various contexts to newly gathered, cultivated or domesticated wild plants; genetically improved domesticates; crops grown in new areas, new uses, or new markets; and genetically improved or altered domesticates (Small, 1999). To clarify the situation we group crops here on the basis of their economic and social importance.

Major traditional crops

These crops are cultivated worldwide in adapted areas with high economic value and are associated with high genetic input. They include many grains, forages, oilseeds, grain legumes, tuber crops, fruits, vegetables and sugar crops. All of these crops were once considered new in certain parts of the world. At the present time many traditional crops are being so genetically altered that some forms may be considered as new crops.

Traditional speciality crops

These are niche crops that, while economically important, have small markets than can be filled by relatively few growers. Included are many fruit, vegetable and spice crops. These speciality crops may easily become new crops when introduced into new areas.

Underutilized crops

These were once more widely grown but are now falling into disuse due to various agronomic, genetic, economic or cultural factors. In general, they are characterized by much less genetic improvement than the major crops, but they are being lost because they are less competitive. Examples include cereals such as oats (*Avena sativa*), pseudocereals such as buckwheat (*Fagopyrum esculentum*), oilseeds such as sesame (*Sesamum indicum*) and safflower (*Carthamus tinctorius*), and grain legumes such as chickpea (*Cicer arietinum*) and faba bean (*Vicia faba*). Rescued underutilized crops are the most likely source of new crops for many parts of the world.

Neglected crops

These are traditionally grown in their centres of origin where they are important for the subsistence of local communities and are maintained by socio-cultural preferences and traditional uses. These crops remain inadequately characterized and, until recently, have been largely ignored by agricultural researchers and genetic conservation. Yet they may represent our most valuable potential resource for the future. In some cases, their lack of exploitation is an historical accident. Examples include the primitive cereals einkorn (*Triticum boeoticum*), emmer (*T. turgidum*) and spelt (*T. spelta*); Andean root and tuber crops such as oca (*Oxalis tuberosa*), ullucu (*Ullucus tuberosus*) and mashua or añu (*Tropaeolum tuberosum*); and the millets such as species of *Panicum*, *Paspalum* and *Digitaria*.

Newly created crops

There are two types of newly created crops: (i) those recently developed from wild species whose virtues are newly discovered or formerly collected (wild-crafted species); and (ii) synthesized crops created from interspecific or intergeneric crosses. Together they represent only a handful of cultivated species and very few are included as new foods. Examples of new food domesticates include kiwifruit (*Actinidia deliciosa* and recently *A. chinensis*) developed in New Zealand from a species previously gathered in China. Kiwifruit is now an important world fruit. Totally new crops from wild species are mainly associated with industrial crops. These include *Limnanthes alba* (meadowfoam), a source of unique seed oils; *Simmondsia chinensis* (jojoba) considered useful as a potential substitute for sperm whale oil as a lubricant, but presently used in the cosmetic industry; and drug-bearing plants such as *Taxus brevifolia*, a source of Taxol®, a valuable anti-carcinogen. Newly synthesized crops include triticale (×*Triticosecale*), developed from an intergeneric cross between wheat and rye (*Secale cereale*) in the last half of the 20th century, and two crops derived from interspecific crosses in *Brassica*: harukan, a heading crucifer, and oo, a fodder rape.

Genetically modified (GM) crops

Crops drastically altered by intensive genetic techniques, either by traditional means (intra- and interspecific hybridization) or from transgene technology, are a relatively new class of crops. It now appears that they may have to be treated as new crops. Thus, high-oil maize which has been achieved by selection for large embryo size has unique properties for animal feeding. Other endosperm mutations of maize such as waxy and high amylose create potential new products.

Transgene technology refers to changes induced by recombinant DNA technology. Gene splicing is now an established technique and literally hundreds of transgenic crops are being evaluated in the USA. Since 1996, three transformed crops, soybean (*Glycine max*), cotton and maize, have had very high rates of farmer adoption. By 1999, herbicide-resistant soybean accounted for 57% of the crop area, *Bt* cotton 55%, and *Bt* maize 33%. There is consumer resistance in Europe where transgenic crops have been derisively termed 'Frankenfoods' by their detractors, and production is essentially banned. The short-term future of GM food is cloudy, but the long-term future is positive. GM food is unlikely to be a problem in Asia in view of the high need for increased production and acceptance of the technology by China.

New Crops and Crop Diversity

Historical review

The agricultural history of most areas, including the USA and European countries, chronicles the rise and fall of introduced crop species. Very few are native; most were obtained from other areas. Yet through a process of introduction, and trial-and-error, agriculture in advanced economies is increasingly dependent on relatively few species. In the USA almost 89% of annual row crops is planted to maize, soybean or wheat. Despite the fear of food shortages the developed world is in fact bedevilled by surpluses and low prices due to the tremendous increases in productivity brought about by modern agriculture (see Fig. 18.1). Cynics might complain that futurists have promised famine but they have not delivered. The economy of overproduction has led to low commodity prices which has resulted in price supports for the most common crops, as well as increasing research inputs for the major crops. This has led to a search for non-food crops to increase agricultural earnings in Europe. The picture in Asia, considered hopeless 33 years ago (*Famine, 1975* by William and Paul Paddock was published in 1967), has dramatically improved. Surpluses of some crops are now even a problem in some areas of India and China but the population increases expected in the next 50 years are troublesome. Africa is the problem continent. The per capita production of food has actually declined in sub-Saharan regions over the last 30 years. The present food crisis is a combination of low technology and political unrest, problems that may be solvable only in the political arena.

A discussion of agriculture on a long-term basis must be by necessity a discussion of technologies and species. In very primitive societies, crops are stable but in developed agricultural societies, the picture is one of change, with crops continually entering and leaving the system. Superimposed on this are the dynamic changes taking place within crop species due to genetic manipulation. Thus, crop diversity varies in different parts of the world. Two examples, New Zealand and the state of Indiana in the USA, demonstrate the key role of crop change in agriculture. Although each differs in terms of crop dynamics, both areas show a loss of diversity over the last 50 years.

New Zealand

This island nation, east of Australia, is an ideal laboratory to study the introduction, persistence and fate of new crops. Crop history since the European incursion in 1840 has been analysed by Halloy (1999) to investigate patterns of crop diversification. During the last 150 years total cultivated area steadily rose and then slowly levelled off. While the number of species increased dramatically, there has been a marked decrease in diversity in the last 50 years, as in the USA and Europe. Crops arise from nothing, persist or become abundant for some time, then in turn, disappear or become extinct, as a result of economic, historical and cultural events (Fig. 18.2).

The early dominance of potato and wheat was related to the need to feed the new population. The prolonged period of dominance of feed grains persisted to provide feed for draught animals (the horse). The rise and dominance of timber crops at present is explained by economic considerations.

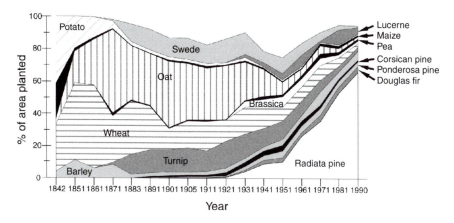

Fig. 18.2. Abundance dynamics (1842–1990) of the 14 highest ranking crops of New Zealand in 1990. Source: Halloy (1999).

The average lifespan of species (defined as covering more than 1% land area) is about 10 years, while the effective lifespan of crop cultivars is about 6 years, 4–10 for wheat for example, comparable to a depreciation of crop cultivars by 7% per year. The dynamic flow of crops was maintained by accessing new germplasm from abroad. Economic necessity has encouraged the introduction of many neglected and gathered fruit species such as the kiwifruit, feijoa (*Feijoa sellowiana*) and babaco (*Carica pentagona*). Kiwifruit was domesticated in New Zealand early in the last century and is one of the best examples of a new crop developed from a wild-crafted crop. It followed a typical boom and bust cycle. The bust was due to world competition which piggybacked on technology developed in New Zealand. However, the competitive position of New Zealand is returning with the success of *Actinidia chinensis*, a yellow-fleshed species marketed as Zespri™ Gold.

Indiana

This state in the Midwest of the USA is in the heart of the Eastern Corn Belt. Fertile soils and summer rains permit the culture of crops without irrigation although drought stress often curtails yields. Analysis of crop patterns in the past century indicates a remarkable shift away from diversity in the major crops (Fig. 18.3) as well as minor crops. The total cultivated area has hardly changed in 100 years and the area devoted to maize has remained steady at about 2.3 Mha, about 46% of total cropland, but the other major crops (wheat, oats, hay) have declined slowly over the century. Soybean, a new crop in the 20th century, has shown a remarkable rise in area and now constitutes 2.0 Mha and is responsible for taking over areas previously planted to other crops. Together maize and soybean now are 87% of total

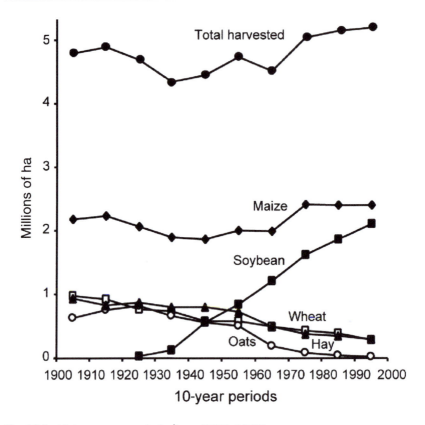

Fig. 18.3. Major crop areas in Indiana (1900–1999).

cultivated area in the state. Typically the two crops are rotated, but continuous corn is also grown.

Within each crop category there has been a dynamic change in cultivar. In maize, hybrid corn was quickly adopted after its introduction in the late 1930s, followed by a change from double-cross to single-cross hybrids. Since 1996 the adoption of genetically transformed (herbicide-resistant) soybean was very rapid, and reached about 70% of planted area in 1999, much higher than the rest of the country. The adoption of *Bt* maize was lower, about 20%, because the risk of European corn borer has not been severe. In Texas where the borer problem is severe, practically all the maize planted was *Bt*.

Why does crop diversity decrease?

What is the reason for this diminution of diversity in our food crops? One would expect that there would be many species among the 350,000 available to us, to have equal or better attributes than the ones we now consume. There are at least four explanations:

1. The crops chosen were not random ones but represent thousands of years of trial and error. They have survived because of unique attributes that cannot be denied. Wheat, an ancient crop of South-West Asia and the Middle East, is a complex interspecific hybrid, adapted to bright sunny weather and cool climates for early growth. Its unique properties are based on a combination of seed proteins (gliadin and glutenin) that makes possible varied bakery products such as bread, pastry and pasta. Maize, an ancient crop of Central America and a C_4 plant, is amazingly productive. Its nutritional deficiencies (low lysine) can be overcome by complementing animal rations with the protein of grain legumes. Rice, a native of China, is especially adapted to grow in wet climates. Potato, adapted to cool climates, has very high potential yields, well-balanced protein, and high versatility in processing and usage.

2. Our major crops have received an increasing amount of grower and scientific attention that has overcome or compensated for many of their deficiencies and increased their adaptation. A number of value-added processing procedures have been developed for crops, increasing their economic importance. Thus, maize grown principally as a source of poultry and swine feed is now widely used as a source of starch, a sweetener replacing cane or beet sugar, and as a source of ethanol. Soybean oil, and the resulting protein-rich cake, has become the basis for a huge chemurgic industry and the source of many products including tofu, soy milk, soy sauce and spun protein for meat substitutes.

3. Our important crops have become part of our social fabric as well as our religious and cultural heritage. We have become addicted to them and in various culinary forms they have become mainstays of our diet. A meal without rice is unacceptable in Asia (and much of South America), as is a meal without bread or potato in Europe or North America. It is very difficult to change basic food habits.

4. Finally, the political influence of the growers of basic food grains have encouraged governments to protect them with subsidies and to support them with basic research funds and marketing assistance. This is true in Japan where rice cultivation is even found in urban areas, an indefensible practice from an economic standpoint. It is true in the USA where maize growers had long been protected by subsidy, and in spite of a planned phased elimination, direct payments are still received by producers plus subsidies for ethanol.

New Crop Candidates

It is extremely difficult to predict the future and those foolish enough to try do so at their peril. However, because of the lag inherent in crop changes, speculation in this area is less risky.

Cereals and pseudocereals

There is no evidence that the main cereal grains (maize, wheat, rice) will be replaced or even be diminished in the near term and it is likely that they will continue to dominate world agriculture. These crops received the bulk of world research in terms of genetic improvement and cultural research and they have been, and are most likely to be further enriched by advances in genetics and biotechnology.

Maize

The improvement of maize in the USA will continue to be derived from two sources: genetic transformation and the selection of unique genotypes for feed and industrial purpose. Thus, high-oil maize will increase in its use in feed for cattle, poultry and swine. Maize has become an important industrial crop in the USA. A dramatic example is the increased use of 'corn sweeteners' (glucose and fructose), now the principal sweetener for carbonated soft drinks in the USA. The incorporation of various starch mutants such as *waxy* (amylopectin) and *high amylose* (amylomaize) shows high promise. The gluten from processed maize is also used in feeds. Incorporation of apomixis in maize allowing for the fixation of heterosis would permit poor farmers to save their own seed and could be a breakthrough for the developing world, particularly Africa (Savidan, 2000).

Rice

The staple cereal in Asia, rice, is increasing in popularity in Africa and South America. The advances in hybrid rice introduced by Chinese scientists and its adoption in India suggest that yield increases can still be expected. The recent development of rice high in β-carotene (a precursor to vitamin A) by molecular genetic techniques is an important breakthrough.

Wheat

It is likely that the demand for bread and durum wheat will increase on a world scale as globalization increases the demand for bread and pasta in traditional rice-eating cultures such as China and other Asian countries. In the 20th century the advances have come from traditional genetic alterations, particularly disease resistance and the short-stature introduced in the Green Revolution. Further advances in the 21st century will probably be from biotechnology.

Sorghum

Research in sorghum has lagged behind maize. Sorghum is not as nutritious a feed as maize because of problems of low digestibility which accounts for the lower world price. Sorghum, however, is better adapted to arid areas than maize.

Millets

The traditional drought-resistant grains of Africa have received little research support but millets may be the best potential new crops for arid regions. Pearl millet (*Pennisetum glaucum*) is the most likely to succeed and has received significant research attention. It is a high quality poultry feed, and forage types have been well accepted in the southern USA.

Ancient grains

Many ancient grains are neglected crops. In the USA there has been a limited interest in speciality wheats for the emerging health-food industry, such as einkorn (*Triticum boeoticum* AA), emmer (*T. turgicum* AABB) and spelt (*T. spelta* AABBDD), as well as tef (*Eragrostis tef*), the dominant cereal of Ethiopia. Demand for these grains is still small. One grower in the state of Washington has developed a small industry for the organic market for a cereal called Kamut®, a relative of durum wheat, but it is unlikely that this could be more than a niche crop.

Pseudograins

These include grain-like crops that are not members of the grass family. Buckwheat, amaranth (*Amaranthus* spp.) and quinoa (*Chenopodium quinoa*) have been targeted as new crops, but their adoption is still marginal. Buckwheat is a favourite grain for noodles (soba) in Japan, accepted in Eastern Europe (kasha), and has a small market in the USA for some products (buckwheat pancakes). Buckwheat is very nutritious and it has a chance to increase in value providing it receives research support and promotion. Amaranth, an ancient grain of the Incas, has received some research support and recent advances in non-shattering types suggest potential for this crop. Quinoa, another ancient Inca grain, has problems because the sapogenins in the seed make the grain bitter unless it is leached, but the unique properties of the starch suggest industrial use.

Legumes

The legume family (*Fabaceae*) is the second most important plant source of human and animal feed after the cereal grasses (*Poaceae*). Legumes can biologically fix nitrogen while seed and foliage of many legumes are rich in protein. Many legumes also contain organic chemicals in sufficient quantity to be economically useful as feedstocks or raw materials but these genetic resources have barely been exploited. Legumes grown for their seeds are known as grain legumes or pulse crops. Soybean, originally consumed as a vegetable in China, and introduced in the USA as a forage crop in the 18th century, has been transformed in the 20th century into the major US oilseed and protein feed source and is the best example of a successful new crop. Recently, soybean has become important in South America, particularly Brazil and Argentina. An edible form of soybean called edamame is popular in Japan and could be transformed into a new vegetable or snack food in the USA with proper promotion.

Other grain legumes such as chickpea, cowpea (*Vigna sinensis*), faba bean, lupins (*Lupinus* spp.), mungbean (*Phaseolus aureus*) and pigeonpea (*Cajanus cajan*) are examples of neglected legumes that could become new crops in various parts of the world. In the USA these crops are being considered to supply markets for ethnic foods or for exports.

Edible and industrial oils

The major world oil-producing crops include soybean, cotton (cottonseed oil is a by-product of cotton grown for fibre), rapeseed (*Brassica napus*), peanut (*Arachis hypogaea*), oil palm (*Elaeis guineensis*) producing both an edible oil (palm kernel) and an industrial oil (palm oil from the fruit), and olive (*Olea europaea*) producing an important edible oil in Mediterranean climates. Minor seed-oil crops include flax (*Linum usitatissimum*), jojoba (*Simmondsia chinesis*) and tung (*Aleuritis fordii*). A number of other crops have been suggested as new oilseeds for industrial use including cuphea (*Cuphea* spp.) to produce the medium-chain fatty acids such as lauric acid and capric acid; crambe (*Crambe abyssinica*), a rich source of erucic acid; meadowfoam (*Limnanthes* spp.) and the money plant (*Lunaria annua*) for long-chain fatty acids; lesquerella (*Lesquerella* spp.) for its hydroxy fatty acids; and vernonia (*Vernonia galamensis*) and Stokes' aster (*Stoksea laevis*) as a source of epoxy fatty acids. However, these new oilseed candidates are truly struggling. Only a few have achieved even minor importance. Jojoba, which produces a liquid wax similar to sperm whale oil, was considered to have tremendous promise as a source of transmission oil because of good lubricity at high temperatures. At present jojoba has found a niche market for use in the cosmetic industry and, at present, prices remain too high for other potential uses. Meadowfoam has achieved limited success as a new crop in Oregon.

Non-wood fibre

There are hundreds of textile fibre crops but the most important have been cotton, flax, ramie (*Boehmeria nivea*), jute (*Cochorus* spp.) and sisal (*Agave* spp.). Cotton, although under pressure from synthetic fibres, has made a resurgence worldwide and remains the most improved crop species producing lint plus oil and meal from seed. Flax as a fibre crop has almost disappeared although flax is still cultivated as an oilseed crop and is making a comeback in the UK. Fibres from the decorticated straw from oilseed cultivars (grown from linseed oil and meal) are used in paper making.

Recently hemp (*Cannabis sativa*) has received great interest in Canada, Germany and The Netherlands where concerted attempts have been made to convert agricultural lands used to produce food crops to non-food crops. Hemp is illegal in the USA because strains high in tetrahydrocannabinol (THC), known as marijuana, are considered a narcotic drug, yet illicit cultivation is wide scale.

In the USA, long-term efforts have been made to promote kenaf (*Hibiscus cannabinus*) as a source of fibre for newsprint and high quality paper. Much of kenaf research has been promoted by the US Department of Agriculture and recently collaborations have been made with private industry. However, despite enormous effort, commercial area in the USA is only about 1000 ha and commercial paper production has not expanded beyond the pilot-plant stage despite a number of uses for bedding and mulch.

The pods of milkweed (*Asclepias syriaca*), once gathered as a fill for life jackets during World War II, have been used as a non-allergenic fill to replace imported duck and goose down in comforters. A company in Nebraska has attempted to cultivate the crop but it has survived more on gathered than on cultivated milkweed.

Industrial crops

A number of important food and non-food crops principally used in a processed form are often classified as industrial crops and are increasingly an important part of world agriculture. These include the sugar crops – tropical sugarcane (*Saccharum officinalis*) and the temperature sugarbeet (*Beta vulgaris*); the beverage crops – coffee (*Coffea* spp.), tea (*Camellia sinensis*) and cacao (*Theobroma cacao*); *Hevea* rubber; and a number of oilseed crops. Many of these crops survive despite historical attempts at replacement. For example, the sugarbeet was a 19th century attempt to replace sugarcane brought about by the English blockade of France during the Napoleonic wars. Recently corn sweeteners have provided additional competition for cane sugar and replacement of cane could be accelerated with new developments in fermentation technology of waste products such as stover.

Hevea rubber, which created a spectacular boom with the introduction of pneumatic tyres, is a relatively new crop to the world although long-gathered in pre-Columbian eras for various uses. Guayule (*Parthenium argentatum*) has long been considered a domestic source for rubber in the USA but has never been commercialized. Recent problems with latex allergies for *Hevea* rubber products have reversed the declining interest in guayule.

Narcotic crops pose a special challenge and despite worldwide condemnation and approbation sadly continue to hold a place in world agriculture. The long-term fate of tobacco is in doubt, at least in the USA, because of increased anti-tobacco sentiment as a result of the social and medical costs, but no one expects tobacco to disappear on a world basis soon because the nicotine addiction is so strong. Marijuana, or THC-rich hemp, continues to bedevil all attempts to eliminate it, and its proponents have questioned the hypocrisy of banning and criminalizing it despite medical benefits (certainly questionable if smoked) in contrast to the support and acceptance of tobacco. Tremendous social problems on both the supply and the demand side of the issue have resulted from the cultivation of coca (*Erythroxylon coca*) in South America for the illegal manufacture of cocaine; poppy (*Papaver somniferum*) in Asia for illegal production of heroin; khat (*Catha edulis*) in the Middle East and North-East Africa; and betel (*Piper betle*) in Asia. The high demand and resultant high prices for these dangerous substances have made it extremely difficult to find socially acceptable alternatives for growers and as a result narco-criminals and narco-politics continue to devastate the producing and consuming communities.

Energy crops

There has been considerable interest in crop plants as a source of renewable fuels (biofuels). On-farm energy production is an old concept in the USA and early in the last century, 27 million draught animals were fuelled by crops produced on 34 Mha of grasslands, but this renewable energy source was replaced by petroleum and tractors. The rise in energy prices in the 1970s encouraged the concept of renewable energy from biomass and extensive research has been conducted on energy crops. These include maize stover (an estimated 200 million dry tonnes of residue remain available in the USA) and switchgrass (*Panicum virgatum*), fermentation of maize seed to make ethanol, and biodiesels from oilseeds. Ethanol production from maize has been subsidized in the USA, and fuel-blends with gasoline in a ratio of 1:10 are sold as gasahol in order to reduce greenhouse gas (GHG) for ecological reasons. A gas additive called methyl tertiary butyl ether (MTBE), used by the industry for the same purpose, has been shown to be hazardous to water supplies and if banned could greatly increase the demand for ethanol. Promising bio-lubricants have been achieved recently with canola oil (*Brassica napus* and *B. rapus*) blends (Johnson, 1999).

Maize stover has the potential to produce 50 Mt of sugar annually, doubling the current sweetener industry. The sugar could also be converted to ethanol. Realization would depend on sustainable harvest of corn stover, the improved conversion of stover to sugars, a commitment to reduce GHG emissions, and increases in the price of petroleum. The will to progress may be as much political as technical.

However, the consensus on biofuels had been that the economic incentives were lacking, that the technology was uncompetitive, and that energy crops were unlikely to be economic in the near term especially as new petroleum reserves were discovered. Furthermore, biofuels would have to compete with other alternative fuels such as wind, solar and photoelectric power. In Brazil, which must import practically all of its petroleum, the production of ethanol from cane sugar to produce energy for cars was carried out on a massive scale, but the programme was non-competitive when gasoline prices decreased. As recently as 1990, 90% of Brazil's automobiles were powered by fuel made by alcohol with generous subsidies buoying up sugar production. Support was dismantled and in 2000 only 1% of automobiles operated on sugar-based fuels. Part of the shift was encouraged by an increase in domestic sources based on offshore drilling in Brazil. A world oil price of US$25 per barrel has been considered too low for renewable sources to be competitive without subsidy unless the technology of conversion to energy (bio-refineries) increases sharply or the price of oil drastically increases. However, a spike in oil prices in the latter half of 2000 has reinvigorated the supporters of biofuels.

Fruits and nuts

The traditional fruit crops have had a close association with humans throughout recorded history. These include pome fruits such as apple (*Malus* × *domestica*) and pear (*Pyrus communis*); stone fruits such as peach (*Prunus persica*), cherry (*P. avium, P. cerasus*), plum (*Prunus* spp.), and apricot (*P. armeniaca*); the berry fruits such as strawberry (*Fragaria* × *ananassa*), brambles (*Rubus* spp.), currants and gooseberries (*Ribes* spp.); vine crops such as grape (*Vitis* spp.); subtropical fruits such as avocado (*Persea americana*), citrus (*Citrus* spp.), date (*Phoenix dactylifera*), fig (*Ficus carica*), persimmon (*Diospyros kaki*) and pomegranate (*Punica granatum*); and tropical fruits such as banana and plantains (*Musa* spp.), mango (*Mangifera indica*) and pineapple (*Ananas comosus*). The European incursion in the New World and the age of exploration resulted in an exchange of fruit species that is still going on. Mango, an ancient tropical fruit in India, is increasing in popularity in Europe and North America while some fruits that were once well known have almost disappeared such as the quince (*Cydonia oblonga*) and the medlar (*Mespilus germanica*). Fruits and nuts continue to be introduced and developed in various locations and a few have achieved true success. These include: (i) the successful introduction of pistachio (*Pistacia vera*) to California; (ii) kiwifruit, a vine crop formerly gathered in China, transformed into a new crop in New Zealand and spread to temperate regions worldwide; (iii) the oriental pear or nashi (*Pyrus pyrifolia*), now widely cultivated outside of Asia; (iv) *Vaccinium* species including blueberry and cranberry, domesticated in the 20th century in the USA but moving to Europe and South America; (v) macadamia (*Macademia ternifolia*), a nut native to Australia domesticated in Hawaii; and (vi) various cacti including *Opuntia* (cactus peak), and *Hylocereus* and *Cereus* (pitaya) native to South America but now cultivated in other parts of the world including Taiwan and Israel.

Vegetables

Vegetables are herbaceous crops, usually annuals but some perennials, that are consumed fresh or processed. Vegetables have an important place in the human diet and each culture has selected a great variety of them, but even here the trend is towards a loss of diversity. The important vegetable crops incluce various crucifers (*Brassica* species such as cabbage, cauliflower, broccoli, collards); solanaceous fruit crops such as tomato (*Lycopersicon esculentum*), aubergine (*Solanum melongena*), sweet and hot peppers (*Capsicum* spp.); cucurbits such as cucumber (*Cucumis sativus*) and various melons; various root and tuber crops such as radish (*Raphanus sativus*), potato, sweetpotato (*Ipomoea batatas*), cassava (*Manihot esculenta*); various leafy salad crops and greens such as lettuce (*Lactuca sativa*), endive (*Cichorium endivia*) and spinach (*Spinacia oleracea*); and edible beans (*Phaseolus communis*) and peas (*Pisum sativum*). Domestic vegetables continue to be exchanged. These include various cabbages from Asia (*Brassica campestris*, *B. pekinensis* and *B. chinensis*); solanaceous fruits from South America such as chayote (*Sechium edule*), husk tomato (*Physalis* spp.) and naranjilla (*Solanum quitoense*); and speciality crops from Europe such as arugula (*Eruca sativa*), artichoke (*Cynara scolymus*) and asparagus (*Asparagus officinalis*). In the USA and Europe, immigrants have introduced their ethnic foods, and as a result large supermarkets have greatly increased their offerings of new fruits and vegetables in order to attract consumer interest. There have been attempts to introduce the cultivation of these vegetables where feasible, and many have become successful speciality crops but often relatively few producers satisfy local demand. In the USA most of these crops can be easily produced in California and Florida.

Parallel to this development is the genetic alteration of standard crops that have created changes that transform traditional crops to 'new' crops. Successful examples include supersweet corn with replacement of *shrunken 2* for the *su* (*sugary*) gene; sugar snap peas with edible sweet, thick pods; the slow-ripening tomato using heterozygotes at the *rin* and *nor* locus; and the hybrid sweet peppers with brightly coloured fruit.

Aromatics, culinary herbs and medicinals

There has been continual interest in the aromatic and pungent properties of plants from the beginning of human history. Biblical sources give evidence for the ancient spice trade and indicate a very early interaction between the Mediterranean cultures and the Orient. Spices became the source of great wealth and the spice-rich Orient was the impetus for the Age of Exploration. While the relative economic importance of spices is decreasing, these crops are still silent partners to cooks the world over. The medicinal value of herbs was recognized early and most modern medicines, although now chemically synthesized, were originally derived from plant sources. The current interest in alternative medicines has sponsored a revival of herbal medicine. The growth and culture of these ancient plants provide a continual source of new crops. For example, the market for Asian ginseng (*Panax ginseng*), highly prized in Asian cultures, has long encouraged the collection of American ginseng (*Panax quinquefolius*) for export to Asia. This species is now cultivated in both Canada and the USA but the industry lately has been threatened by low prices due to Chinese production. Other popular herbal medicinals include species of *Echinacea*, goldenseal (*Hydrasius canadensis*), St John's wort (*Hypericum* spp.) and saw palmetto (*Serenoa serrulata*). Although the botanicals industry is plagued by wildly fluctuating prices, the demand for herbal supplements is increasing with probably the greatest consumption in Germany. Some new anti-cancer plants such as *Taxus brevifolia* and *Vinca minor* have been introduced as new crops but the cultivation of these crops is presently in very few hands.

Ornamentals

The ornamental industry is characterized by great diversity and there are more ornamental species cultivated than all other agricultural and horticultural crops combined. Furthermore, novelty is an important attribute of ornamentals. Nevertheless, standard ornamental crops continue to constitute an important part of the market, but the trend to new offerings is increasing. Many of these new offerings for greenhouse-produced crops, however, are not really new but have been in the trade for a century. Sources of new ornamentals include minor outdoor-growth groups, new cultivars of ornamental field-grown plants, woody and landscaping plants, ornamental cultivars of field crops, plants found in botanical gardens, and wild plants in their native habitat (Halevy, 1999).

The Challenge: a Strategy for New Crop Development

A legitimate case can be made for expanding crop diversity and for reversing the trend away from diversity in many parts of world agriculture (Janick *et al.*, 1996). New crop advocates suggest that successful new introductions offer alternative means of increasing farm income by diversifying products, hedging risks, expanding markets, increasing exports, decreasing imports, improving human and livestock diets, and creating new industries based on renewable agricultural resources. Diversification could spur economic development in rural areas by creating local, rural-based industries such as processing and packaging, and by providing general economic stability. Furthermore, an expansion of alternative crops could serve the strategic interests of nations by providing domestic sources for imported materials and by providing substitutes for petroleum-based products. Diversification would also serve as a form of world food security and would make agronomic sense because reliance on few species poses special hazards and risks due to biotic hazards. The southern maize blight epidemic of 1973 arose because the common male sterile (T) cytoplasm of practically all hybrids grown in the USA was susceptible to an outbreak of a new strain of *Helminthosporium maydis*, a fungal pathogen resulting in a billion dollar loss in a single year. Finally, the use of new species is also important for potential sources of new products of industrial compounds (meadowfoam), new foods (kiwifruit) and new medicinals (Taxol® from *Taxus*). There are, of course, extremely successful examples of new food crops developed from underutilized species of which soybean and canola (low erucic acid, low glucosinolate rapeseed) are the best examples. Soybean has contributed more than $US500 billion to the US economy from 1925 to 1985 and canola has become almost a billion dollar annual crop for Canada by virtue of its healthfulness as a cooking oil based on a significant fraction of long-chain monoenoic fatty acids.

The long time required for the genetic improvement of wild species, and the high risk involved, makes it unlikely that a rescreening of wild germplasm would be a profitable activity for uncovering new food crops. Only in industrial or medicinal crops, where a focused screening can uncover new and valuable molecules, would a search through wild species make sense. Problems in this area relate to tensions between the countries where these plants are found and collectors who attempt to exploit these resources for cultivation in the developed world. Because wild medicinal plants cannot be patented, drug companies had little incentive to conduct botanical explorations, although the increased interest in new botanicals may spur development in this area. However, it seems clear that this effort will require cooperation between the public and private sector and an equitable arrangement between all the parties concerned.

The reinvestigation of neglected and underutilized crops presents a better opportunity for food crops. Recent efforts with pearl millet (*Pennisetum glaucum*) in the USA suggest that a number of grains could have wider appeal, especially for special situations such as arid areas or for double cropping in short seasons. Furthermore, the globalization of our economy has increased interest in ethnic foods, opening up the market for new products.

Many neglected and underutilized crops are locally well adapted and constitute an important part of the local diet, culture and economy; require relatively low inputs; and contribute to high agricultural sustainability. However, traditional agricultural research in developed countries has hitherto paid little attention to, or ignored these crops. Consequently they have attracted little research funding despite the fact that they are adapted to a wide range of growing conditions, contribute to food security, especially under stress conditions, and are important for nutritional well-balanced diets. Although these traditional crops are often low yielding and cannot compete economically with improved cultivars of major crops, many of these crop species have the potential to become economically viable. A major factor hampering the development of these traditional crops is the lack of genetic improvement and, at times, a narrow genetic diversity for important agronomic traits. Further constraints are the lack of knowledge on the taxonomy, reproductive biology, and the genetics of agronomic and quality traits. However, because these crops represent the greatest resource for meeting new food demands in this century, research in this area would appear to be a logical role for public centre research. The development of the Consultative Group on International Agricultural Research (CGIAR) Centers and other associated groups have carried out research efforts in this area. This includes the International Potato Center (CIP) which supports effort in tuber crops, the International Crops Research Institute for the Semi-arid Tropics (ICRISAT) and the International Centre for Agricultural Research in Dry Areas (ICARDA) which carry out research on crops of dry areas and the semiarid tropics such as sorghum, pearl millet, pigeonpea, chickpea and lentils; and the Asian Vegetable Research and Development Center (AVRDC) which carries out work on tropical vegetable improvement. Unfortunately, funding for publicly supported research, both nationally and internationally is no longer increasing and, in many cases is declining at the same time that the cost of doing research is soaring. This has prevented the development of a serious, long-term, world strategic plan for this type of research effort. With a shortage of funds, international research efforts had in the past emphasized those major food crops that lead to food security: rice, wheat, maize, sorghum, banana and plantain.

At the present time there are two competing strategies for meeting the food needs of the future. One is to increase food diversity by exploiting the potential in underutilized and neglected crops. The bottleneck is the need for genetic improvement, which requires a long-term effort. Unfortunately there are no financial incentives in either the public or private sector to accomplish this feat. Only emphasis on world cooperation will be able to maximize this effort.

The other competing strategy is to seek further improvement of our present major crops emphasizing the new technology of molecular biology now fortified by genomics. For example, it has been successfully demonstrated that advances in biotechnology make it possible to engineer fatty acid modification. Thus, it is possible to convert the fatty acids of adapted crops such as soybean and canola (rapeseed) to mimic the fatty acids of other lower yielding oilseeds. Clearly the present incentive to do this is powerful and present protection of intellectual property rights through patents will encourage the private sector to pursue this goal. However, it is unclear at present whether the molecular–genetic approach or the new crop strategy will prevail for oil crop improvement.

The present success of genetically transformed crops in the USA (*Bt* maize and cotton, and herbicide-resistant soybean) provides an incentive for this approach. However, because of the enormous expense of this endeavour, the multinational research companies are reluctant to move outside of any but the most important crops. Thus, the trend towards reducing genetic diversity in agriculture is constantly being reinforced. Although the recent controversy over genetically modified foods has also put a damper on enthusiasm for this technology in Europe, research continues unabated. Because of the decision of multinational research organizations to emphasize grower-adapted technology, the consumer has not seen the virtue of this technology. However, this may change as the research emphasis may turn to the biotechnology to introduce value-added characters apparent to the consumer (such as quality or increased nutrition) as well as to other exciting uses such as plant-derived vaccines and pharmaceuticals. Vociferous critics have suggested that transgene technology will go the way of nuclear energy, but supporters continue to be optimistic for the long term.

The coming controversy will be to decide which strategy leads to a more productive and sustainable agriculture. It should not be overlooked that molecular biology may also contribute to the genetic improvement of underutilized and neglected species. Because many of these crops are not inherently productive, traditional plant breeding is still essential. All avenues need to be pursued. The molecular biological approach cannot be ignored because the tide of history is on its side. A way must be found to pursue both options. The only way to do this is to foster real cooperation between the public and private sector, between national and international research organizations, and among universities and researchers. The challenge of increasing food resources to meet a doubling of population before the end of the century depends on such an approach.

References

Allad, R.W. (1999) *Principles of Plant Breeding*, 2nd edn. John Wiley & Sons, New York.

Halevy, A.H. (1999) New flower crops. In: Janick, J. (ed.) *Perspectives on New Crops and New Uses*. American Society for Horticultural Science Press, Alexandria, Virginia, pp. 407–409

Halloy, S.R.P. (1999) The dynamic contribution of new crops to the agricultural economy: is it predictable? In: Janick, J. (ed.) *Perspectives on New Crops and New Uses*. American Society for Horticultural Science Press, Alexandria, Virginia, pp. 53–59.

Janick, J. (1999) The search for new food resources. *Plant Biotechnology* 16, 27–32.

Janick, J., Blase, M.B., Johnson, D.L., Jolliff, G.D. and Myers, R.L. (1996) *Diversifying*

US Crop Production. CAST Issue Paper 6, Council of Agricultural Science and Technology, Ames, Iowa.

Johnson, D.L. (1999) High performance 4-cycle lubricants from canola. In: Janick, J. (ed.) *Perspectives on New Crops and New Uses.* American Society for Horticultural Science Press, Alexandria, Virginia, pp. 247–250.

Savidan, Y. (2000) Apomixis: genetics and breeding. *Horticultural Reviews* 18, 13–86.

Small, E. (1999) New crops for Canadian agriculture. In: Janick, J. (ed.) *Perspectives on New Crops and New Uses.* American Society for Horticultural Science Press, Alexandria, Virginia, pp. 15–52.

Warren, G.F. (1998) Spectacular increase in crop yields in the United States in the twentieth century. *Weed Technology* 12, 752–760.

Plant Biotechnology: Methods, Goals and Achievements

19

U. Sonnewald[1] and K. Herbers[2]

[1]*Institut für Pflanzengenetik und Kulturpflanzenforschung, Leibnitz-Institut Corrensstrasse 3, D-06466 Gatersleben, Germany;* [2]*SunGene GmbH & CoKGaA, Corrensstrasse 3, D-06466 Gatersleben, Germany*

Introduction

Tremendous progress in plant molecular biology over the last two decades has opened up ample opportunities to improve crop plants in a way not foreseen a few years ago. Starting with the first transgenic tobacco plant in 1983, plant transformation has become a more or less routine technology for most crop plants. The invention of marker genes conferring antibiotic or herbicide resistance allowed for the efficient selection of transformed cells and in the case of herbicide resistance, even the creation of the first commercially relevant plants. Now the first transgenic plants engineered to be herbicide or insect resistant are outcompeting conventional crops and have started to revolutionize pest-management strategies leading to a major rethinking of the chemical industry. For the sake of simplicity the first generation of transgenic crop plants is characterized by the introduction of single-gene traits. This includes either the expression of additional cellular activities or the suppression of endogenous functions. Suppression of endogenous functions has been achieved mainly by two strategies, homology-dependent gene silencing or expression of antisense transcripts. In addition, limited success has been reported by using the ribozyme technology. To alter more complex traits comprehensive knowledge concerning the metabolic pathways, the underlying regulatory networks and the availability of respective genes are prerequisites. Based on worldwide efforts to study genome function almost any gene of interest is or will soon be available, making complex pathway modulations feasible. Thus, identification of gene function to identify suitable candidate genes will be the major challenge of the future. Main objectives of plant biotechnology are attempts to engineer metabolic pathways for the production of tailor-made plant polymers or low molecular weight compounds, increased resistance towards biotic and abiotic stresses, improved food quality, and the production of novel polypeptides for pharmaceutical or technical use. For predictable manipulations cis-regulatory elements and sub-cellular targeting sequences to direct foreign gene products towards desired tissues and sub-cellular compartments are essential pre-requisites. In this chapter we will summarize available tools to create transgenic plants and give representative examples for the use of transgenic plants in biotechnology.

Plant Transformation

Gene transfer into plants requires the availability of suitable gene delivery, selection and regeneration methods. The method of choice to engineer desirable genes into plants is *Agrobacterium*-mediated gene transfer. The current knowledge concerning transfer of DNA from *Agrobacterium tumefaciens* into plant cells as well as the history of *A. tumefaciens* has recently been summarized by Zupan *et al.* (2000) and will not be discussed in detail here. Until recently it was believed that *Agrobacterium* transformation would be limited to dicotyledonous plants. Therefore, direct gene transfer systems have been developed allowing efficient transformation of cereals (summarized in Christou, 1996). Direct gene transfer methods include polyethylene glycol (PEG)-mediated DNA uptake into protoplasts, electroporation of cells or protoplasts, particle bombardment of cells or tissues, and microinjection of individual cells or protoplasts. The latter technology has recently been extended for transient plastome transformation by direct injection of plasmid DNA into chloroplasts using a galinstan expansion femtosyringe (Knoblauch *et al.*, 1999). Plastome transformation has a number of advantages compared with nuclear transformation. These include higher expression levels, no gene silencing, no position effect due to homologous recombination, the possibility to express polycistronic RNAs, and no pollen transmission of foreign DNA (for recent review see Heifetz, 2000). Compared with *Agrobacterium*-mediated gene transfer, direct gene transfer methods possess several disadvantages: transfer of DNA is usually limited to relatively small DNA fragments, integration of a high number of gene copies frequently occurs, and many of the gene copies are subject to rearrangements. To overcome these limitations protocols for *Agrobacterium*-mediated gene transfer were successfully developed for rice (Hiei *et al.*, 1994), maize (Ishida *et al.*, 1996) and wheat (Cheng *et al.*, 1997). Thus, the use of *Agrobacterium*-mediated gene transfer is no longer restricted to dicots.

   Due to the relatively rare event of transformation, it is necessary to select the transformed cell harbouring the desired transgene. Therefore, the gene of interest is usually linked to a selectable marker gene. The most widely used marker genes confer antibiotic or herbicide resistance. Although efficient for the selection in tissue culture both marker systems possess undesirable characteristics. Due to consumer concerns the spread of antibiotic resistance genes should be limited and an increasing number of genetically modified plants resistant to herbicides will ultimately lead to the accelerated appearance of resistant weeds rendering the use of the respective chemical inefficient. Thus, development of alternative selection systems is required. To this end a selection system based on the non-toxic sugars, xylose and mannose, has been established (Fig. 19.1a). Both sugars cannot be metabolized unless the converting enzymes, xylose isomerase and mannose-6-phosphate isomerase, respectively, have been transformed into the plant cell (Bojsen, 1993; Haldrup *et al.*, 1998). Recent findings of xylose isomerase genes in plants (Kaneko *et al.*, 1998), however, question the universal use of the xylose selection system in plants. Following a similar idea a selection marker based on detoxification of 2-deoxyglucose has been developed (Sonnewald and Ebneth, 1997). 2-Deoxyglucose is a glucose analogue which, following hexokinase-mediated phosphorylation to 2-deoxyglucose-6-phosphate, severely impairs plant growth due to multiple effects on metabolism (Fig. 19.1b). Detoxification is achieved by overexpression of a specific yeast-derived 2-deoxyglucose-6-phosphate phosphatase. Another approach is the use of visible reporter systems such as expression of green-fluorescent protein (Sheen *et al.*, 1995) or enhanced anthocyanin biosynthesis due to over-expression of the maize *Lc* transcription activator (Goldsbrough *et al.*, 1996). Disadvantages of visible marker systems are that a large number of putative transformands have to be analysed for marker gene expression and, in the case of altered pigment formation, regenerated plants are phenotypically different from control plants. Recently, inducible expression of the isopentenyl transferase gene from *A. tumefaciens* has been utilized to select transgenic tobacco and lettuce plants based on the ability of transformed cells to develop shoots in the absence of cytokinin (Kunkel *et al.*, 1999). Although development of alternative selection systems will be important to meet consumers expectations and to allow multiple transformation of elite lines, the ultimate goal is the production of marker-free transgenic crops. This can be achieved by either co-transformation and subsequent outcrossing of selectable marker genes or by the use of site-specific excision systems allowing the removal of unwanted marker genes after the selection procedure (Sugita *et al.*, 2000).

Transgene expression

Depending on the specific question, the introduced foreign gene must be expressed in a predictable and desired manner. Nevertheless, the most commonly used promoter up to now is the constitutive 35S cauliflower mosaic virus (CaMV) promoter (Franck *et al.*, 1980). This promoter has been successfully used to engineer herbicide tolerant and pathogen resistant plants, but is of limited use to study metabolic pathways. In the past, numerous promoters with different expression patterns in transgenic plants have been

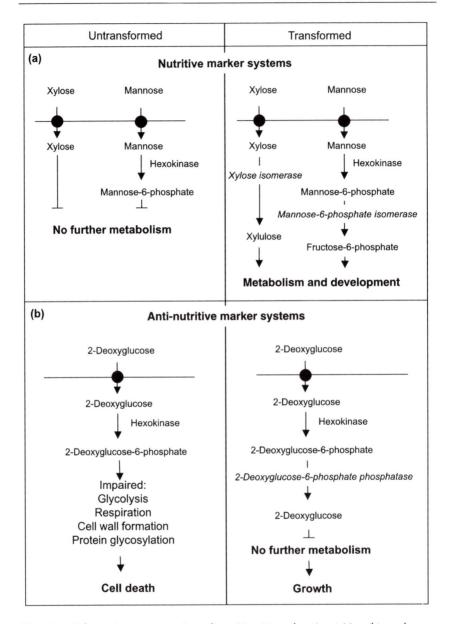

Fig. 19.1. Schematic representation of nutritive (a) and anti-nutritive (b) marker systems developed for plant transformation.

described. To date organ-specific promoters have been reported for different storage sink tissues like seeds, potato tubers and fruits (for review see Edwards and Coruzzi, 1990) and metabolic sink tissues such as meristems (Ito *et al.*, 1994). In addition, leaf expression has been studied in great detail. The most extensively analysed leaf-specific genes are the genes encoding the small subunit of ribulose-1,5-bisphosphate carboxylase-oxygenase (Fluhr *et al.*, 1986), the chlorophyll a/b binding protein (Simpson *et al.*, 1985) and the 10 kDa protein of the oxygen-evolving complex of photosystem II (Stockhaus *et al.*, 1989). Promoters of all three genes confer light-inducible expression in chloroplast-containing cells, i.e. leaf mesophyll cells, guard cells and companion cells (Edwards and Coruzzi, 1990). The list of available promoter sequences is continuously growing including regulatory sequences allowing guard cell-specific (Müller-Röber *et al.*, 1994), mesophyll-specific (Stockhaus *et al.*, 1994), epidermis cell-specific (Mandaci and Dobres, 1997), phloem-specific (Schmülling *et al.*, 1989) and anther-specific (Mariani *et al.*, 1990) expression of reporter genes. Most of the promoters currently used have been isolated following a detailed expression analysis of cDNA clones isolated. To accelerate promoter identification, tagging strategies for isolation of genomic sequences directing transgene expression have been followed in *Arabidopsis*, tobacco, potato (Lindsey *et al.*, 1993) and rice (Jeon *et al.*, 2000). To this end promoterless beta-glucuronidase (GUS) reporter gene constructs were randomly introduced into the plant genome via *Agrobacterium*-mediated gene transfer. This allowed the random fusion of the GUS gene with regulatory sequences of native genes leading to expression of the fusion gene in a number of different cell types. Following histochemical characterization of GUS expression, desired genomic fragments have to be cloned and their expression verified in transgenic plants. Although promoters conferring cell- and tissue-specific expression are versatile tools for biotechnological application, chemical regulation of transgene expression offers the opportunity to switch synthesis of gene products on and off. This feature is especially important if toxic substances or pharmaceutical products are to be synthesized. Various chemical-inducible systems based on activation or inactivation of target genes have been described (for recent review see Zuo and Chua, 2000). In plants several plant promoters exist which are inducible by endogenous signal metabolites. Best studied cases are abscisic acid- (Marcotte *et al.*, 1989), ethylene- (Ohme-Takagi and Shinishi, 1995) and salicylic acid- (Uknes *et al.*, 1993) inducible genes. The use of these promoters is limited, since: (i) the naturally occurring levels of signal metabolites and the responsiveness of the target tissue/cell may vary; (ii) the signal metabolites effect more than the expression of the gene of interest; and (iii) the inducible genes described so far are under complex control including environmental and developmental factors. Therefore, synthetic promoters which do not respond to naturally occurring plant products have been engineered. These include chimeric promoter systems which allow the activation

(Gatz *et al.*, 1991; Wilde *et al.*, 1992; Caddick *et al.*, 1998; Martinez *et al.*, 1999) or repression (Wilde *et al.*, 1994) of gene expression.

Sub-cellular targeting of foreign proteins

Specific metabolic functions of eukaryotic cells are organized in specialized sub-cellular compartments. As a consequence targeting of foreign proteins into desired sub-cellular compartments might be required, if metabolic processes are to be manipulated. In addition production of foreign proteins may require sequestration of the newly synthesized polypeptide in desired compartments stabilizing the protein and allowing its accumulation. The transport of polypeptides to their appropriate destination depends on the presence of specific targeting signals (Fig. 19.2). Secreted or vacuolar proteins are synthesized as pre-proteins with an N-terminal hydrophobic signal peptide, allowing the entry of the nascent polypeptide into the lumen of the endoplasmic reticulum (Vitale and Denecke, 1999). Whereas secretion of proteins in higher plants does not need further information, vacuolar targeting requires additional targeting signals (for review see Nakamura and Matsuoka, 1993). Nuclear encoded mitochondrial and plastidic polypeptides are synthesized as precursor proteins in the cytosol containing N-terminal transit peptides. Fusion of mitochondrial (Boutry *et al.*, 1987) and plastidic (Van den Broeck *et al.*, 1985) transit peptides to foreign proteins enabled the targeting of the polypeptides to the respective organelles. Fusion experiments revealed that the C-terminal tripeptide SKL is necessary and sufficient for targeting of a reporter gene product (chloramphenicol acetyltransferase) to plant peroxisomes (Banjoko and Trelease, 1995). Targeting of nuclear encoded polypeptides into the cytosol does not require sorting information.

Goals of Plant Biotechnology

In general goals of plant biotechnology are not much different from classical breeding goals. They can be divided into attempts to optimize input and output traits. Input traits refer to increased resistance towards abiotic and biotic stresses, strategies to increase crop yield and to improve post-harvest characteristics. In addition, transgenic plants have been modified to become resistant to broad-spectrum herbicides. Although controversially discussed, herbicide-resistant crops allow the use of environmentally and toxicologically acceptable herbicides, not possible otherwise. Attempts to improve output traits include production of foreign proteins for pharmaceutical and technical use, production of endogenous or novel polymers for food and non-food applications as well as the synthesis of low molecular weight compounds including vitamins, essential amino acids and pharmaceutically relevant secondary plant products.

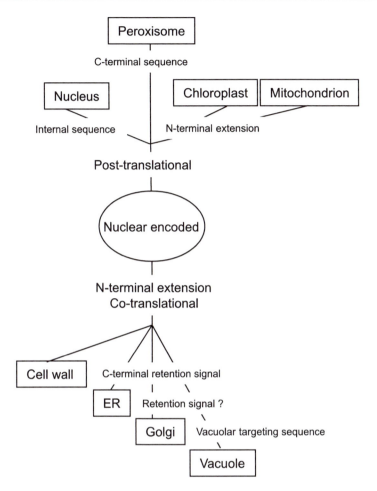

Fig. 19.2. Intracellular protein transport of nuclear encoded polypeptides in plant cells.

Improved stress resistance

Under natural growth conditions plants continuously experience changing environmental challenges (Fig. 19.3). Due to their immobility they are forced to develop strategies to cope with these stresses. Amongst others drought, salinity and freezing are the main limitations for plant productivity. In all three cases, water uptake of cells is impaired which leads to osmotic stress. To withstand these conditions, plant cells have evolved a variety of defence strategies including the production of osmoprotectants, the strengthening of reactive oxygen species (ROS) detoxification systems, and the alteration of membrane lipids. Osmoprotectants, or compatible osmolytes, belong to a

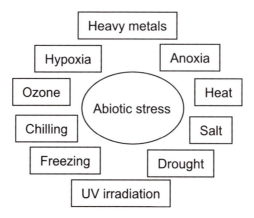

Fig. 19.3. Different environmental stress factors leading to abiotic stress.

diverse group of substances including ions, proteins, amino acids, sugars, polyols and polyamines (summarized in Bohnert and Jensen, 1996). Commonly abiotic stress leads to injury of membranes perturbing membrane-bound processes including photosynthesis. As a consequence ROS accumulate, resulting in further membrane and macromolecule damage. The main approaches to engineering transgenic plants for improved stress tolerance include synthesis of osmoprotectants, increased content of desaturated fatty acids in membranes and improved scavenging of ROS (summarized in Bohnert and Sheveleva, 1998; Holmberg and Bülow, 1998). In addition, attempts are being made to understand the signal transduction pathways underlying stress responses of model plants to engineer crops for a better performance under stress conditions (Warren, 1998). To this end mutant screens are developed allowing selection of stress-tolerant plants. One example is the *eskimo1* mutation which enables *Arabidopsis thaliana* plants to survive freezing conditions without prior cold acclimation (Xin and Browse, 1998). Another example is the overexpression of the transcription factor CBF1 which leads to the coordinate expression of a whole group of cold-induced genes and enhanced freezing tolerance in transgenic *Arabidopsis thaliana* plants (Jaglo-Ottosen *et al.*, 1998). Interestingly, CBF1-controlled genes are not up-regulated in the *eskimo1* mutant, indicating that different mechanisms can lead to enhanced freezing tolerance.

Besides diverse abiotic challenges, plants encounter numerous diseases caused by different groups of pathogens including insects, fungi, bacteria, nematodes and viruses. Insecticidal plants have been created by the expression of *Bacillus thuringiensis* (*Bt*) δ-endotoxins, of proteinase inhibitors directed against diverse proteinases, α-amylase, lectins, chitinases, peroxidase, virus RNA (*Helicovera armigera* stunt virus) and polyphenol oxidases (Schuler *et al.*, 1998). After extensive modifications of the coding region, including an increased GC content, the use of plant-preferred codons, the removal of improper splice sites and polyadenylation signals, expression of *Bt* toxins has become the most effective strategy for insect control. Therefore, the use of *Bt* toxins is expanded to a vast variety of different plant species as transformation technology is improving.

To obtain crop plants resistant to microbial diseases, several strategies have been followed. First attempts to achieve virus resistance were based on genes encoding viral coat proteins (for review see Miller and Hemenway, 1998). Later additional viral sequences including replicases, movement proteins, and defective interfering RNAs and DNAs have been utilized to obtain resistant plants. Since this strategy is based on viral sequences the resistance obtained has been called 'pathogen-derived resistance'. In addition alternative strategies for virus resistance have been developed including expression of RIPs (ribosome-inactivating proteins) and ribonucleases specific for double-stranded RNA molecules, overproduction of salicylic acid for constitutive expression of defence functions (Verberne *et al.*, 2000), and expression of specific resistance genes. Besides their anti-viral function, overexpression of RIPs and overproduction of salicylic acid results in a limited bacterial and fungal resistance. In addition, fungal resistance has been obtained by expression of specific resistance genes, elevated levels of phytoalexins and expression of lysozyme (for summary see Salmeron and Vernooij, 1998; Herbers and Sonnewald, 1999).

Increased yield

The yield of crop plants is dependent on the integrated functioning of specialized plant organs. Leaves reduce carbon dioxide to carbohydrates during photosynthesis, roots are essential for anchoring and supplying water and ions while storage organs guarantee sexual or vegetative propagation. During development and dependent on environmental conditions metabolic requirements may differ, yet synthesis and allocation of metabolites throughout the plant has to occur coordinately at all times. In addition, biosynthetic pathways of primary and secondary plant products within a single cell are complex, highly regulated networks through which metabolite flux is directed according to environmental and developmental necessities. This is referred to as metabolic partitioning. Based on this complexity success in redirecting carbon flow to improve crop performance and yield has been limited. To increase the rate of sucrose biosynthesis and thereby the availability of photoassimilates for sink development different strategies can be envisaged. It seems reasonable to assume that increasing the activity of rate-limiting enzymes within the sucrose biosynthetic pathway might be one of them. To test this hypothesis attempts to elevate the activity of sucrose-phosphate synthase (SPS) have been reported. Although contradictory in part, the results indicate that elevated SPS protein levels do not always lead to increased enzyme activity and a higher rate of sucrose biosynthesis. The reason for this discrepancy resides in the post-translational down-regulation of SPS activity by reversible protein phosphorylation. Thus an increased yield of SPS overexpressing GMPs has not been observed.

Alternatively, transport and/or utilization of assimilates might be important determinants of crop yield (for review see Herbers and Sonnewald, 1998). Several attempts have been reported to improve sink strength by introducing novel pathways such as fructan biosynthesis (Hellwege *et al.*, 2000) or by optimizing sucrose hydrolysis (Sonnewald *et al.*, 1997). To date, none of the strategies has led to significant increases in crop yield.

By introducing an *Arabidopsis* phytochrome B gene into transgenic potato plants, Thiele *et al.* (1999) reported on increased tuber yield in greenhouse experiments. *PhyB* overexpression resulted in pleiotropic effects including semi-dwarfism, decreased apical dominance, anthocyanin accumulation, delayed senescence, delayed tuber induction and a higher number of smaller but thicker leaves. According to the authors increased yield may be explained by a higher photosynthetic capacity due to a delayed onset of senescence. Whether this strategy leads to increased yield under agricultural conditions remains to be shown.

Hybrid seed production

The discovery that crossing inbred lines of corn produces plants with a dramatically increased yield (*c.* 20%) led to the implementation of hybrid seeds. Production of hybrid seeds requires controlled fertilization, which in the case of corn is quite easy. However, for most other self-pollinating plants it is laborious and expensive. Self-pollination can only be prevented by removing the male organs or by inhibiting pollen development. To this end mechanical (de-tasselling), chemical (the use of gametocides) or genetic (self-incompatibility, cytoplasmic male sterility) methods have been developed. Although effective in a number of crop plants, the available methods cannot be applied to all crops and have serious handicaps in restoration or in implementation, under different environmental conditions. Therefore, genetically engineered artificial male sterility systems are a promising alternative. Artificial systems rely on two strategies: (i) pollen-specific destruction of cellular integrity; and (ii) inhibition of cellular functions required for pollen development. Destruction of cellular integrity can be achieved via production of enzymes (proteases, lipases, RNases), toxic substances (RIPs, diphtheria toxin A), growth regulators (auxin, cytokinin), and the activation of herbicide–glucuronide conjugates or pro-herbicides. In addition to the sterility-causing agent, a restorer function is required. The most effective strategy up to now is the expression of an RNase under control of an anther-specific promoter (Mariani *et al.*, 1990). Restoration is achieved by crossing with plants engineered to express a specific RNase inhibitor. In addition to the methods described above, pollen development can be inhibited by the manipulation of carbon metabolism and/or uptake.

Post harvest performance

Ripening of climacteric fruits, such as tomatoes, apples and bananas, is thought to involve ethylene. The process of ripening is associated with alterations in colour, texture, flavour, aroma and susceptibility to pathogens. Softening of the fruit is accompanied by cell wall degrading processes breaking down the rigid structure of the fruit. Biochemical and molecular changes associated with fruit ripening have been extensively studied in tomato (for review, see Gray *et al.*, 1994). Genes encoding enzymes involved in cell wall degradation, colour change and ethylene synthesis have been cloned and transgenic plants with deficiencies in different aspects of ripening have been created. Strategies to alter fruit ripening concentrated on the inhibition of ethylene biosynthesis and pectin degradation.

In addition to fruit softening, the development of brown discoloration of fruits and vegetables reduces consumer acceptability and is therefore a major issue for producers and the food industry. In potato tubers, mechanical injury during harvest or storage leads to the formation of 'blackspots' caused by the formation of polyphenolic compounds. The enzyme polyphenol oxidase (PPO) is thought to be responsible for the initial steps of polyphenol accumulation in damaged tissue. In agreement with this hypothesis antisense inhibition of PPO in transgenic potato plants leads to the inhibition of enzymatic browning in harvested potato tubers (Bachem *et al.*, 1994).

Although potato tubers represent a carbohydrate resource for many diets and are used as basis for a variety of processed products, their industrial application is limited since sprouting of stored tubers has to be inhibited. To prevent sprouting, tubers are either treated with dormancy prolonging chemicals or are kept at low temperatures. Low temperature treatment of tubers leads to the accumulation of reducing sugars which is unsuitable for the potato crisping and chipping industry due to the occurrence of the Maillard reaction between reducing sugars and amino acids during frying. The cause of cold-induced sweetening is thought to be an imbalance between starch breakdown and glycolytic activity. Consequently sucrose is formed via sucrose-phosphate synthase which is subsequently hydrolysed by invertases, yielding the reducing sugars glucose and fructose. To inhibit hexose formation, Greiner *et al.* (1999) expressed an invertase inhibitor in transgenic potato plants. As a consequence invertase activity was inhibited and hexose accumulation strongly decreased.

An alternative strategy to delay sprouting relies on the inhibition of assimilate supply of the developing sprout. In this context it could be shown that removal of cytosolic pyrophosphate delayed sprouting of transgenic

Fig. 19.4. Influence of *E. coli* pyrophosphatase on potato tuber sprouting (for detailed information see Hajirezaei and Sonnewald, 1999). Following 12 months of storage at room temperature wild-type tubers (wt) show extensive sprout growth whereas sprout growth of transgenic tubers is strongly delayed (line 1) or absent (line 2 and 3).

potato tubers dramatically (Hajirezaei and Sonnewald, 1999; Fig. 19.4). Sprouting of wild type tubers started after 4 months of storage at room temperature, whereas transgenic tubers did not sprout even after a prolonged storage period of 2 years. The reason was speculated to be a reduced sucrose export and the inhibition of sucrose utilization (Hajirezaei and Sonnewald, 1999). Sucrose export is most likely inhibited due to a reduced sucrose uptake of companion cells based on limited energy supply as a consequence of cytosolic PP$_i$ depletion. This assumption is supported by the finding that removal of PP$_i$ in phloem cells of transgenic tobacco plants resulted in photoassimilate accumulation in leaves and reduced growth of tobacco plants (Lerchl *et al.*, 1995).

Production of proteins

Transgenic plants represent an attractive and cost-efficient alternative to conventional systems for the production of bio-molecules. Obvious advantages of the use of transgenic plants as bioreactors can be summarized as follows: (i) due to their phototrophy and autotrophy plants are able to produce their biomass using sunlight and inorganic substances, resulting in low energy costs for production; (ii) the production is flexible and can be done in bulk quantities; (iii) in contrast to microbial systems plants carry out post-translational protein modifications which can be essential for enzyme activity; (iv) enzymes used in the food, feed or paper industry do not necessarily need to be purified in the case where the plant material is directly used; and (v) human pathogens or toxins do not reduce the quality of the product. However, for phyto-farming to be commercially attractive a number of prerequisites need to be fulfilled, namely: (i) the polypeptides need to be produced in considerable amounts; (ii) they must be stable in the plants; (iii) enzymes must stay functional during harvest and subsequent enrichment/ purification procedures; (iv) if purification is needed the procedure must be easy and cheap; and (v) plant growth must be unaffected by the expression of the heterologous enzyme. Numerous industrial enzymes, vaccines and antibodies have been produced in transgenic plants (for recent reviews see Herbers and Sonnewald, 1999; Doran, 2000; Walmsley and Arntzen, 2000).

Endogenous polymers

Starch is the major reserve carbohydrate in vascular plants and consists essentially of linear chain (amylose) and branched chain (amylopectin) glucans. Depending on the plant species amylose makes up between 10 and 25% of starch. Chain length of the linear regions of amylopectin is grouped into low (5–30 Glc units) and high (30–100 Glc units) molecular weight chains. Some of the glucans may be phosphorylated, most notably in the case of potato-tuber starch. At present, starch is extracted commercially from a limited number of sources (mainly maize and potato). Physico-chemical properties (such as gelatinization temperature, retrogradation and viscosity), important for technical uses, are influenced by granule size and size homogeneity, the amylose to amylopectin ratio, the distribution of low and high molecular weight glucose chains, and the lipid, ash and phosphorus content. Therefore, molecular manipulations concentrate on the manipulation of starch content (overexpression of deregulated ADP-glucose pyrophosphorylase, decreased starch breakdown), chain length, amylose to amylopectin ratio and the phosphate content (summarized in Slattery *et al.*, 2000).

Fats and oils exist in the form of triglycerides, in which fatty acid molecules are linked by ester bonds to the three hydroxyl groups of glycerol. Carbon chain length may vary from 8 to 20 but most commonly 16 and 18 carbon fatty acids are found in higher plants. More than 200 types of fatty acids are known to be produced in plants, but only palmitic, stearic, oleic, linoleic and linolenic acids are commercially exploited. Approximately 90% of vegetable oil produced is used for human consumption whereas only 10% is used for non-food applications. The major non-food market for vegetable oils is the production of soaps, detergents and other surfactants. Depending on the specific use, chain length and saturation of fatty acids are important properties. For example, medium-chain fatty acids have ideal properties as surfactants. Currently, major commercial sources of medium-chain fatty acids are coconut and palm tree oils. Chain length is determined by the action of specific thioesterases hydrolysing fatty acid-ACPs. Thioesterases of most crop plants hydrolyse 16:0 and 18:1-ACPs; however, thioesterases of several other plant species hydrolyse medium-chain fatty acyl-ACPs; leading to the accumulation of medium-chain fatty acids in their seed oils. The pivotal role of thioesterases was demonstrated in transgenic *Arabidopsis* and rapeseed plants expressing a lauryl-ACP specific thioesterase from the California bay tree (Voelker *et al.*, 1992). The oil isolated from seeds of the transgenic plants contained 25–40% lauric acid providing an economic alternative to coconut and palm tree oils. Dehesh *et al.* (1996) isolated a cDNA clone encoding acyl-ACP thioesterase from *Cuphea hookeriana*. Seeds of *C. hookeriana* accumulate mainly caprylate (8:0) and caprate (10:0). Introduced into transgenic rapeseed plants expression of the *C. hookeriana* acyl-ACP thioesterase led to the accumulation of 8:0 and 10:0 fatty acids. Accumulation of the novel fatty acids was accompanied by a decrease of linoleate (18:2) and linolenate (18:3). In addition to thioesterases, the activity of acyl-ACP desaturases has been altered in transgenic plants with the aim of changing the level of unsaturation (summarized in Napier *et al.*, 1999). Also attempts to increase the level of polyunsaturated fatty acids (PUFAs) are of special interest since their health-promoting properties are widely accepted.

Novel polymers

In addition to the manipulation of endogenous plant products discussed above, attempts to redirect carbon flow towards the production of novel bio-molecules have been reported. One of the first examples of the synthesis of a novel biopolymer in transgenic plants was the production of poly-3-hydroxybutyrate (PHB) in *Arabidopsis* plants (summarized by Poirier, 1999).

Furthermore, attempts to produce fructans in non-fructan-storing plants by expression of bacterial or plant enzymes have been reported (Ebskamp *et al.*, 1994; Sevenier *et al.*, 1998). Plant fructans are polymers of fructose (DP of 5–60 fructose units) carrying a D-glucosyl residue at the end of the chain attached via a β-2,1-linkage. They are widely distributed in the plant kingdom and can be classified into three main types: inulin, phlein and branched inulin. Besides plants, many microorganisms have been described to produce fructans. However, chain length of bacterial fructan is usually much larger (DP of more than 100,000 fructose units) than in plants.

Low molecular weight sugars

In addition to attempts described above, the production of cyclodextrins, trehalose and palatinose in plants has been envisaged. Cyclodextrins (CDs) are cyclic oligosaccharides of six, seven or eight α-1,4-linked glucose molecules. CDs are used as pharmaceutical delivery systems, for flavour and odour enhancement and the removal of undesired compounds from foods. CDs are formed by the enzyme cyclodextrin glycosyltransferase (CGT) using hydrolysed starch as substrate. CGT is exclusively found in some bacteria which are commercially used for the production of CDs. Using a chimeric CGT gene from *Klebsiella pneumoniae* allowing the targeting of the polypeptide into amyloplasts, Oakes *et al.* (1991) attempted to engineer potato tubers to produce CDs. Analysis of tubers harvested from the transgenic plants revealed that 0.001–0.01% of the starch was converted to CDs.

Trehalose, a non-reducing disaccharide consisting of two glucose units, is present in a variety of microorganisms and animals. It has been described as a stress protectant, stabilizing proteins and membranes during desiccation and heat stress. Commercially, trehalose is used as a food additive improving quality and flavour of dried and processed food as well as a stabilizer of vaccines, hormones and blood components. Due to high production costs, trehalose is not used as a routine food additive. In yeast and *E. coli*, trehalose biosynthesis requires two enzyme activities: trehalose-6-phosphate synthase (TPS), converting glucose-6-phosphate and UDP-glucose into trehalose-6-phosphate; and trehalose-6-phosphate phosphatase (TPP), converting trehalose-6-phosphate to trehalose and inorganic phosphate. To engineer plants for trehalose synthesis both genes have been expressed in transgenic plants. However, due to activity of an endogenous trehalase only low levels of trehalose accumulated in transgenic plants (for a recent review see Goodijn and van Dun, 1999). Palatinose (isomaltulose, 6-*O*-α-D-glucopyranosyl-D-fructose) is a structural isomer of sucrose with very similar physico-chemical properties. Due to its non-cariogenicity and low caloric value it is an ideal sugar substitute for use in food production. Palatinose is produced on an industrial scale using immobilized bacterial cells. To investigate the potential of transgenic plants as a production facility the *palI* gene from

Erwinia rhapontici encoding a sucrose isomerase catalysing the conversion of sucrose into palatinose was introduced into tobacco plants. Expression of the *palI* gene within the apoplast of transgenic plants led to high-level accumulation of palatinose. As a consequence plant growth and development was dramatically impaired suggesting that palatinose itself is no substrate for plant metabolism (Börnke and Sonnewald, unpublished; Fig. 19.5).

Fig. 19.5. Leaf phenotype of palatinose-accumulating tobacco plants (Börnke and Sonnewald, unpublished).

Vitamins

Vitamin E includes eight naturally occurring fat-soluble nutrients called tocopherols. α-Tocopherol has the highest biological activity and the highest molar concentration of lipid-soluble antioxidant in man. Unfortunately, plant oils, the main dietary source of tocopherols, typically contain α-tocopherol as a minor component and high levels of its biosynthetic precursor, γ-tocopherol. In plants γ-tocopherol is converted into α-tocopherol via the action of the enzyme γ-tocopherol methyltransferase. To improve α-tocopherol content in plant oils DellaPena and co-workers overexpressed this methyltransferase in *Arabidopsis* seeds. Total tocopherol content in seeds of transgenic *Arabidopsis* plants was unaltered; however, the composition was shifted in favour of α-tocopherol (Shintani and DellaPenna, 1998).

Another example for improved vitamin content in transgenic plants is the so called 'golden rice'. Rice endosperm is naturally poor in several essential nutrients, including provitamin A. To increase the level of provitamin A in transgenic rice Potrykus and co-workers introduced the entire β-carotene biosynthetic pathway into rice endosperm (Ye *et al.*, 2000).

Challenges

During the last two decades the major challenge of agricultural biotechnology was the isolation of potentially useful genes and the development of transformation methods allowing the introduction and controlled expression of those genes in appropriate model or commercially relevant crop plants. This led to the creation of first generation genetically modified crop plants, being herbicide- and/or insect-resistant, which made their way to the market.

In the future, second generation genetically modified crop plants will express not only input but, in addition, output traits. Thus, the challenge will be to create the tools of public and regulatory acceptance that will allow the generation of crop plants with controlled expression of several target genes.

Due to tremendous achievements in genomic research complete genome sequences of model organisms have been deciphered. Consequently, the necessity for targeted gene isolation continuously decreases whereas high throughput post-genomic technologies, essential to determine the function of those genes, are challenges of the future. These technologies will not necessarily lead to genetically modified plants but will become invaluable tools for marker-assisted breeding, and molecular kits for food and feed safety. This in turn will help to improve public acceptance of genetically modified crops.

References

Bachem, C.W.B., Speckmann, G.J., van der Linde, P.C.G., Verheggen, F.T.M., Hunt, M.D., Steffens, J.C. and Zabeau, M. (1994) Antisense expression of polyphenol oxidase genes inhibits enzymatic browning in potato. *Biotechnology* 12, 1101–1105.

Banjoko, A. and Trelease, R.N. (1995) Development and application of an *in vivo* plant peroxisome import system. *Plant Physiology* 107, 1201–1208.

Bohnert, H.J. and Jensen, R.G. (1996) Strategies for engineering water-stress tolerance in plants. *TIBTECH* 14, 89–97.

Bohnert, H.J. and Sheveleva, E. (1998) Plant stress adaptations – making metabolism move. *Current Opinion in Plant Biotechnology* 1, 267–274.

Bojsen, K. (1993) A positive selection system for transformed eukaryotic cells based on mannose or xylose utilization. *WO 94/20627*.

Boutry, M., Nagy, F., Poulson, C., Aoyagi, K. and Chua, N.H. (1987) Targeting of bacterial chloramphenicol acetyltransferase to mitochondria in transgenic plants. *Nature* 328, 340–342.

Caddick, M.X., Greenland, A.J., Jepson, J., Krause, K.-P., Qu, N., Riddell, K., Salter, M.G., Schuch, W., Sonnewald, U. and Tomsett, A.B. (1998) An ethanol inducible gene switch for plants used to manipulate carbon metabolism. *Nature Biotechnology* 16, 177–180.

Cheng, M., Fry, J.E., Pang, S., Zhou, H., Hironaka, C.M., Duncan, D.R., Conner, T.W. and Wang, Y. (1997) Genetic transformation of wheat mediated by *Agrobacterium tumefaciens*. *Plant Physiology* 115, 971–980.

Christou, P. (1996) Transformation technology. *Trends in Plant Sciences* 1, 423–431.

Dehesh, K., Jones, A., Knutzon, D.S. and Voelker, T.A. (1996) Production of high levels of 8:0 and 10:0 fatty acids in transgenic canola by overexpression of Ch FatB2, a thioesterase cDNA from *Cuphea hookeriana*. *Plant Journal* 9, 167–172.

Doran, P.M. (2000) Foreign protein production in plant tissue culture. *Current Opinion in Biotechnology* 11, 199–204.

Ebskamp, M.J., van der Meer, I.M., Spronk, B.A., Weisbeek, P.J. and Smeekens, S.C. (1994) Accumulation of fructose polymers in transgenic tobacco. *Biotechnology* 12, 272–275.

Edwards, J.W. and Coruzzi, G.M. (1990) Cell-specific gene expression in plants. *Annual Review of Plant Physiology and Plant Molecular Biology* 24, 275–303.

Fluhr, R., Kuhlemeier, C., Nagy, F. and Chua, N.H. (1986) Organ-specific and light-induced expression of plant genes. *Science* 232, 1106–1112.

Franck, A., Guilley, H., Jonard, G., Richards, K. and Hirth, L. (1980) Nucleotide sequence of cauliflower mosaic virus DNA. *Cell* 21, 285–294.

Gatz, C., Kaiser, A. and Wendenburg, R. (1991) Regulation of a modified CaMV 35S promoter by the Tn10-encoded Tet repressor in transgenic tobacco. *Molecular and General Genetics* 227, 229–237.

Goldsbrough, A.P., Tong, Y. and Yoder, J.I. (1996) *Lc* as a non-destructive visual reporter and transposition excision marker gene for tomato. *Plant Journal* 9, 927–933.

Goodijn, O.J.M. and van Dun, K. (1999) Trehalose metabolism in plants. *Trends in Plant Science* 4, 315–319.

Gray, J.E., Picton, S., Giovannoni, J.J. and Grierson, D. (1994) The use of transgenic and naturally occurring mutants to understand and manipulate tomato fruit ripening. *Plant Cell and Environment* 17, 557–571.

Greiner, S., Rausch, T., Sonnewald, U. and Herbers, K. (1999) Ectopic expression of a tobacco invertase inhibitor homolog prevents cold-induced sweetening of potato tubers. *Nature Biotechnology* 17, 708–711.

Hajirezaei, M. and Sonnewald, U. (1999) Inhibition of potato tuber sprouting: low levels of cytosolic pyrophosphate lead to non-sprouting tubers harvested from transgenic potato plants. *Potato Research* 42, 353–372.

Haldrup, A., Petersen, S.G. and Okkels, F.T. (1998) The xylose isomerase gene from

Theromanaerobacterium thermosulfurogenes allows effective selection of transgenic plant cells using D-xylose as the selection agent. *Plant Molecular Biology* 37, 287–296.

Heifetz, P.B. (2000) Genetic engineering of chloroplasts. *Biochemie* 82, 655–666.

Hellwege, E.M., Czapla, S., Jahnke, A., Willmitzer, L. and Heyer, A.G. (2000) Transgenic potato (*Solanum tuberosum*) tubers synthesize the full spectrum of inulin molecules naturally occurring in globe artichoke (*Cynara scolymus*) roots. *Proceedings of the National Academy of Sciences USA* 97, 8699–8704.

Herbers, K. and Sonnewald, U. (1998) Molecular determinants of sink strength. *Current Opinion in Plant Biology* 1, 207–216.

Herbers, K. and Sonnewald, U. (1999) Production of new/modified proteins in transgenic plants. *Current Opinion in Biotechnology* 10, 163–168.

Hiei, Y., Ohta, S., Komari, T. and Kumasho, T. (1994) Efficient transformation of rice (*Oryza sativa*) mediated by *Agrobacterium* and sequence analysis of the boundaries of the T-DNA. *Plant Journal* 6, 271–282.

Holmberg, N. and Bülow, L. (1998) Improving stress tolerance in plants by gene transfer. *Trends in Plant Science* 3, 61–66.

Ishida, Y., Saito, H., Ohta, S., Hiei, Y., Komari, T. and Kumashiro, T. (1996) High efficiency transformation of maize (*Zea mays* L.) mediated by *Agrobacterium tumefaciens*. *Nature Biotechnology* 14, 745–750.

Ito, M., Sato, T., Fukuda, H. and Komamine, A. (1994) Meristem-specific gene expression directed by the promoter of the S-phase-specific gene, *cyc07*, in transgenic *Arabidopsis*. *Plant Molecular Biology* 24, 863–878.

Jaglo-Ottosen, K.R., Gilmour, S.J., Zarka, D.G., Schabenberger, O. and Thomashow, M.F. (1998) *Arabidopsis* CBF1 overexpression induces *COR* genes and enhances freezing tolerance. *Science* 280, 104–106.

Jeon, J.S., Lee, S., Jung, K.H., Jun, S.H., Jeong, D.H., Lee, J., Kim, C., Jang, S., Yang, K., Nam, J., An, K., Han, M.J., Sung, R.J., Choi, H.S., Yu, J.H., Choi, J.H., Cho, S.Y., Cha, S.S., Kim, S.I. and An, G. (2000) T-DNA insertional mutagenesis for functional genomics in rice. *Plant Journal* 22, 561–570.

Kaneko, T., Kotani, H., Nakamura, Y., Sato, S., Asamizu, E., Miyajima, N. and Tabata, S. (1998) Structural analysis of *Arabidopsis thaliana* chromosome 5. V. Sequence features of the regions of 1,381,565 bp covered by twenty one physically assigned P1 and TAC clones. *DNA Research* 5, 131–145.

Knoblauch, M., Hibberd, J.M., Gray, J.C. and van Bel, A.J. (1999) A galinstan expansion femtosyringe for microinjection of eukaryotic organelles and prokaryotes. *Nature Biotechnology* 17, 906–909.

Kunkel, T., Niu, Q.-W., Chan, Y.-S. and Chua, N.-H. (1999) Inducible isopentenyl transferase as a high-efficiency marker for plant transformation. *Nature Biotechnology* 17, 916–919.

Lerchl, J., Geigenberger, P., Stitt, M. and Sonnewald, U. (1995) Inhibition of long-distance sucrose transport by inorganic pyrophosphatase can be complemented by phloem-specific expression of cytosolic yeast-derived invertase in transgenic plants. *The Plant Cell* 7, 259–270.

Lindsey, K., Wei, W., Clarke, M.C., McArdle, H.F., Rooke, L.M. and Topping, J.F. (1993) Tagging genomic sequences that direct transgene expression by activation of a promoter trap in plants. *Transgenic Research* 2, 33–47.

Mandaci, S. and Dobres, M.S. (1997) A promoter directing epidermal expression in transgenic alfalfa. *Plant Molecular Biology* 34, 961–965.

Marcotte, W.R. Jr, Russell, S.H. and Quatrano (1989) Abscisic acid-responsive sequence from *Em* gene of wheat. *Plant Cell* 1, 969–976.

Mariani, C., De Beuckeleer, M., Truettner, J., Leemans, J. and Goldberg, R.B. (1990) Induction of male sterility in plants by a chimaeric ribonuclease gene. *Nature* 347, 737–741.

Martinez, A., Sparks, C., Drayton, P., Thompson, J., Greenland, A. and Jepson, I. (1999) Creation of ecdysone receptor chimeras in plants for controlled regulation of gene expression. *Molecular and General Genetics* 26, 546–552.

Miller, E.D. and Hemenway, C. (1998) History of coat protein-mediated protection. *Methods in Molecular Biology* 81, 25–28.

Müller-Röber, B., la Cognata, U., Sonnewald, U. and Willmitzer, L. (1994) A truncated version of an ADP-glucose pyrophosphorylase promoter from potato specifies guard cell-selective expression in transgenic plants. *Plant Cell* 6, 601–612.

Nakamura, K. and Matsuoka, K. (1993) Protein targeting to the vacuole in plant cells. *Plant Physiology* 101, 1–5.

Napier, J.A., Michaelson, L.V. and Stobart, A.K. (1999) Plant desaturases: harvesting the fat of the land. *Current Opinion in Plant Biology* 2, 123–127.

Oakes, J.V., Shewmaker, C.K. and Stalker, D.M. (1991) Production of cyclodextrins, a novel carbohydrate, in tubers of transgenic potato plants. *Biotechnology* 9, 982–986.

Ohme-Takagi, M. and Shinishi, H. (1995) Ethylene-inducible DNA binding proteins that interact with an ethylene-responsive element. *Plant Cell* 7, 173–182.

Poirier, Y. (1999) Production of new polymeric compounds in plants. *Current Opinion in Biotechnology* 10, 181–185.

Salmeron, J.M. and Vernooij, B. (1998) Transgenic approaches to microbial disease resistance in crop plants. *Current Opinion in Plant Biology* 1, 347–352.

Schmülling, T., Schell, J. and Spena, A. (1989) Promoters of the *rolA, B,* and *C* genes of *Agrobacterium rhizogenes* are differentially regulated in plants. *Plant Cell* 1, 665–670.

Schuler, T.H., Poppy, G.M., Kerry, B.R. and Denholm, I. (1998) Insect-resistant transgenic plants. *TIBTECH* 16, 168–174.

Sevenier, R., Hall, R.D., van der Meer, I.M., Hakkert, H.J., van Tunen, A.J. and Koops, A.J. (1998) High level fructan accumulation in transgenic sugar beet. *Nature Biotechnology* 16, 843–846.

Sheen, J., Hwang, S., Niwa, S., Kobayashi, H. and Galbraith, D.W. (1995) Green-fluorescent protein as a new vital marker in plant cells. *Plant Journal* 8, 777–784.

Shintani, D. and DellaPenna, D. (1998) Elevating the vitamin E content of plants through metabolic engineering. *Science* 282, 2098–2100.

Simpson, J., Van Montagu, M. and Herrera-Estrella, L. (1985) Light-inducible and tissue-specific expression of a chimeric gene under control of the 5′ flanking sequence of a pea chlorophyll a/b-binding protein gene. *EMBO Journal* 4, 2723–2729.

Slattery, C.J., Kavakli, I.H. and Okita, T.W. (2000) Engineering starch for increased quantity and quality. *Trends in Plant Science* 5, 291–298.

Sonnewald, U. and Ebneth, M. (1997) 2-Deoxyglucose-6-phosphate phosphatase as a selectable marker in plants. *WO1998EP0002069*.

Sonnewald, U., Hajirezaei, M., Koßmann, J., Heyer, A., Trethewey, R.N. and

Willmitzer, L. (1997) Increased potato tuber size resulting from apoplastic expression of a yeast invertase. *Nature Biotechnology* 15, 794–797.

Stockhaus, J., Schell, J. and Willmitzer, L. (1989) Identification of enhancer elements in the upstream region of the nuclear photosynthetic gene *ST-LS1*. *Plant Cell* 1, 805–813.

Stockhaus, J., Poetsch, W., Steinmüller, K. and Westhoff, P. (1994) Evolution of C₄ phosphoenolpyruvate carboxylase promoter of the C_4 dicot *Flaveria trinervia*: an expression analysis in the C_3 plant tobacco. *Molecular and General Genetics* 245, 286–293.

Sugita, K., Kasahara, T., Matsunaga, E. and Ebinuma, H. (2000) A transformation vector for the production of marker-free transgenic plants containing a single transgene at high frequency. *Plant Journal* 22, 461–469.

Thiele, A., Herold, M., Lenk, I., Quail, P.H. and Gatz, C. (1999) Heterologous expression of *Arabidopsis* phytochrome B in transgenic potato influences photosynthetic performance and tuber development. *Plant Physiology* 120, 73–83.

Uknes, S., Dincher, S., Friedrich, L., Negrotto, D., Williams, S., Thompson-Taylor, H., Potter, S., Ward, E. and Ryals, J. (1993) Regulation of pathogenesis-related protein-1a gene expression in tobacco. *Plant Cell* 5, 159–169.

Van den Broeck, G., Timko, M.P., Kausch, A.P., Cashmore, A.R., Van Montagu, M. and Herrera-Estrella, L. (1985) Targeting of foreign proteins to chloroplasts by fusion to the transit peptide from the small subunit of ribulose-1,5-bisphosphate carboxylase. *Nature* 313, 358–363.

Verberne, M.C., Verpoorte, R., Bol, J.F., Mercado-Blanco, J. and Linthorst, H.J. (2000) Overproduction of salicylic acid in plants by bacterial transgenes enhances pathogen resistance. *Nature Biotechnology* 18, 779–783.

Vitale, A. and Denecke, J. (1999) The endoplasmic reticulum – gateway of the secretory pathway. *Plant Cell* 11, 615–628.

Voelker, T.A., Worrell, A.C., Anderson, L., Bleibaum, J., Fan, C., Hawkins, D.J., Radke, S.E. and Davies, H.M. (1992) Fatty acid biosynthesis redirected to medium chains in transgenic oilseed plants. *Science* 257, 72–74.

Walmsley, A.M. and Arntzen, C.J. (2000) Plants for delivery of edible vaccines. *Current Opinion in Biotechnology* 11, 126–129.

Warren, G.J. (1998) Cold stress: manipulating freezing tolerance in plants. *Current Biology* 8, R514–516.

Wilde, R.J., Shufflebottom, D., Cooke, S., Jasinska, I., Merryweather, A., Beri, R., Brammar, W.J., Bevan, M. and Schuch, W. (1992) Control of gene expression in tobacco cells using a bacterial operator-repressor system. *EMBO Journal* 11, 1251–1259.

Wilde, R.J., Cooke, S.E., Brammar, W.J. and Schuch, W. (1994) Control of gene expression in plant cells using a 434:VP16 chimeric protein. *Plant Molecular Biology* 24, 381–388.

Xin, Z. and Browse, J. (1998) *Eskimo1* mutants of *Arabidopsis* are constitutively freezing-tolerant. *Proceedings of the National Academy of Sciences USA* 95, 7799–7804.

Ye, X., Al-Babili, S., Klöti, A., Zhang, J., Lucca, P., Beyer, P. and Potrykus, I. (2000) Engineering the provitamin A (β-carotene) biosynthetic pathway into (carotenoid-free) rice endosperm. *Science* 287, 303–305.

Zuo, J. and Chua, N.H. (2000) Chemical-inducible systems for regulated expression of plant genes. *Current Opinion in Biotechnology* 11, 146–151.

Zupan, J., Muth, T.R., Draper, O. and Zambryski, P. (2000) The transfer of DNA from *Agrobacterium tumefaciens* into plants: a feast of fundamental insights. *Plant Journal* 23, 11–28.

Transgenic Plants for Sustainable Crop Production

20

B. Keller and E. Hütter Carabias

Institute of Plant Biology, University of Zürich, Zollikerstrasse 107, CH-8008 Zürich, Switzerland

Introduction

Crops have been genetically modified for a variety of purposes: to be resistant to fungal, bacterial and viral diseases and insect pests; to tolerate herbicides; and to grow better under restricting environmental conditions such as drought, high salt or metal-containing soils. Increasingly, the improvement of quality traits such as nutritional content and suitability for industrial processing is also of importance. Conventional breeding has very similar goals such as to develop plant varieties with higher yield and/or better nutritional qualities, disease and pest resistance and lower cultivation costs. For example, the genes of *Bacillus thuringiensis* (*Bt*) transferred to maize confer the same trait (insect resistance) as some specific insect-resistance genes used in classical breeding. Such genes were used to breed for resistance to the leafhopper in rice or the hessian fly in wheat and were introduced into these crops using traditional breeding methods. The advantage of the biotechnological methods compared with conventional breeding is the targeted isolation of a gene from the donor organism and its subsequent, relatively rapid, transfer into the genome of the recipient plant. The introduced gene will be stable in the following generations. Compared with selective breeding through hybridization, in which whole genomes are mixed and where it can take many years to breed a plant with a specific trait, gene technology is more precise and, if the relevant genes are available, can save considerable time in the development of a new cultivar.

At the current stage of the technology, the integration of transgenes into the plant genome cannot be targeted. This random integration could cause some problems due to insertional inactivation of a gene at the integration site or unwanted activation of neighbouring genes. These new risks have to be compared with the problems putatively associated with conventional plant breeding where two whole genomes are mixed. It has also to be considered that plant genomes are not naturally stable entities: active transposons (jumping genes) exist in many plant species and can basically cause the same unwanted effects as integration of transgenes. Transposition can inactivate genes and induce rearrangements and recombination events resulting in unpredictable positional effects. However, no safety problems related to transposons have ever been reported. The sequence analysis of large genome fragments in plants with larger genomes has also revealed a remarkable level of ancient transposon integration events with gene fragments spread in the genome and complex rearrangements at the loci studied.

Although transgenic crops and traditionally bred varieties are very similar in many aspects, the impact of the new technology on the environment and the economy has to be assessed carefully for industrialized as well as for developing countries. In addition, ethical aspects have to be specifically considered. In our opinion this is necessary because of a general lack of experience with the new technology, and not because there is evidence that the genetic modification of plants is inherently less safe than conventional breeding.

Importance and Prospects of Transgenic Plants in Industrial and Developing Countries

There has been a 23-fold increase in the global area planted with transgenic crops from 1996 to 1999. These crops are mostly grown in countries of the north. The main transgenic crops in today's agriculture are soybean (54%), maize (28%), cotton (9%) and oilseed rape (9%). Minor transgenic crops are potato, squash and papaya (www.isaaa.org). Global area planted with transgenic crops is expected to continue to grow but started to plateau in 2000 reflecting the unprecedented high adoption rates and the high percentage of principal crops already planted with transgenics in the USA, Argentina and Canada. The major issues modulating adoption after the year 2000 are public acceptance, which drives market demand, regulation and commodity prices.

In developing countries, most of the transgenic varieties currently grown are cash crops. A number of important crops, including most of those cultivated in developing countries for human consumption, have received little attention in biotechnological research so far. These species include cassava, sweetpotato, banana, grain legumes and many others (Conway and Toenniessen, 1999).

The traits modified by genetic engineering can be divided into three broad categories: agronomic improvement, food modification and industrial exploitation (Table 20.1).

Biotechnology research for and in the developing countries has focused

Table 20.1. Important crop traits modified by genetic engineering.

Agronomic traits	• weed control
	• insect pest resistance
	• resistance to diseases (fungi, viruses, bacteria, nematodes)
	• tolerance to environmental stresses (e.g. heat, cold, drought, salt, Al, heavy metals)
	• increased nitrogen fixation
Quality traits	• enhanced nutritional quality of food crops (e.g. vitamins, minerals, proteins)
	• delayed ripening of fruits
	• changes of colour, flavour and texture, as well as modification of oil, starch and protein composition elimination of toxic or anti-nutritional factors
Industrial uses	• high-value chemicals
	• modified and speciality oils
	• recombinant or engineered proteins including enzymes
	• production of pharmaceuticals (e.g. antibodies, vaccines)
	• renewable non-food products (e.g. fuel, plastics)
	• bioremediation

largely on traits for disease and pest resistance. Now the focus has widened to include improved nutritional quality and adaptation of plants to environmental stresses. One recent development is the successful introduction of genes into rice that result in the production of the vitamin A precursor β-carotene in the rice grain, which is converted to vitamin A in the human body (Ye *et al.*, 2000). Other transgenes include ferritin, phytase and a cystein-rich metallothionin-like gene leading to higher amounts of iron in rice and better intestinal resorption (e.g. Goto *et al.*, 1999). Genes that confer tolerance to high concentrations of aluminium (found in many tropical soils) or enhance phosphorus uptake have been added to different plant species (De la Fuente *et al.*, 1997; López-Bucio *et al.*, 2000).

Commercial applications of transgenic plants

On a global scale, the area planted with transgenic plants increased from about 2 million hectares in 1996 to almost 40 Mha in 1999. This area was planted with eight transgenic crops with a high market value such as herbicide-tolerant soybean, corn, cotton and rapeseed, *Bt* corn, *Bt* cotton, *Bt* potatoes, virus-resistant crops (squash, papaya) and an increasing acreage of crops with combined traits (Tables 20.2 and 20.3).

Although cash crops are dominating transgenic varieties, field tests with transgenic breeding lines have also been carried out with trees (birch, eucalypt, poplar, spruce, pine, walnut), vegetables (broccoli, cabbage, carrot,

Table 20.2. Global area of commercially grown transgenic crops from 1996 to 1999 (million hectares) (www.transgen.de, December 1999; www.isaaa.org).

Crop	Trait*	1996	1997	1998	1999
Soybean	HT, oil profile altered	0.5	5.2	14.4	21.6
Maize	IR, IR + HT	0.3	3.2	8.3	11.1
Cotton	IR, HT, IR + HT	0.8	1.3	2.5	3.7
Canola/rapeseed	HT, HT + male sterility	0.1	1.3	2.4	3.4
Potato	IR, VR	<0.1	<0.1	<0.1	<0.1
Squash	VR	0	0	0	<0.1
Papaya	VR	0	0	0	<0.1
Tomato	IR, VR, fruit ripening altered	0	<0.1	<0.1	<0.1
Total		1.7	11	27.8	39.9

* HT, herbicide tolerance; IR, insect pest resistance; VR, virus resistance.

Table 20.3. Genetically modified traits of commercially grown crops, 1996 to 1999 (% of acreage) (www.isaaa.org).

Traits	1996	1997*	1998	1999
Herbicide tolerance	23	54	71	71
Insect resistance	37	31	28	22
Herbicide tolerance and insect resistance	0	0.1	1	7
Virus resistance	40	14	<1	<1

* Plus 1% quality traits.

cauliflower, chicory, cucumber, aubergine, lettuce, melon, pea, tomato), fruit plants (apple, blueberry, cranberry, grape, kiwi, papaya, plum, raspberry, strawberry), flowers (carnation, chrysanthemum, gerbera, gladiolus, petunia), industrial crops (flax, sugarbeet, sugarcane, sunflower), food crops (barley, wheat, rice, sweetpotato, potato) as well as fodder crops like lucerne and lupin and others (mustard, pepper, bent grass or tobacco) (www.transgen.de, December 1999).

In 1999, 82% of the transgenic crops were grown in industrial countries, and 18% in developing countries (Table 20.4) with most of the area in Argentina (17% of the total area in developing countries), and the rest in China, South Africa and Mexico (www.isaaa.org). In several other developing countries field trials are conducted (www.isaaa.org<http://www.isaaa.org>).

Table 20.4. Transgenic crops commercially grown in industrial and developing countries (million hectares) (www.isaaa.org).

Countries	1996	1997	1998	1999
Industrial countries	1	8.2	23.4	32.8
Developing countries	0.7	2.8	4.4	7.1

Trends in development

The most promising trends in development go towards improved nutrient uptake from the soil, chloroplast transformation to avoid transgene spread through pollen flow, alternative marker genes (other than antibiotic resistance genes) and apomixis. Apomixis circumvents meiosis, and thereby avoids segregation and recombination of genetic information and preserves the maternal genotype by parthenogenetic embryo development. Heterozygosity and epistasis would be maintained, while hybrid varieties of most crops lose heterosis after the first generation due to segregation and recombination. Research on genetic engineering of apomixis is still in an early phase (van Dijk and van Damme, 2000). Another important development is based on the molecular identification of genes involved in nutrient solubilization by extruding enzymes, organic acids and other compounds altering the chemistry of the rhizosphere. This work has great potential for the genetic enhancement of nutrient extraction from soil. Significant progress has been made by Hirsch and Sussman (1999) by the identification of genes encoding proteins directly involved in transport of nutrients.

Transgenic breeding still suffers from the low number of genes available for the targeted modification of specific traits. The genome sequencing projects as well as the knowledge on the expressed part of the genome of important crop species offer an excellent opportunity for the identification of agronomically important genes. The systematic approach to mutate all genes by transposon or T-DNA mutagenesis in maize, rice and *Arabidopsis* will give information on the function of many genes in those genomes. The sequencing of the *Arabidopsis* genome is now complete and the public rice genome project is rapidly advancing. Large numbers of expressed sequence tags (EST) from many other plants are known already.

A controversial trend in development is the gene use restriction technology (GURT), i.e. the production of transgenic plants that produce lethal proteins in the seeds at the time of maturity. The seeds harvested from these genetically engineered plants are sterile and hence cannot be kept for next season's sowing (Steinbrecher and Mooney, 1998; Kangasniemi, 1999). This innovation utilizes a set of introduced genes which enable an external signal (i.e. an applied chemical) to trigger a set of specific molecular events within the developing plant. Examples of how the technology might be used include: preventing re-germination of seeds or gene escape (the sterile seed application); controlling the flowering behaviour of the plant; and aiding the plant's pest defence mechanisms. The technology has only been developed thus far in a model system, tobacco, which is not intended for commercial deployment. The components of the system have not yet been assembled into a single plant intended for commercialization.

Ecological, Economic and Ethical Aspects of Using Transgenic Plants in Cropping Systems

Transgenic plants contain genes from other plant species, bacteria, viruses, fungi or rarely animals. New varieties selected by conventional breeding also contain 'novel' genes which were transferred from a variety of the same crop plant or a closely related species. The genomes of both transgenic and conventionally bred plants are not absolutely stable, as mutations and transposition events can occur. Assessing the risks of transgenic crops should always include a comparison with the potential risks of traditionally bred cultivars. Basically, risk assessment should focus on the transgene introduced and not the method by which it was transferred to the plant. Nevertheless, at the present early stage of gene technology undesired physiological effects caused by random integration of the transgenes as well as minor but undesired side effects cannot be excluded and justify a closer inspection. The experience with conventional plant breeding where whole genomes are mixed and DNA from related species has been introduced without problems can help in the risk assessment of transgenic plants. It is likely that the problems, if any, encountered with transgenic varieties will be very similar to those known from classical varieties. In general, few problems have occurred in classical plant breeding, indicating that plants react in a very benign way to the introduction of new genetic material. It is unlikely that transgenic plants will behave differently.

The recent debate on the effects of transgenic crops concerns ecological, economic and ethical aspects but also pays attention to so-called secondary effects. These include changes of agricultural practice, the quantity and quality of applied pesticides and fertilizers, as well as the extension of cultivated areas and the consequences of the consumption of genetically modified organisms on human and animal health. Most results on possible effects of transgenic crops reflect either processes observed in laboratory, greenhouse or field trials or estimations based on models or historical data and only very few mid- or long-term studies are available yet.

Ecological aspects

The spectrum of potential environmental impacts of transgenic plants depends on the characteristics of the crop and on the environment in which it is to be introduced. Therefore, a 'case-by-case' and 'step-by-step' approach is a suitable basis for the risk assessment of transgenic plants. The following positive aspects are frequently discussed in the assessment of ecological implications of transgenic crops. First, pest- and herbicide-resistant plants can reduce the number and the amount of pesticide applications, leading to less damage of the environment by pesticides. Second, the new varieties might result in higher yield, increasing food production and making land use more efficient. Herbicide-tolerant crops are compatible with reduced or no-tillage methods in weed control, which help to preserve topsoil exposed to wind and water erosion. On the other hand, putative negative impacts have also to be considered. These include impacts on biodiversity, vertical gene transfer and effects on non-target organisms.

Positive ecological effects due to reduction in pesticide application

A review of the data available on pesticide use in the transgenic plants of major importance reveals conflicting results, ranging from a decrease, to no change to an increase, depending on the pest pressure varying from year to year and on the type of the previously applied pesticides.

The few data available on herbicide use in glyphosate (Roundup) tolerant soybean show a decrease in herbicide use (for a complete list of references see www.unizh.ch/botinst). In many cases, one glyphosate application substituted for two or more active ingredients applied in traditional soybean varieties. This accounts for the decline in the number of herbicide applications in US soybean production between 1995 and 1998 (Gianessi and Carpenter, 2000). Glyphosate is a non-selective, broad-spectrum systemic herbicide with no residual soil activity, a low toxicity and therefore a better ecotoxicological profile than other herbicides.

Two surveys of *Bt* corn (1996 and 1997) showed a decrease in insecticide use (for a complete list of references see www.unizh.ch/botinst). The development of the corn borer occurs mainly in the stem and therefore the timing of insecticide application is critical, as it has to occur before the caterpillar bores into the corn stalk and is protected from insecticides. It is estimated that less than 5% of the field corn acreage in the US corn belt was treated with insecticides against the corn borer prior to the introduction of *Bt* corn varieties. Insecticide use against the corn borer is much higher in states like Nebraska, Kansas, Colorado and Texas (40%), where the spider mite is an annual problem and corn fields are scouted more frequently than elsewhere, detecting also the presence of the corn borer (Gianessi and Carpenter, 1999). On the other hand, losses due to the European corn borer are unpredictable and vary each year depending on the infestation levels. The question remains open whether *Bt* corn is useful to reduce insecticide applications. As indicated below, considerations on increasing yield by reducing the losses are probably more relevant for *Bt* corn.

In 1995, the year before *Bt* cotton varieties were introduced, an average of 3.4 (0.5–7) insecticide applications were made to control bollworm (*Helicoverpa zea*) and budworm (*Heliothis virescens*) across all cotton-producing states (Gianessi and Carpenter, 1999). In 1997, *Bt* varieties were treated between two to four times, whereas conventional varieties were treated six to eight times. Estimation of reduced insecticide use from 1995 to 1999 in *Bt* cotton (for a complete list of references see www.unizh.ch/botinst) has to consider that in many cotton-producing areas, boll weevil eradication programmes have been pursued. This means that in these areas, the return of beneficial insects that naturally control bollworm/budworm should reduce the number of treatments needed for control of these pests. On the other hand, insecticides that were being used to control bollworm/budworm were also controlling other secondary pests. With a reduction in the number of treatments towards tobacco budworm, cotton bollworm and pink bollworm (*Pectinophora gossypiella*), populations of these secondary pests have increased (e.g. Stewart *et al.*, 1998). Therefore, it remains an open question if *Bt* cotton leads to a reduction in insecticide use in the long term.

Bt potatoes controlling Colorado potato beetle (*Leptinotarsa decemlineata*) are only grown on a small area (approximately 4% of US potato acreage in 1999). There are no clear data available on the impact of these crops on pesticide use. *Bt* potatoes are not resistant to several other insect pests that need to be controlled in potato fields. An application of a systemic insecticide provides control against Colorado potato beetle, aphids and other foliar feeders, while the *Bt* potato controls the Colorado potato beetle exclusively. Therefore, most growers would still need to apply insecticides in *Bt* potatoes.

Yield increases

Several studies indicate that *Bt* corn and *Bt* cotton give higher yields (see below, economic aspects). Whether this yield increase in the countries of the north will translate into a reduction of the acreage planted with a particular crop or whether it will lead to a decline of prices remains an open question.

Loss of biodiversity and genetic diversity

There is concern that transgenic crops will lead to a loss of biodiversity and of genetic diversity within crop species. There is an immense reduction of biological diversity due to deforestation and the conversion of native land to agriculture. In this situation, the possible loss of biodiversity due to cultivation of transgenic crop varieties is not significant and can be prevented by careful preservation of landraces eventually displaced by transgenic varieties. Protection of agrobiodiversity has to be specifically considered in cropping systems rich in biodiversity like agroforestry systems and intercropping and legume-based rotations which are practised in developing countries today. Large-scale cultivation of only a few varieties of transgenic crops might harm genetic biodiversity in these countries severely. On the other hand, every yield improvement on existing agricultural land can save land that is rich in biodiversity from being brought into cultivation.

Vertical gene transfer and introgression

The possibility of an escape of transgenes from cultivated crops into wild relatives by pollen flow is considered to be the most serious environmental risk of transgenic plants. But interbreeding is not unique to transgenic crops and also occurs with regard to non-transgenic resistance traits. It occurs naturally as soon as the main conditions for interbreeding are met: (i) the wild relative must be in the range of the crop pollen and must flower at the time crop pollen is available; (ii) fertilization must occur in the wild relative and viable seeds must be produced; (iii) the seeds must survive and germinate; and (iv) the progeny of the hybrid seeds must be fertile or survive vegetatively. There is no evidence that pollen from genetically modified plants behaves differently from pollen of non-transgenic plants and cross-pollination between crop plants and wild relatives can be predicted by the data on outcrossing of classically bred varieties.

Twelve out of 13 of the world's most important food crops hybridize with their wild relatives in some geographical areas of their agricultural distribution (Ellstrand *et al.*, 1999). If new alleles, e.g. herbicide tolerance, disease or insect resistance from transgenic plants persist in wild populations, gene flow might lead to significant evolutionary change in the recipient populations. This would mainly be expected if the transgene confers a selective advantage to the wild plant, e.g. by creating a more aggressive weed. Increased weediness resulting from gene flow from traditionally improved crops to weedy relatives has been well documented (Ellstrand *et al.*, 1999). Another potential problem with natural hybridization between domesticated plants and their wild relatives is the increased risk of extinction of wild taxa. There has been almost no research on this topic.

Horizontal gene transfer

Non-sexual transmission of recombinant DNA from the transgenic crop to microorganisms, called horizontal gene transfer, is theoretically possible but has so far not been proven in the field (e.g. Ammann *et al.*, 1996). Horizontal gene transfer would only be of importance for the environment if the transferred DNA was expressed in the transformed microorganisms and resulted in a higher competitiveness.

Effects on non-target organisms

Transgenic plants with insecticidal properties may have direct or indirect negative impacts on non-target organisms such as beneficial insects, second-

ary pests or so-called indifferent organisms (i.e. organisms not in direct interaction with the transgenic plant). Insect predators or parasitoids which prey
on crop pests, or insects feeding on pollen from transgenic crops ingest insecticidal compounds and might be affected. The data available show conflicting results (for more information see Schuler *et al.*, 1999).

Development of resistant pest and weed populations

There is considerable experience with the use of monogenic disease and pest
resistance in classical resistance breeding. The use of specific fungal disease
resistance genes in cereals, to give just one example, has frequently resulted
in the rapid breakdown of these genes. The rapid change of resistance source
and the development of virulent pathogen races produced the famous boom-
and-bust cycles of new varieties and pathogen populations. This use of genetic resources is not sustainable and should be avoided, in classical as well
as molecular breeding. The widespread use of transgenic disease- or pest-
resistant crop varieties based on single gene resistances may cause or accelerate the development of resistance of the target pest or pathogen strains.
Permanent expression of a toxin produced by an insect-resistant plant creates a high selection pressure. Several insect populations, including the European corn borer, have developed resistance against pesticides and also
against *Bacillus thuringiensis* formulations (Tabashnik, 1994). To date, no
resistance of target pests to *Bt* expressed in transgenic corn, cotton or potato
has been reported. After large-scale commercial planting of transgenic crops,
different glyphosate-resistant weeds have emerged in herbicide-resistant
soybean (Owen, 1997; Benbrook, 1999).

Viral disease resistance and concerns about new viruses

Virus resistance was one of the first targets for genetic modification.
Transgenic plants resistant to viral pathogens were transformed with genes
encoding viral coat proteins. Exchange of these genes with other viral patho-
gens may be possible, creating new virus strains with unknown properties.
Expert opinions vary widely as to whether this could be a significant problem
(Robinson, 1996; Nuffield, 1999).

Economic aspects

Estimating costs and returns

The relevant costs for comparing genetically engineered with traditional
crop varieties include the cost of seed, the full costs of pest control, including
pesticide material, pesticide applications, pest scouting, and any alternative
(e.g. mechanical) pest control. Gross returns have to be estimated as the
value of production using the actual crop yield multiplied by a state-average
harvest-period price for each commodity. The value of production less total

seed and weed or pest control costs represents the relevant net returns for comparing transgenic crops with traditional crops (Fernandez-Cornejo and McBride, 2000). Because factors influencing cost-effectiveness of transgenic plants, such as the price of the product or the infestation level of a pest, vary from year to year and from location to location, the economics of transgenic plants will only become clear in the long term.

From the eight commercially cultivated transgenic crops in 1999, data on net returns are available for herbicide-tolerant soybean, cotton and corn and *Bt* cotton and *Bt* corn. Adoption of herbicide-tolerant soybean led to conflicting results in net returns, while two studies on the adoption of herbicide-tolerant cotton and corn showed the same net return. *Bt* cotton and *Bt* corn led to significantly increased net returns except in years with low pest pressure (for a complete list of references see www.unizh.ch/botinst).

The introduction of the competitively priced Roundup-Ready™ weed control systems led to reductions in the prices farmers paid for herbicides. For example, the price of glyphosate was reduced in 1998 by 22%. The result of lower priced Roundup treatments in comparison with competing herbicides and the lowering of the price for key herbicides including glyphosate meant that soybean growers spent significantly less on herbicides in 1998 than in 1995 (Gianessi and Carpenter, 2000). An increasing number of farmers need two applications of Roundup and one application of at least one additional active ingredient. There are little or no herbicide cost-savings on such farms relative to growers planting conventional varieties and using integrated weed management. Many price regulations, volume discounts and product guarantees have made it very difficult to compare the actual cost of herbicide-based systems (Benbrook, 1999).

Ethical aspects

The assessment of ethical aspects of transgenic crops has to be carried out with interdisciplinary approaches. Environmental, economic, political and social criteria need to be assessed by the corresponding experts and the ethical assessment can then be carried out based on the professional statements.

The three basic principles for the analysis of ethical aspects and sustainability of transgenic crops are: (i) the conservation of present conditions including knowledge; (ii) the conservation of options; and (iii) the conservation of access (Weiss, 1989). Each generation should leave to its successors the same or better conditions keeping a variety of choices, including such that are not foreseeable at the time. One such choice is whether traditional cropping systems will still function after the planting of transgenic crops. Conservation of options is defined as conserving the diversity of the resource base and not the preservation of the status quo. This can be accomplished in part by new technological developments such as transgenic crops tolerant to environmental constraints that represent substitutes for existing resources (traditional varieties), as well as by conservation of existing resources such as soil and water. The principle of conservation of access aims at balancing justice requirements between and within generations. For example, a new technology aiming at environmental conservation benefiting future generations must not be at the expense of the poorest people in developing countries today. In the context of transgenic crops, seeds must be accessible and affordable for people in developing countries and further developments in biotechnology must be conducted in accordance with the needs of developing countries. As the commercial production of genetically engineered crop varieties requires many different patents and licences, e.g. for the vectors, the genes used in the plant transformation systems and selectable markers, patents on these materials may create major problems and potentially significant additional expenses for the public-sector breeding programmes that produce seed for poor farmers. Intellectual property will most probably reduce the possibility of research institutions in developing countries commercializing transgenic plants if the patents or licences are not distributed for free to these institutions.

The consequences of transgenic crops on floral and faunal biodiversity and on single species are of major importance for the analysis of the environmental impact of new crop varieties. Therefore, a careful benefit and risk assessment has to be conducted, including comparison with traditional pest management practices or traditionally bred varieties.

Prospects and Risks of Improving Sustainability and Food Security

Environmental, economic and social consequences of a transgenic variety could vary substantially by crop and country or region. Therefore, the development of transgenic plants must be based on profound knowledge of regional characteristics, needs and constraints to guarantee sustainability and food security in the long term.

Sustainability in crop production depends on ecological and socio-economic as well as ethical factors. Indicators for ecological sustainability in plant production include agricultural productivity, soil erosion, groundwater quantity and quality, surface-water quality and the conservation of non-renewable resources such as soil fertility and biodiversity. Indicators for socio-economic sustainability are the regeneration, conservation and improvement of life quality in the sense of providing sufficient quantities of nutritionally and hygienically valuable food, i.e. food security, and the conservation and improvement of the arable land (Schulte, 1998) while the indicators of ethical sustainability are the three principles mentioned above: conservation of present conditions, of options and of access.

In developing countries, the needs for biotechnology are quite different from those of industrial countries. For example, enhanced food production and adaptation of plants to environmental constraints are the main goals of biotechnology in developing countries. Continuous improvement of crop yields by growing high-yielding crop varieties or by reintroducing agriculture into deserted areas would save land and thereby protect the last undisturbed habitats. Agriculture in industrial countries is characterized by overproduction and is driven by market and environmental arguments. The debate about biotechnology for developing countries should not be whether or not they need biotechnology, but how biotechnology can be applied in sustainable ways that contribute to improved agriculture and to the social and economic welfare of the people in the long term.

Socio-cultural, economic and geographical factors all play an important role in food supply for developing countries. Given the complexity of factors involved, the approaches to guarantee food security have to be very diverse (Leisinger, 1999). Transgenic breeding is one of several technical options that can contribute to increased food production. However, without simultaneous other efforts, e.g. higher political priority for rural areas and the support of indigenous and traditional technologies, transgenic crops can only benefit a small minority of underprivileged people (Schreiber, 1999).

The question of sustainability has to be discussed case by case: for example, the development of salt-tolerant plants will only be a useful tool for sustainability if these crops can be grown on salt-affected dryland and can be irrigated with seawater or water of marginal quality. If such crops would be planted on soils with high levels of salt due to intensive plant production, the problem would only be aggravated by continued irrigation. The development of plant varieties more efficient in the use of phosphorus (P) represents a way to achieve a more sustainable agriculture. Phosphorus is one of the most important nutrients limiting agricultural production worldwide. To reduce P deficiencies and ensure plant productivity, nearly 30 million tons of P fertilizer are used every year. Up to 80% of these 30 million tons are lost because P becomes immobilized and unavailable for plant uptake. Furthermore, plant varieties with an increased P acquisition capacity should be of great importance for subsistence farmers who cannot afford commercial P fertilizer (López-Bucio *et al.*, 2000). Aluminium-tolerant crops also represent a way towards more sustainable agriculture particularly in tropical areas. Aluminium is found in soluble form in many acidic soils that comprise about 40% of the total arable land. Such soils are frequently found in the tropics. Aluminium is toxic for many crops and it is the major limiting factor for plant productivity on these soils.

Referring to the trends in development discussed above, gene use restriction technology would clearly disadvantage poor farmers, who at present keep seeds from year to year and are not able to buy new seed every year. A similar situation of restricted access to better varieties has been created by the use of F1 hybrids which are planted only in parts of the developing world (Nuffield, 1999). On the other hand, apomixis technology, if developed in a form which is applicable to the major staple crops, will be very valuable to local breeders and for the maintenance of within-crop genetic diversity. Hopefully, this technology, once developed, will be widely available, particularly to breeders in developing countries.

Vertical gene transfer might be the biggest risk for the sustainable use of transgenic crops. If we assume that non-target plants acquire insect-resistance genes from transgenic crops, insect populations feeding on previously non-toxic wild plants might be seriously damaged. Besides these direct effects, the acquisition of insect resistance or herbicide tolerance by wild plants could change their population dynamics and increase their invasive potential. Such considerations also apply to resistances against viral and fungal diseases. This is an even more serious issue for developing countries where control of invasive plants is a major problem for subsistence farmers and may have implications for ecosystems of global importance. For many of the important crop species, the risk for vertical gene transfer is higher in the developing than in industrialized countries for two reasons: (i) wild relatives are often common in developing countries as the crop plants originate there; and (ii) cultivated land is far more mixed with uncultivated land than in industrialized countries. Vertical gene transfer might also be a problem for preservation of the biodiversity in the centres of origin of the particular crop species. The assessment of the extent of gene flow between transgenic crops and wild relatives has to be examined on a case-by-case basis in different geographical regions.

CGIAR (Consultative Group on International Agricultural Research), ISAAA (International Service for the Acquisition of Agri-Biotech Applications), the Rockefeller Foundation, ISNAR (International Service for National Agricultural Research) and many other organizations and smaller institutions support knowledge and technology transfer to developing countries. This support should allow these countries to develop their own varieties adapted to local conditions and practices. Donor agencies considering investing in agricultural biotechnology research in developing countries should carefully determine in collaborative efforts what are the most pressing needs and current opportunities and what research is already going on. Within the framework of the Life Sciences Programme, UNESCO (United Nations Education, Scientific and Cultural Organization) supports three programmes for education and training in plant molecular biology, plant biotechnology and aquatic biotechnology in several developing regions of the world. The goal of these training centres is to ensure that the application of biotechnology is beneficial for the developing countries (Brink et al., 1999).

The following concerns are most frequently expressed in the discussion about the sustainable use of transgenic plants in developing countries (Leisinger, 1999), but are also of major importance for non-transgenic crops:

- The prosperity gap between north and south might widen, e.g. by the substitution of tropical agricultural export products with genetically modified products, or by the exploitation of indigenous genetic resources of the south without appropriate compensation by the north.

- Most of the world's biodiversity and centres of genetic diversity for important crop plants can be found in developing countries: the farmers might increasingly use a small number of more productive genetically modified varieties instead of the many thousands of traditional local varieties previously used, resulting in a loss of biodiversity.

Not all developing countries have yet implemented laws and regulations to handle the safe introduction of genetically modified products. Field tests or the commercial introduction of genetically modified plants in such countries could pose a greater risk as the local environment has not been appropriately considered during risk assessment of the same or similar transgenic plants in countries with an established regulatory system.

Conclusions

Genetically modified plants are the products of a very new technology and it is not yet possible to draw general conclusions about its application as these are often made in the heated political debate. The present commercial, agricultural use of transgenic plants is mainly based on two traits: herbicide resistance and *Bt*-based insect resistance. No general conclusion can be drawn about the usefulness of gene technology in plant breeding based on such a minimal set of traits. In the future, it is likely that gene technology will allow increases in crop yields per hectare, as this is possibly already the case for *Bt* corn. This would not only increase food production, but also reduce the land required for farming and contribute to the preservation of forests and other habitats. In addition, it will be possible to grow crops again in areas where agriculture was given up due to environmental constraints. These applications will be of significant benefit for developing countries. The use of transgenic crops will also contribute to a sustainable agriculture if pesticide use can be decreased and loss of topsoil can be reduced. Economic and environmental conditions vary widely from year to year and from place to place. Thus, an accurate assessment of the yield changes and net returns requires 10 years or more of commercial cultivation of transgenic cultivars. The recent figures discussed in this chapter only reflect experiences from the first 2 to 3 years of commercial application.

The potential benefits of plant biotechnology for developing countries are unlikely to be realized unless the technology and knowledge gap between industrialized and developing countries is bridged and the transgenic seeds are provided to all farmers free or at nominal cost. This will require public investment by national governments and donors, in collaboration with the private sector. Research as well as the subsequent distribution of seed and technical advice has to be promoted. Breeding programmes will also need to include staple foods such as cassava, upland rice, African corn, sorghum and millet.

Challenges

The molecular understanding of physiological processes in plant development and interactions with the environment is only in its beginnings. The complete sequences of the two model genomes *Arabidopsis* and rice will soon be available and there is a constant improvement of our tools for functional analysis of a large number of genes. Thus, the possibilities for improving key agronomical traits will grow rapidly in the future. The first genetically modified varieties used in commercial agriculture are no more than prototypes and will be succeeded by improved products. We suggest that, in order to continue the development of gene technology in agriculture, the risks must be carefully analysed and the plants monitored to detect possible problems. We are convinced that the new and powerful technology has to be further developed for the benefit of the next generations.

References

Ammann, K., Jacot, Y. and Rufener Al Mazyad, P. (1996) Field release of transgenic crops in Switzerland – an ecological risk assessment of vertical gene flow. In: Schulte, E. and Käppeli, O. (eds) *Gentechnisch veränderte krankheits- und schädlingsresistente Nutzpflanzen – Eine Option für die Landwirtschaft?, Band 1*. BATS, Basel, pp. 101–157.

Benbrook, C. (1999) Evidence of the magnitude and consequences of the Roundup Ready Soybean yield drag from university-based varietal trials in 1998. *AgBioTech InfoNet Technical Paper 1, July 13*.

Brink, J.A., Prior, B. and DaSilva, E.J. (1999) Developing biotechnology around the world. *Nature Biotechnology* 17, 434–436.

Conway, G. and Toenniessen, G. (1999) Feeding the world in the twenty-first century. *Nature* 402(6761), C55–C58.

De la Fuente, J.M., Ramirez-Rodriguez, V., Cabrera-Ponce, J.L. and Herrera-Estrella, L. (1997) Aluminium tolerance in transgenic plants by alteration of citrate synthesis. *Science* 276, 1566–1568.

Ellstrand, N.C., Prentice, H.C. and Hancock, J.F. (1999) Gene flow and introgression from domesticated plants into their wild relatives. *Annual Review of Ecological Systematics* 30, 539–563.

Fernandez-Cornejo, J. and McBride, W.D. (2000) Genetically engineered crops for pest management in US agriculture: farm-level effects. *Agricultural Economic Report No. 786*, Economic Research Service/USDA, Washington, DC.

Gianessi, L.P. and Carpenter, J.E. (1999) Agricultural biotechnology: insect control benefits. National Center for Food and Agricultural Policy NCFAP, Washington, DC.

Gianessi, L.P. and Carpenter, J.E. (2000) *Agricultural Biotechnology: Benefits of Transgenic Soybeans*. National Center for Food and Agricultural Policy (NCFAP), Washington, DC.

Goto, F., Yoshihara, R., Shigemoto, N., Toki, S. and Takaiwa, F. (1999) Iron fortification of rice seed by the soybean ferritin gene. *Nature Biotechnology* 17, 282–286.

Hirsch, R.E. and Sussmann, M.R. (1999) Improving nutrient capture from soil by the genetic manipulation of crop plants. *Trends in Biotechnology* 17, 356–361.

Kangasniemi, J. (1999) Are exclusion technologies inclusive? 'Terminator Gene' controversy raises hopes and fears. *Diversity* 15, 19–21.

Leisinger, K.M. (1999) Biotechnology and food security. *Current Science* 76, 488–500.

López-Bucio, J., Martínez de la Vega, O., Guevara-García, A. and Herrera-Estrella, L. (2000) Enhanced phosphorus uptake in transgenic tobacco plants that overproduce citrate. *Nature Biotechnology* 18, 450–453.

Nuffield, Council on Bioethics (1999) Genetically modified crops: the ethical and social issues. Nuffield Council on Bioethics, London. http://www.nuffieldfounda tion.org

Owen, M. (1997) North American developments in herbicide tolerant crops. In: British Crop Protection Council (ed.) *The 1997 Brighton Crop Protection Conference: Weeds*, Vol. 3. Brighton, pp. 955–963.

Robinson, D.J. (1996) Environmental risk assessment of releases of transgenic plants containing virus-derived inserts. *Transgenic Research* 5, 359–362.

Schreiber, H.-P. (1999) Socioethical and sociopolitical reflections on the application of gene technology in developing countries. In: Hohn, T. and Leisinger, K.M. (eds) *Biotechnology of Food Crops in Developing Countries*. Springer Verlag, Vienna, pp. 35–37.

Schuler, T.H., Poppy, G.M., Kerry, B.R., and Denholm, I. (1999) Potential side effects of insect-resistant transgenic plants on arthropod natural enemies. *Trends in Biotechnology* 17, 210–216.

Schulte, E. (1998) Kriterien für eine nachhaltige Landwirtschaft. *BioTeCH Forum* 4, 8–9.

Steinbrecher, R.A. and Mooney, P.R. (1998) Terminator technology. *The Ecologist* 28, 276–279.

Stewart, S., Reed, J., Luttrell, R. and Harris, F.A. (1998) Cotton insect control strategy project: comparing *Bt* and conventional cotton management and plant bug control stragegies at five locations in Mississippi, 1995–1997. In: National Cotton Council (ed.) *1998 Proceedings Beltwide Cotton Conference*, Vol. 2. San Diego, California, pp. 1199–1203.

Tabashnik, B.E. (1994) Evolution of resistance to *Bacillus thuringiensis*. *Annual Review of Entomology* 39, 47–79.

Van Dijk, P. and van Damme, J. (2000) Apomixis technology and the paradox of sex. *Trends in Plant Science* 5, 81–84.

Weiss, E.B. (1989) *In Fairness to Future Generations: International Law, Common Patrimony, and Intergenerational Equity*. The United Nations University, Tokyo.

Ye, X., Al-Babili, S., Klöti, A., Zhang, J., Lucca, P., Beyer, P. and Potrykus, I. (2000) Engineering the provitamin A (β-carotene) biosynthetic pathway into (carotenoid-free) rice endosperm. *Science* 287, 303–305.

A complete bibliography can be found at www.unizh.ch/botinst

21

Crop Science: Scientific and Ethical Challenges to Meet Human Needs[1]

L.O. Fresco

Assistant Director-General, Agriculture Department , Food and Agriculture Organization of the United Nations, Rome, Italy

The First International Crop Science Congress was held in 1992. Many of you sitting here today were present then. Looking around the room you may recognize several of your neighbours. They may have reached more senior positions and of course their publications lists are longer. Fortunately, there are also a few more women and there is a new generation of post-doctorates among us. But overall, not much has changed. As individuals, we humans are slow to develop.

Yet, in those 8 years, the world around us has transformed itself at what seems an unprecedented pace, be it in terms of population and income growth or access to modern information technology. And, more importantly for our purpose today, the world of science has changed as well. Obviously, we may congratulate science on the major progress made in the area of crop research, as witnessed by the fact that average crop yields increased fourfold and total crop harvest increased sixfold, with clear implications for food security in the course of the last century. Agricultural research, although still under-funded, can be considered as one of the most profitable investments, something which is not irrelevant given today's mood. Nevertheless, the share of agriculture in total lending has fallen from US$19 billion in the 1970s to US$10 billion in the 1990s (FAO, 1996a).

The public at large, however, equates biological agricultural science with the failure to protect human health, the destruction of the countryside and above all the creation of 'Frankenstein food'. Charles, Prince of Wales (1996), who is widely credited with generating this term, questions man's right 'to experiment, Frankenstein-like, with the very stuff of life'. Popular

---

[1] Presented at the Opening Ceremony of the Third International Crop Science Congress, Hamburg, 17 August 2000.

perception has it that the world of agricultural science has isolated itself from the man in the street (or the woman in the field), and is seeking to impose its ideas on the planet, rather than to understand public needs. These views are not new but have quickly become more vigorous.

In his opening address to the Iowa Congress in 1992, Vernon Ruttan mentioned three waves of social concerns: about potential limits on growth, about pollution and about environmental quality and human health at large (Ruttan, 1992). These are still valid, and so are his recommendations, but the picture today is definitely more complex. While agricultural science never took place in a vacuum, it now has to stand up to intense public scrutiny on how it conducts its business, what subjects it addresses and what its effects are. So it is above all the agricultural scientist who should take to heart what Sir Julian Huxley, eminent scientist, writer and international public servant – a rare combination even in his time – wrote in 1957:

> It is as if man had been appointed managing director of the biggest business of all, the business of evolution . . . whether he is conscious of what he is doing or not, he is in point of fact determining the future direction of evolution on this earth. That is his inescapable destiny, and the sooner he realizes it and starts believing in it, the better for all concerned.

I propose to analyse a number of current and emerging trends in the world around us from an ethical perspective in order to define a selected number of scientific challenges for crop science and technology.

Some World Agricultural Trends Seen from an Ethical Perspective

The basic trend: the uneven distribution of food

Food production has increased dramatically over the last 35 years: in spite of a 70% increase in world population, per capita supply has increased by almost 20% (FAO, 1996b). In developing countries, population has almost doubled while per capita supply grew by almost 30%. As a result, the percentage of hungry people was halved from 36% in 1970 to 18% in 1995–1997. In absolute numbers the decline is less spectacular: 790 million people in developing countries and 34 million in developed countries are still undernourished. At the 1996 World Food Summit all countries committed themselves to bringing down the world's hungry to 415 million. However, our latest assessment indicates that 580 million individuals would still be undernourished by 2015, with 23 countries having more than 25% of the population undernourished (FAO, 2000). Notwithstanding urbanization, the majority of those suffering from food insecurity are in rural areas. Cereals will remain the principal source of food supplies, accounting for about half of daily calorie intakes. Around half of the increase will be for food, and about 44% for animal feed. Feed use, especially in developing countries, will be the most dynamic element driving the world cereals economy.

The unbalanced availability of food is mirrored by the uneven application of improved production technologies. The inequalities refer to effective demand, productive capacity and economics. For example, the production of 1 tonne of rice costs approximately US$1000 in Japan, US$300 in Italy and Niger, and US$80 in the Philippines. A detailed analysis is required to understand how these differences can be attributed to the cost of labour, fertilizers, technologies, or biophysical potential.

Although, to a large extent, the uneven application of technology is caused by factors outside the realm of science, there is no reason for complacency. Scientists do bear a part of the responsibility for the selective applicability of technologies to more favourable ecological circumstances and as a result their uneven application. The ethical questions here relate to what crops are addressed, what type of production systems are targeted and what means are utilized in crop improvement.

The true global trend: globalization

In recent years, the world has expanded beyond the global village to the global market, with ever increasing mobility of capital, labour and goods. Globalization is not only a question of size, but also of kind: it is inextricably linked to privatization. It sets afoot major economic restructuring in both developed and developing countries, and has greatly changed the balance of public and private sectors, also in science. Privatization of knowledge through intellectual property – particularly in the high-technology area – affects the public sector.

Globalization also results in concentration. For the seed and agrochemical industries, for example, it has been estimated that the world's ten top agrochemical industries account for about 85% of the global agrochemical market. In 1998, just four companies controlled 69% of the seed market in the USA (Hayenga, 1998). What will the effects of the development of such huge conglomerates be on the direction of scientific research, in particular in view of the evolving needs of small farmers in developing countries? In some cases, concentration also means reducing product diversity: recently, the world's largest vegetable seed corporation announced that it would eliminate 25% of its product line as a cost-cutting measure.

In many resource-poor countries, agricultural production for export is seen as one of the motors for development. This implies control over the various phases of production and a dependable export certification programme to meet the food safety and other sanitary and phytosanitary regulations of importing countries. Although the World Trade Organisation's Sanitary and Phytosanitary (SPS) agreement has set disciplines to sanitary and phytosanitary measures, this has not yet resulted in less regulation and easier market access. Harmonization, through the setting of international standards, is a process that still needs considerable efforts before it realizes its full potential for trade facilitation.

Whatever its potential benefits, globalization also exacerbates the existing differences among countries and regions and calls for specific strategies to be developed according to different needs. It has offered opportunities to the poorer countries, and driven rapid development and local capital accumulation.

The emerging response: diversification of diets and crops

Partly in response to globalization, but mostly as a result of rising incomes and urbanization, the food demand is becoming more diversified and more quality oriented. This trend appears also in developing countries, albeit in some cases with a considerable time lag. As affluence increases further, we also observe consumer demand trends for food produced with technologies that are regarded as being environmentally sustainable, especially organic agriculture. But diversity is not just for the urban rich. There is convincing evidence that the low intake of micronutrients by the poor has detrimental health effects (Ames, 1998). For the poor too, diversity in food production and consumption is essential.

Finally, new crops for food and industrial purposes may open up rural employment opportunities. The widespread adoption, recently, of sunflower as an oilseed in the Loess plateau of China is but one example of the potential for new crops. Similarly, the production and marketing of local traditional crops and varieties may satisfy urban demand and become a motor of rural development as in the case of indigenous leafy vegetables being grown for urban markets in eastern and southern Africa.

Diversification of crops and products and increasing nutritional qualities demand a sophisticated scientific approach. Here, the ethical issue is again how choices are made and priorities are set with respect to the needs of specific underprivileged target groups.

The background trend: agricultural services to society

The diversification of diets and products does not imply diversification of cropping systems *per se*. Intensive monoculture is still very much the norm in market economies. However, there is growing recognition that agriculture provides more services to society than just producing calories or dollars per hectare. By nature, agriculture deals with common goods and public concerns. Next to its economic services, which also include employment, the agricultural sector is increasingly held responsible for environmental services such as the preservation of watersheds, the protection of agricultural biodiversity, the sequestration of carbon, possibly the production of renewable energy and so on. Moreover, balanced national development implies maintaining rural livelihoods and traditions to keep remote areas liveable. As a result, rural activities such as nature conservation and agro-tourism set new standards and limits for agricultural production systems. In fact, the term sustainability has thus acquired an even broader meaning than before. Such multipurpose land-use systems also herald an era of decentralized and participatory decision making involving all stakeholders. The ethical and technical question for the scientist is how to provide an objective scientific basis, including indicators of environmental, economic and social impact, to allow adequate decision making on balanced agricultural growth.

The overwhelming trend: the information revolution

Information technology is shaping agricultural science and its application. Developing countries are quickly taking advantage of this situation. But we know that on the internet, scientific excellence and nonsense exist side by side. While information technology may be a great trans-boundary equalizer, the need for reliable sources of scientific data is growing. The ethical issue here is whether scientists, in the private and public sectors alike, are sufficiently sharing their results, including their doubts and failures.

The Scientific and Technological Challenges

Much is at stake for crop science: in order to contribute to poverty alleviation, sufficient food and a balanced diet for an increasing world population, intensive cropping systems need to be developed that have a beneficial or at least non-harmful effect on the environment and provide a multitude of services to society. In order to meet this challenge, crop scientists must define more clearly their moral obligations in an increasingly privatized world where the general public has become resolutely biased against agricultural science. Let me highlight a few specific scientific issues in this respect.

Responsible land and water use

The optimal use of natural resources, in particular land and water, is the basis for land use intensification. Today, the per capita world cropland has decreased to only 0.27 ha, while in China it has dropped to 0.08 ha per capita (Lal, 1989). There are estimates that over 25 billion tonnes of topsoil are lost yearly (FAO, 1996b), mainly due to deforestation and overgrazing. Erosion and cropping system-induced soil chemical changes all require research aimed at a better matching of crop species and cultivation systems with specific environments.

Throughout the world average per capita water availability has dropped from 16,000 to 7000 cubic metres between 1950 and 2000 as a result of population growth (FAO, 1994). A case in point is Morocco with a projected availability of only 780 m^3 per person in 2025, while irrigation currently represents 92% of all water use in the country (FAO, 1996b). The World Water Forum recently stressed the need for increased efficiency in agricultural water use. A maize crop producing about 8 t ha^{-1} of grain consumes more than 5 million litres per hectare of water by evapotranspiration during the growing season, requiring approximately 1000 mm of rainfall or 10 million litres of irrigation water (Pimentel *et al.*, 1997). Balancing the requirements of agriculture with those of the population and industry forces crop science to re-examine crop yield performance. If crop yield is expressed per unit of water rather than per unit of land, and if realistic water pricing is applied, this may drive a significant shift towards crops showing a high return per unit of water. The expected shift from rice to wheat in China in the next 25 years for instance will have important implications in terms of water saving (FAO, 2000). Many Low Income Food Deficit Countries face increasing water shortages. More research is needed, not only on physiological mechanisms to increase water-use efficiency, but also on simple techniques such as water harvesting. Such techniques can both reduce risk and increase yields in dry areas. We agree with the Hamburg Declaration that water-saving strategies in irrigated cropping systems and better adaptation of crops to limited water availability deserve attention as such.

But there is more. Within an integrated land and water approach, the logical complement to improving water availability to crops is the development of new lines that are drought resistant or, at least, drought tolerant. The revolution in molecular genetics has now made it possible, at least in theory, to target quantitative trait loci and thus increase the efficiency of breeding for some traditionally intractable agronomic problems such as drought resistance and improved root systems.

Harnessing diversity

Only nine plant species alone provide more than 75% of all human food. A mere three plant species (rice, wheat and maize) provide more than half the dietary energy of the world's population (FAO, 1999). This is largely the effect of a selective focus of agricultural science, which has hitherto neglected the domestication of a multitude of species suited to different ecological niches. FAO recognizes that food security calls for continuing work on the genetic improvement of the main crops, especially to increase their adaptability to the wide diversity of agroecological conditions. However, I also want to emphasize the need to explore a wider range of species that are already adapted to different and marginal ecologies.

Domestication of new crops may be time consuming, but there seems to be a lot of scope in the improvement of locally important minor crops, which in many regions make a major contribution to the diet, but which attract limited R&D resources. There is a key role for international public science in this respect.

Whatever the species, a promising research line seems to be for higher net photosynthetic rates. Again, new molecular technologies may increase the possibilities of adapting C_4 species to new environments and of transferring the C_4 metabolism to important C_3 species.

Food insecure populations would also benefit from a focus on multipurpose species. Chinese scientists have demonstrated the huge potential of sweet sorghum (*Sorghum bicolor*) in terms of grain and biomass production for human food, sugar production, animal feed and bio-energy (Li Dajue, 1997). Similarly, insufficient attention has been given to biological nitrogen fixation (BNF). Although it is recognized that possibilities exist for enhancing BNF performance of crop legumes, for increasing the role of legumes in cropping systems, and for transferring BNF capabilities to non-leguminous crop species (especially the major cereals), research has been fragmented, sporadic and under-funded (Boddey *et al.*, 1997).

Furthermore, there is considerable scope for taking a fresh look at perennial crops versus annuals, because of their reduced fertilizer requirements, protection against soil erosion and their provision of shelter for useful species. Let me cite just one example of a perennial species with a high potential for irrigated arid lands: the date palm (*Phoenix dactylifera*). Recently its cultivation has been introduced successfully in arid regions of southern Africa.

Let me also mention the potential for crop diversification strategies that promote mitigation (or reduction) in the emission of greenhouse gases and favour carbon storage. Also, advances in agricultural ecology have led to cropping systems that harness a wider range of natural resources and thereby divert solar energy to non-harvested species. FAO's IPM programme has demonstrated the benefits of enhancing associated biological diversity and promoting the role of non-harvested species within production ecosystems that are critical to ecosystem functioning, even when actual crop diversity is low.

Sustainability: more empirical and integrated approaches

Farming systems may be deemed sustainable and efficient if organic matter, nutrient levels, soil structure, erosion and ease of root penetration are all at acceptable levels, if nutrient cycling rates are acceptable and if this is proven by sustainable crop yields that are fully satisfactory, as percentages of the yield potentials, and by acceptable rates of off-site losses (Tinker, 2000). This implies that research institutions must become more closely involved with real farms on a long-term basis in order to ensure that advances in crop science can be adapted to the scale and reality of production, especially in food-insecure regions. Pragmatism in crop research planning could help to offset the risks of fleeing into modelling and neglecting the empirical and applied work that really needs to be done in the field: the high prestige work tends to be behind the computer and in the laboratory. Moreover, crop science has become too atomized and specialized to be able to perform the integrating function needed to achieve sustainability. Neither curricula nor career development currently allow much opportunity for the creation of a new generation of empirical agronomists with sustainable systems design in their terms of reference.

Genetically modified organisms

The most forceful public questions are being asked about both the sharing of benefits and the perceived negative effects on human health and the earth's environment of the uncontrolled application of genetically modified crops. FAO's position is that we must use every means at our disposal to improve food security subject to careful assessments being made. It is undeniable that one of the promising technical ingredients for agricultural development is biotechnology, because it will allow a more precise adaptation of genotypes to environmental conditions, nutritional and dietary needs and market preferences. But is biotechnology increasing the amount of food in the world and is more food accessible to the hungry because of such innovation?

Public concern and scientific ethic have raised awareness about the potential risks associated with the transfer of genes in the absence of a better understanding of their functional characteristics. State-of-the-art knowledge is not yet adequate to predict either the unsought and adverse effects from the inserted gene itself, or the way in which the inserted gene may alter expression of existing genes in a plant or animal. It should be emphasized that no human health problems resulting from biotechnology have been documented to date. However, the lack of evidence of adverse effects is not the same as knowledge that genetic modification is safe. Internationally, food safety is addressed in the FAO/WHO Codex Alimentarius and the effects of living modified organisms on biodiversity are addressed in the Cartagena Protocol. International and national regulation and risk assessment are relatively new, and public trust in these processes is low.

Classical plant breeding and biotechnology depend on naturally occurring genes. For this reason, the maintenance of biodiversity is a global concern. Agricultural biodiversity differs qualitatively from wild biodiversity, and its specific features and problems need specific solutions. The most important genes are held within agricultural systems, at intra- and interspecific levels, which have been evolving in a truly international endeavour since the Neolithic Period. No country can now do without resources from elsewhere. A number of tropical or sub-tropical countries, which are poor from the economic point of view, are in the centres of origin of crop plants, and rich in agricultural genetic diversity. International cooperation for the management of plant genetic resources for food and agriculture as a global common good is therefore not an option, but a necessity, which FAO Member Countries are pursuing through the negotiation of the revised International Undertaking on Plant Genetic Resources. An important step in this direction has been the unanimous recognition by FAO Member Countries of Farmers' Rights as the complement of Plant Breeders' Rights. The current challenge before us is to make this concept operative.

Last but not least, although the developed world has created the necessary legal frameworks, there are no economic or legal mechanisms to compensate or provide incentives for the developers of the raw material, the genetic resources themselves. Agricultural biotechnologies also require a solution to the problem of access to the enabling technologies, which are subject now to patent protection, as are the final products.

Transparency of information and decision making

At the basis of public concern is a feeling of being left out, or worse, of not being told the truth. With the wisdom of hindsight we may surmise that science, and particularly agricultural biotechnology, could have done much more to get the public on its side by being open and communicative about its pioneering work in molecular biology and genetic engineering as it was developing. A recent opinion poll has indicated that over 80% of British and German citizens are against consuming genetically modified food. It would appear that *post factum* attempts to overcome consumers' doubts through such instruments as information campaigns and consensus conferences have not helped much in allaying suspicions which have, by now, become entrenched.

The current debate over labelling of foods containing ingredients from genetically modified organisms has highlighted the role of transparency. Food processors and retailers, as actors in the food supply chain, are close to consumers and very sensitive to consumer opinion. Food industry leaders recognize that public confidence is essential to the success of any product and they are very aware of the public's ambivalence towards genetic engineering. Some leading food companies have excluded genetically engineered ingredients from their products because they are wary of consumer rejection. On the other hand, others are lobbying strongly against the public pressure to segregate genetically modified (GM) foods, with the net result that polarization over the issue of labelling continues to perturb governments, public administrations and industry throughout the world.

There can be no doubt that scientists have an absolute moral responsibility to provide objective, peer-reviewed information to the public and to refrain from publicizing immature, insufficiently tested results.

FAO and the Need for Concerted International Efforts

I have referred several times to FAO and the international public sector. Beyond taking up issues of common interest for which no immediate and remunerative markets exist, let me highlight two related areas where the existence of an intergovernmental forum is essential to further science for food security, sustainable land use and protection of the environment. One is the area of sanitary and phytosanitary regulations, including food standards. In this era of globalization, international regulatory mechanisms need to be developed to maximize the potentials and minimize the risks. Despite divided opinion, FAO is confident that consensus on GM food standards can be achieved. We also believe that the conclusion of the Cartagena Protocol is a large step towards facilitating the protection of the environment. Other international agreements, like the International Plant Protection Convention, may play a role in the sustainable use of GMOs. FAO has just established an international 'ethics committee' adding the input of philosophers and religious representatives to that of scientists to investigate human factors related to GM agriculture so that strategies can be developed to use this tool in the fight against hunger and malnutrition whilst taking all the necessary precautions to protect human health and the environment.

The second area is information sharing. A massive amount of information is quickly available in all parts of the world through the Web and scientific journals. However, there are few mechanisms to enhance the access of poor countries to information and to help decision-makers to sift through the significance and applicability of bewildering and often contradictory data. We are concerned that information resulting from scientific research is not adequately shared and spread, and that the information tools used are not always the most appropriate ones. We are actively promoting databases to remedy the situation. A visit to FAO's website (www.fao.org) can turn up several examples of databases that could be useful for those dealing with crop production.

We concur with the proposal in the 'Hamburg Declaration' – that DNA sequences of plant genomes should be released to a public database – as was also suggested to the CGIAR (TAC, 2000). In the context of the negotiations on plant genetic resources such proposals could perhaps be regarded as a kind of tax in kind whereby institutions and corporations would effectively devote some of their resources to respond to world food problems and open questions about equity.

Finally

In the 8 years since Ames, we have grown older and perhaps also sadder and wiser. To be a crop scientist today is not easy, especially when we see young and brilliant minds lured away into computer science or business administration and when public opinion considers that we are interfering with evolution. However, all of us have a responsibility toward the weak and the poor even if, in our rapidly globalizing and unequal world, this is not self-evident. Crop scientists need to look beyond their sub-disciplines and support policy and regulatory measures to protect international public goods, such as water, soil nutrients and genetic diversity.

We must distinguish between the emerging global economy – which is a reality – and global society – which has yet to be built. It is perhaps in pursuit of the latter objective that the moral responsibility of science and individual scientists, whether they work in the public sector or private companies, needs to be emphasized. As global markets are not matched by global governance, many trans-boundary issues remain poorly solved or managed. Such issues are phytosanitary standards and risk analysis, optimal use of the earth's land and water resources, and the mitigation and contributing role of agriculture in global climatic change. The trend to privatization may leave major areas of concern to humanity unaddressed, as they have no market value or one that cannot be quantified. While we are more aware of the need to manage international public goods responsibly, the political tools to do so are weak, and, in the globalized economy, small countries, small companies and small farmers have very small voices. Scientists have moral responsibilities to speak for the weak, because they sometimes best understand the likely results of not doing so.

But the first concern must be to regain the credibility and public acceptance of agriculture science. To return once more to Dr Julian Huxley: what should drive us, as scientists, as international civil servants, is, in his words:

> ... curiosity, initiative, originality, and the ruthless application of honesty that count in research – much more than feats of logic and memory alone.

This is, I believe, the ultimate challenge for the crop scientist: to put these moral qualities at the service of the problems of development and food security.

Acknowledgements

The author wishes to acknowledge the contribution and efforts of the many FAO colleagues who collaborated in the preparation of this paper. Special thanks go to Alison Hodder and Caterina Batello, Agricultural Officers, Crop and Grassland Service.

References

Ames, B. (1998) Micronutrients prevent cancer and delay ageing. *Toxicology Letters* 102–103, 5–18.

Boddey, R.M., de Moraes SA, J.C., Bruno, J., Alves, R. and Urquiaga, S. (1997) The contribution of biological nitrogen fixation for sustainable agricultural systems in the tropics. *Soil Biology and Biochemistry* 129, 787–799.

FAO (1994) Water policies and agriculture. Special Chapter of *The State of Food and Agriculture 1993*, FAO.

FAO (1996a) Investment in agriculture: evolution and prospects. *Technical Background Documents of the World Food Summit, Rome, Italy, 13–17 November 1996*, Vol. 2, Document 10. FAO.

FAO (1996b) Food for all. *World Food Summit, Rome, Italy, 13–17 November 1996*. FAO.

FAO (1999) *1994–96 Food Balance Sheets*. FAO.

FAO (2000) *Agriculture: Towards 2015/30*. Technical Interim Report. April 2000, Rome. FAO.

Hayenga, M. (1998) Structural change in the biotech seed and chemical industrial complex. *AgBioForum* 1, 2, Autumn.

Lal, R. (1989) Land degradation and its impact on food and other resources. In: Pimentel, D. (ed.) *Food and Nature Resources*. Academic Press, San Diego, pp. 85–140.

Li Dajue (1997) Developing sweet sorghum to accept the challenge problems on food, energy and environment in 21st Century. In: Li Dajue (ed.) *Proceedings of the First International Sweet Sorghum Conference*. FAO/CAS, Beijing, September 1997, 19–34.

Pimentel, D., Houser, J., Preiss, E., White, O., Fang, H., Mesnick, L., Barsky, T., Tariche, S., Schreck, J. and Alpert, S. (1997) Water resources: agriculture, the environment, and society. *Bioscience* 47, 97–106.

Prince of Wales (1996) *The 1996 Lady Eve Balfour Memorial Lecture*. 50th Anniversary of the Soil Association.

Ruttan, V.W. (1992) Research to meet crop production needs: into the 21st Century. In: Buxton, D.R. (eds) *International Crop Science I, Proceedings of the First International Crop Science Congress, Ames, Iowa. July 1992*. Crop Science Society of America, Madison, Wisconsin, 3–11.

TAC (Technical Advisory Committee) (2000) *2010 Vision and Strategy Papers*. Synthesis of Selected Strategic Documents. Special TAC Meeting on CGIAR Vision and Strategy, January 2000.

Tinker, P.B. (2000) Future NRM Work. (Unpubl.). Think piece for Special TAC Meeting on CGIAR Vision and Strategy, January 2000 (*cf.* TAC, 2000).

Declaration of Hamburg

22

Edited by J.H.J. Spiertz

President, International Crop Science Congress

Preamble

The concern of crop scientists about the role of society and science in meeting the demand for food of a growing population in a sustainable way was expressed by launching the Declaration of Hamburg during the 3rd International Crop Science Congress on August 22, 2000. This Declaration aims at an on-going discussion about our professional and ethical obligations for meeting human needs and maintaining natural resources.

At present, agriculture is able to feed the majority of humans. But scientists are becoming very concerned that this will not be the case when we have 8 billion people on the planet in about 20 years time. Failure to feed 8 billion people in a sustainable way will lead to enormous environmental damage, social dislocation and reduced economic growth that will affect the whole world.

Scientists, gathered at the 3rd International Crop Science Congress, alerted the society and policy makers to take action on the following concerns:

- The lack of awareness on the gravity of food security and poverty issues for the next 20 years on the global level.
- The urgency of protecting genetic resources and biodiversity.
- The scarcity and degradation of natural resources, such as land and water.

Strengthening agricultural research and education at national and international level is a prerequisite to fulfil future human needs. We believe there are grounds for cautious hope in our ability to feed ourselves via improved education, modern technological developments and, most importantly, a shared appreciation of the problem. Failure of the world's agricultural scientists to communicate this message would be an abandonment of one of their most important professional and ethical obligations.

The 3rd International Crop Science Congress highlighted the fact that sustainable development of plant production and resource conservation is essential for achieving and maintaining food security. This requires a better and more comprehensive insight into ecologically sound crop production processes, especially in fragile environments and resource-poor countries.

The participants at the 3rd International Crop Science Congress decided to foster the necessary dialogue on the potential contributions of new technologies, such as molecular biology and gene technology, to production agriculture and resource use management. Science-based knowledge is essential for a well-informed dialogue and for an effective policy and regulation of new technologies.

Appendix to the Declaration of Hamburg

Concerns and prospects

1. *Crop sciences* provide the key-knowledge base for increasing the production and quality of human food, animal feed and biomass for industrial use and the provision of energy. Crop sciences play a vital role in improving the quality of life for all human beings by producing food and renewable resources in a sustainable and safe way that meets the needs and the standards of the Global Village.

2. *Food security* continues to be a growing concern globally, due to widespread poverty and the need for more and better quality food for a fast growing population, particularly in developing countries. The world's food supply depends in many regions on the *availability of land and water*, which is becoming increasingly scarce as demands for it increase. Water saving strategies and a better adaptation of crops to limited water and nutrient availability in semiarid regions are key issues for research.

3. To overcome the large losses due to *pests and weeds* and to reduce the use of pesticides and herbicides, advanced technologies for integrated crop management and ecologically based sustainable cropping systems should become available.

4. When developing new crops and cropping systems knowledge of *genetic resources* and *plant biotechnology* should be developed and applied with a focus on sustainable and efficient plant production. An integrated and multidisciplinary approach to crop science is needed for innovations and improvements in agricultural production and resource conservation at the field, farm/household, regional and global level.

5. To meet the concerns of society, participation of scientists in the public debate on the potential benefits and risks of the application of *modern technologies* should be encouraged by all public and private research organizations.

6. For the sake of human health, a high priority should be given to improvement of the quality and monitoring of the *safety of human food and animal* *feed* throughout the production, storage, processing and delivery chain of plant produce.

7. To bridge the *gap in knowledge and modern technologies* between developed and developing countries, more financial resources should be made available for education, training, access to public and private research as well as for technology transfer.

8. To strengthen the role of farmers in the *stewardship* of the rural landscape, natural resources and environment, more emphasis should be given in research and education to forms of land use other than agricultural production.

9. To foster a greater understanding of *environmental impacts*, and to promote efficient *resource use*, knowledge and research tools in crop science should be combined with advances in information and communication technologies, such as GIS and DSS.

Hamburg
22 August 2000

Index

Numbers in **bold** indicate tables.

α-tocopherol, 344–345
abiotic stress, 81–95, 334, 335–337
    and biotechnology, 111, 113, 114
    and crop choice, 192
    and crop management, 275
    in developing countries, 102
    downstream reactions, 85–86, 94
    genomic research, 148–150
    and grasslands, 76
    interaction of factors, 106, 122
    long-term developmental changes, 89–
        92, 95
    research, 112–114, 137, 150
    salinity, 81, **82**, 86, **86**
    sensing and signalling, 86–89, 95
    and soil microbial populations, 109
    and species selection, 110–111
    tolerance and genomics, 92–95
    traditional management practices, 107–
        108
    and transgenic plants, **353**
abscisic acid, 121, 142, 333
acidification, 6
Africa
agroforestry, 110
    crop strategies and constraints, 262–
        263
    food crisis, 101
    grain yield predictions, **36, 37**, 38
    land degradation, 67
    loss of agricultural land, 5
    maize research, 365
    maize selected for abiotic stress
        tolerance, 145–146
    millet crops, 311, 317

    non-food crop area, 4
    poor-environment crops, 45
    population increase, **16**, 17
    rice popularity, 316
    soil degradation, 9
    vitamin A deficiency, 58
Africa, East
    crop field weed infestation, 184
    land resources, 1
Africa, North
    cereal imports, **20**
    income growth, **17**
    land resources, **3**, 6, **7**
    water resources, 84
Africa, southern
    arid land date palms, 375
    cereal imports, **20**
    crop field weed infestation, 184
    crop selection for drought stress
        tolerance, 111
    land resources, 1
    savanna tree roots research, 180
Africa, sub-Saharan
    cereal imports, **20**
    food production decline, 312
    income growth, **17**
    integrated pest management, 170
    land resources, **3**, 6, **7**
    late blight disease, **238, 239**
    ley farming, 74
    poor-environment crops, 45
    potato bacterial wilt, 168
    undernutrition, 21
    water resources, 119, **120**

Africa, West
    groundwater use by crops and trees, 182
    root research in agroforestry, 179–180
agricultural diversification, 11
agricultural labour productivity, 244
agricultural landscape history, 213–214, 226
agricultural productivity, 25, 26–27
    and food security, 28–29
agricultural research
    funding, 19, 33, 242, 369
    and household food security, **22**, 23
    marginal lands, 139
    priorities, 29–30, 41–49
Agrobacterium tumefaciens, 58, 61
    isopentenyl gene, 149
    mediated gene transfer, 162, 330, 331, 333
agrochemicals and reduced biodiversity, 28, 29
agroecosystems and biodiversity, 232–233
agroforestry, 108, 110, 128–129, 132, 358
    and crop pest reduction, 184, 185
    erosion control, 199–200
    future research, 188–189
    integrated pest management, 187
    light interception, 180
    microclimate changes, 186–187
    pests and diseases, 182–185
    pests and plant diversity, 185–186
    soil conservation, 201
    tree pruning, 181–182, 188–189
agroforestry in tropics
    definition, 176
    as green technology, 175
    resource capture, 177–178
    root research, 179–180
    tree canopy microclimate modification, 178
agrotechnical innovation, 267–270
air pollution, 34
alley cropping, 175, **177**, 179, 180, 188, 199
    and pest control, 186
    tree pruning advantage, 181–182, 188–189
aluminium tolerance, 353, 363
America, Central
    land resources, **3**, 6, **7**
    maize growing, 44
    savanna grass and tree interaction, 178
America, Latin
    cereal imports, **20**
    food crisis, 101
    land resources, 6, **7**
America, North
    agricultural landscape history, 214
    cereal exports, 19
    poor-environment crops, 45
    weed diversity, 221
America, South
    grassland improvement, 73

land resources, **3**
maize growing, 44
poor-environment crops, 45
and rice popularity, 316
savanna grass and tree interaction, 178
Andes
    conservation and biodiversity, 231–242
    crop rotation, 164–165
    ecoregion definition, 234–235
    late potato blight, 157, 168, 170
    oca tuber weevils, 170–171
    potato disease management, 167–170
    root and tuber crops, 233–234, 240–241
animal feed, 371
    consumption, 47
    and disease, 53–54
    safety, 383
animal production
    in developed countries, 67–68
    and grassland, 66, 77–78
    intensive methods, **69**
antibiotic resistance, 330–331
antioxidants, 142
aphids, 251–252
    and species dispersal, 249
apomixis technology, 355, 363–364
Arabidopsis thaliana, 87–89, 90, 91–94, 336, 338, 342, 345
    genome, 83, 87, 149, 355, 366
Argentina
    government policy and crop system changes, 273
    soybean production, 318
    transgenic crops, 352
arid environments
    crop selection, 111
    plant adaptations, 90–92
aromatic crops, 322–323
arthropods, 251
Asia
    agricultural land shortage, 103–105
    Farmer Field Schools, 166–167
    food crisis, 101
    grain yield predictions, 40
    land degradation, 67
    land resources, 1
    non-food crop area, 4
    paddy field arthropod diversity, 221
    pigeon pea and millet crops, 111
    poor-environment crops, 45
    population increase, **16**, 17
    and rice popularity, 316
    rice yields, 285–287
    soil degradation, 9
    water resources, 119, **120**, 137
    yield potentials, 42
Asia, Central, water resources, 84
Asia, East
    cereal imports, **20**
    grassland improvement, 73–74
    income growth, **17**, 19
    land resources, **3**, 6, **7**

milk products demand, 67
undernutrition, 21
Asia, South
    grassland improvement, 73–74
    income growth, **17**
    land resources, **3**, 6, **7**
    late blight disease, **238, 239**
    milk products demand, 67
    poor-environment crops, 45
    rainfed lowland rice, 46
    undernutrition, 21
    vitamin A deficiency, 58
    yield growth rate decline, 19
Asia, South-East
    brown planthopper, 166–167
    cereal imports, **20**
    Imperata cylindrica control tree fallows,
        183
    income growth, **17**
    land resources, 6, **7**
    late blight disease, **238, 239**
    monogenic resistance breeding, 158
    poor-environment crops, 45
Asia, South-West, late blight disease, **238,
    239**
Asia, West
    cereal imports, **20**
    income growth, **17**
    yield growth rate decline, 19
Asian Vegetable Research and Development
    Center, 325
Assessment of Soil Degradation in South and
    SE Asia, 9
Australia
    agroforestry and ecosystems, 179
    cereal exports, 19
    crop diversification and wheat growing,
        209
    crop diversification in wheat zone,
        202–203, 209
    crop transpiration efficiency, **123**
    dryland cropping systems, 124–125
    grain yield predictions, **36, 37,** 38
    ley farming, 70
    poor-environment crops, 45
    water resources, 84
    wheat root zone drainage, **128**
    zinc-deficient soil, 59
Australia, South
    wheat growing economics, 200, 202
Australia, south-east, and salinization, 199–
    200
Australia, Western, crop and pasture
    legumes, 204, 206–207

β-carotene, 345, 353
    supplementation, 55
    and transgenic rice, 58–59, 62, 283,
        316
Bacillus thuringiensis (*Bt*), 158, 159, 187,
    222, 336, 337, 351, 360, 365

Bangladesh
    land resources, 3
    paddy field weed diversity, 221
    rainfed lowland rice, 46
barley (*Hordeum vulgare*), 202, 203–204,
    205, 235
    domestication of species, 217
beans (*Phaseolus vulgaris*)
    and intercrops, 198
    and iron and zinc deficiency, 60
    and nematodes, 185, 186
beverage crops, 319
biocides, 224
biodiversity, 28, 29, 34, 48, 66–67, 210,
    376–377
    in agroecosystems, 213–226, 232–233
    dynamics, 247–252
    and farm crops, 196–197
    functional, 214
    in grasslands, 71–72
    and human planning, 214
    and integrated pest management, 187
    and landscape planning, 256–257
    and monocultures, 192, 194
    and pest control, 185–186
    sustainability in agriculture, 222–225
    threatened by land clearance, 11
    traditional knowledge and
        sustainability, 233
    tropical rainforest species, 215
biodiversity decrease, 308
biodiversity loss and transgenic plants, 358–
    359, 362
biodynamic farming, 102
biofuels, 4–5, 41–42, 68, 320–321, 373
    see also oilseeds
biogeochemical barriers, 252–253
biogeochemistry, 257
biological control of rice pests, 167
biological nitrogen fixation (BNF), 375
biomass, 180
    in ley farming, 70
    and radiation, 44
    as soil enricher, 11–12
biophysical constraints on production, **47,**
    48
biota
    natural and semi-natural, 214, 215–
        216, **223**
    planned, 214, 217–219, **223**
    unplanned, 219–222, **223,** 226
biotechnology, 41, 329–345
    and abiotic stress mitigation, 111, 113,
        114
    compared with classical breeding, 351–
        352
    and crop production, 208
    and food needs, 170
    and pest control, 158
    and plant breeding, 161–162
biotic stress, 155–171, 334, 335–337
    and crop choice, 192–193
    and grasslands, 76

biotic stress *continued*
  monogenic resistance, 158–160
  resistance breeding, 157–158
birds
  decline and food sources, 251
  and habitat disturbance, 216, 220
  and habitat patches, 220
  and island biogeography, 248
  population decline, 223–224, 225
Bolivia
  Farmer Field Schools for pest
      management, 170
  potato genetic resources rescue, 237
  potato late blight, 157
  subsistence farming, 234–235
Brazil
  biofuel use, 5
  ethanol production, 321
  grain yield predictions, **36, 37**, 38
  grassland improvement, 73
  soybean production, 318
breeding techniques and salinity, **82**
buckwheat (*Fagopyrum esculentum*), 310,
    317
butterflies and field boundaries, 251

Canada
  herbicide-resistant canola, 222
  productivity increases, 45
  transgenic crops, 352
canola *see* rapeseed
carabids
  and field boundaries, 251
  and species dispersal, 249
carbon
  isotopic signature in wheat leaves, 122,
      123
  sequestration, 11–12
carbon dioxide
  fluxes, 67, 76, 78
  and global warming, 283, 293–294,
      298–299, 300
Caribbean
  income growth, **17**
  population increase, **16**
carotenoids, 53, 54, 55, **56**
  see also β-carotene
cassava and iron and zinc deficiency, 60
cDNA sequencing, 158, 160, 333, 342
cereals, 316–317, 370–371
  combined with legumes, **195**
  and crop diversification, 209
  import figures, **20**
  prices, 27
  world crop comparisons, **307, 308**
  yield increases, 18
  yield potential research, 43–44
chemical fertilizers, 10, 12
chemical pesticides, 68, 188
  and biodiversity of pest species, 222,
      224
chickpea (*Cicer arietinum*), 202, 310, 318

China
  change from rice to wheat crops, 374
  crop surpluses, 312
  genetically modified organisms
      adoption, 311
  grain yield predictions, **36, 37**, 38
  hybrid rice yields, 289
  late blight disease, 168, **238, 239**
  loss of agricultural land, 3
  northern plain water table decline, 119
  population increase, **16**
  production of sunflowers, 372
  soybean, 317–318
  sweet sorghum development, 375
  wheat demand, 316
  zinc-deficient soil, 59
cholesterol absorption, 55, **56**, 57
climate as abiotic stress, 105
climate change, 67, 76–77, 236
  and rainfall distribution, 84
  and yield levels, 48
  see also global warming
co-operation for landscape management,
    255, 257
Common Agricultural Policy, 243–244
Communal Seed Banks, 236, 237
compaction control, 139
conjugated linoleic acid, 70
conservation headlands, 249
conservation of agrobiodiversity and farmers'
    co-operation, 233
conservation tillage, 42
Consultative Group on International
    Agricultural Research, 325, 364
corn *see* maize
corridors, 247, 248–249, 250–251, 255
cotton, 318
  Bt transgenic, 311, 326, 352, 353,
      354, 358, 361
  crop rotation, 165
  and nematodes, 185
  with tree shading, 178
cowpea (*Vigna sinensis*), 318
  and water stress, 102
crop boundary planting, **177**
crop diversification, 41, 191–210, 274, 275,
    312, 323–326
  in Australian wheat zone, 202–203
  and conservation, 208
  decrease, 314–315
  equipment, 200
  and farmers' skill, 201
  and natural disasters, 240
  new varieties, 315–323, 374–375
  optimal, 203, **205**, 208, 242
  and transgenic crops, 219
crop ecology, 192–193
crop management, 275
crop prices and yield increases, 18
crop production
  and biodiversity, 232–233
  environmental impact, 383
  genetic erosion, 218

global changes, 372
and profitability, 192, 193, 194
and resource protection, 200
species choice for water utilization, 103
species selection for abiotic stress
adaptation, 110–111
surpluses and price slump, 312
value-added processing, 315
water resources, 102–103
crop science
and cropping systems development,
274–276
environmental aspects, 373
moral responsibility for research, 379
crop yield, 334
and climate, 105
and diversification, **197**
increases, **309**, 337–338, 358, 363
investment in research, 299–300
and water use, 131–132, 374
cropping systems, 261–277
adoption, 266–268, 275–276
co-operation for change, 274
crop science contribution, 274–276
decision rules, 265
decision support systems (DSSs), 270,
272–273, 277
definition, 261
future change, 276–277
government policy and change, 273
international concerns, 261–262
new targets and criteria
implementation, 268
planning for farm resources, 263–265
prototype modelling, 268, **269**, 270
crops
development of new markets, 323–326
historical review, 311–314
newly created, 310–311, 324
pest inhibition by trees, 184–185
prehistoric domestication, 308
residues, 112
roots compared with trees, 180–182
rotation, 164–165, 197–199, 202,
204, 217, 235, 254–255,
358–359
speciality, 310
traditional species, 309–310
underutilized and neglected species,
310, 324, 325
volume comparisons, **307, 308**
culinary herbs, 322–323
cultivar blends, 194
cyclodextrins, 343
cystein-rich metallothionein-like gene, 353

decision support systems (DSSs), 270, 272–
273, 277, 383
deep drainage, 124–129
and perennial vegetation, 128–129
and root research, 179–180

deforestation, 373
and fragmentation, 246
dehydration
plant growth and development
adjustment, 89, 90–92
stress adaptation, 94
destructive biota, 214
developed countries
growth in cereal production, 19
income growth, **17**
intercrops, 198
population increase, **16**
transgenic crops, **355**
developing countries
access to biotechnology research, 362
education and funding requirement,
383
genetic resource wealth, 377
grassland management, 73–76
growth in cereal production, 19
income growth, **17**
and intercrops, 198
population increase, **16**
sources of genetic diversity, 365
transgenic crops, **355**, 362–365
development strategy and food security, 25–
26
diet diversification, 17–18
DNA markers, 146–147
domestication of species, 217, 375
drought, 12, 40, 46, **47**, 81, 84
risk assessment, 138
drought stress tolerance, 110–111, 374
in maize populations, **141**
plant breeding, 121–123, 130
in rice, 140
and transgenic research, 123, 149
dryland cropping systems, 124–125, 132
Dunaliella salina, 93

ecological compensation areas, 225–226
ecology and agroforestry, 175–177
economics
and biodiversity, 225–226
and crop choice, 192–194, **195**, 200
and food security, 33
and integrated pest management, 187
transgenic vs. traditional crops, 360–
361
and water competition, 103
ecotone, 247
Ecuador
improved potato population breeding,
239
subsistence farming, 234–235
education for change, 15, 165–170, 239
einkorn (*Triticum boeticum*), 217, 310, 317
emmer (*Triticum turgicum*), 217, 310, 317
energy crops
see biofuels; oilseeds
environmental degradation, 28–29

environmental policy
  and grassland, 66–67
environmental stress adaptation
    mechanisms, 83, 110–111
  and crop production success, 113–114
  and genomic research, 92–95
  immediate (downstream) reactions, 85–86
  long-term developmental changes, 89–92
  stress sensing and signalling, 86–89
  variety selection, 143–150
environmental stress factors interaction, 106
Environmentally Sensitive Areas, 71–72
equipment for crop diversification, 200
erosion, 27, 29, 103, 107, 252, 253, 255, 273–274, 373
eskimo 1 mutation gene, 336
ethanol, 315, 320, 321
ethics and food production, 356, 361–362, 372, 373, 378
ethylene inducible genes, 333
Europe
  forest ecosystems, 215
  vegetable crops, 322
  weed diversity, 221
Europe, Eastern
  government policy and crop system changes, 273
  ley farming, 70
Europe, North, yield productivity growth, 45
European Union
  cereal exports, 19
  grassland policy, 66
European Union Agri-Environmental Measures, 71–72
evapotranspiration, 121–123, 125–127, 374
expressed sequence tags (ESTs), 83, 92–94, 149, 355

farm size, 244–245
  and crop choice, 194
farm subsidies, 28, 67, 71–72 315, 273
farmer education
  for pest management, 165–170
farmer field schools
  and potato disease, 239
farmers' participation in research, 30, 75–76
farmyard manure, 107
  and wheat production, **108**
fatty acids, 325, 342
faba bean (*Vicia faba*), 235, 310, 318
feed grain prices, 18, 47–48
ferritin in transgenic plants, 353
fertigation, 113, 139
fertilizers, 18, 27, 38, 113, 125–126, **127**
  experimentation for wheat production, **108**
  and field boundaries, 249
  and grassland nutrition, 66
  and increased phytochemicals, 62

  and iron and zinc deficiency, 59–60
  nitrogen, potassium and rice, 289, 295, 301
  phosphorus, 363
  and pollution, 273
  and reduced unit costs, 68
  and rice disease control, 166
  and tropical soil leaching, 107
fibre production, 4, 68, 318–319
field beans and agroforestry research, 179
field boundaries, 247, 249–251
  and mechanization, 244–245
  species composition, 250
field peas (*Pisum sativum*), 202, 206–207
  and intercrops, 198
financial resources and crop system change, 266–267
flavenoids, 54, 55, **56**, 62
flax (*Linum usitatissimum*), 318–319, 375
  mimicry by weeds, 221
flooding, 46, 81
flower crops, 4
  transgenic varieties, 354
flowering time and abiotic stress adaptation, 89, 90, 91
fodder crops, 354
food aid, 27–28
food demand diversification, 372
food fortification and supplementation, 58
food labelling, 377
food prices, 16, 26–27, 38, 312
  subsidies, 28
  and undernutrition, 25
food production
  and biodiversity, 226
  uneven distribution, 370–371
food safety, 383
food security, 30, 381, 382
  and biodiversity utilization, 242
  and biotechnology, 362–365
  and crop diversification, 374–375
  and decreasing resources, 28029
  definition, 1
  and land degradation, 10, 12
  and pests and diseases, 155
  policy, 23–24
  and research, 43–44
food storage, 25–26, 139
food surpluses, 41–42
  predictions, 40–41
forage systems research priorities, 46–48
foreign proteins subcellular targeting, 334, **335**
forest ecosystems, 215
forest gardens, 102, 110
forest land clearance, 28, 29
forest trees and regeneration, 215–216
France
  grain yield predictions, **36**
  nitrogen fertilization decision rules, **271**
  prototype cropping system, 268, **269**
fructans, 343

fruit crops, 321–322
    world crop comparisons, **307, 308**
fungicides
    and carcinogens, 170
    and late blight disease, 238
    potato disease management, 167–168
Future Harvest Centers, 231, 233

Gambia millet growing, **263**
gene banks, 233, **234**
gene transfer, horizontal and vertical, 359,
    364
gene use restriction technology (GURT),
    355–356, 363–364
genetic conservation of crop germplasm, 208
genetic control of disease, 157–158, 159,
    238, 239–240, 336–337, 351
genetic diversity
    and crop production, 208
    reduction in modern farming, 217–
        218, 218–219
    and species survival, 215–216
    and transgenic plants, 358–359, 362,
        365
genetic drift, 216
genetic engineering
    and biodiversity, 214
    and genetic erosion, 218–219
    and potato varieties, 163
genetic erosion, 236
genetic mapping, 146–147
genetic resources, 377, 382
    and crop production, 232–233
genetically modified organisms (GMOs), 12
    and increased phytochemicals, 62
    and iron and zinc deficiency, 61
    national safety controls, 365
    and pest resistance, 222
    public resistance, 19, 159–160, 311,
        326, 331, 352, 369–370,
        376, 377
    regulatory mechanisms, 378
    vegetable changes, 322
    see also transgenic crops
genetics
    and breeding for target environment,
        143–150
    stress tolerance mechanism study, 83,
        139–143
genomics, 158, 160–161, 162
    and abiotic stress, 92–94, 148–150
geographic information systems (GIS), 138,
    150, 157, 245, 256, 383
Global Assessment of Land Degradation, 8–9
global food demand and supply predictions,
    34–39
global food security, 40–41
Global Forum of Agricultural Research, 30
global warming, 102, 105, 283, 293–294,
    298–299, 300
    see also climate change
globalization of food market, 371–372, 379

glucosinolates, 53, 54, 55, **56,** 62
glycophytes, 86, 87, 93
glyphosate, 357, 361
golden rice *see* β-carotene
government investment in biotechnology,
    365
grain storage, 139
grain yield gap predictions, 36, 38, 41
grassland
    area statistics and definition, 65–66
    biodiversity, 71–72
    and climate change, 76–77
    conversion to cropping, 67
    in developed countries, 67–73
    in developing countries, 73–76
    and pollution, 66–67
    and soil erosion, 253
grassland research, 12, 78
    priorities, 46–48
    and traditional practices, 75–76
grazing, 68
    and fatty acid reduction, 70
green manures, 112
green revolution, 102, 107, 119, 266
greenhouse gases, 76–77, 78, 252, 299,
    320
    legislation, 4

habitat
    biodiversity, 232–233
    connectivity, 215, 246–247, 248–249
    disturbance, 220
    fragmentation, 214, 215–216, 223,
        225, 246, 248–249
    heterogeneity, 215
    patch dynamics, 219–220
    recreation and conservation, 225, 226
halophytes, 86, 87, 93
health value of crops, 55
heavy metals, 81, 353, 363
hedgerows, 247, 248–249, 250–251
    and mechanization, 244–245
herbicide resistance, 330–331, 334, 357–
        358, 359
herbicide tolerance, **354**
herbicides, 198
    and field boundaries, 249
    flora and fauna decline, 251
    plant genetic resistance, 222, 224
herbivores as pests, 189
heterogeneity, 224
    definition, 245–246
    and scale, 247
heterogeneity of landscape, 254–255
    and sustainable cropping systems, 243–
        257
heterosis, 140
home-gardens and genetic diversity, 217
horizontal gene transfer, 359
household food security, 8, 10, 19–28
    and food aid, 27–28
    and food prices, 16

household food security *continued*
 and policy actions, 23–24
hunger, 21, 25–26

income growth and food demand, 16, 17.18
India
 crop surpluses, 312
 genetic erosion of rice, 218
 grain yield predictions, **36, 37**
 late blight disease, **238, 239**
 natural vegetation survival, 215
 pest damage to trees, 184
 population increase, **16**
 Punjab water table decline, 119
 rainfall utilization and agroforestry,
  181
 rainfed lowland rice, 46, **47**
 semiarid millet and agroforestry, 180
 water resources, 84, 102
 zinc deficient soil, 59
Indonesia
 grain yield predictions, **36, 37,** 38–39
 rainfed lowland rice, 46
 rice production and pesticides, 166,
  167
industrial-usage crop improvements, **353**
information technology and research, 49,
  373, 378
infrastructure and household food security,
  21, 23
insecticide use and *Bt* transgenic plants,
  357–358
insects, 46, **47**
 and field boundaries, 251
 flight paths, 186
 and island biogeography, 248
 resistant plants, 123, **354**
integrated crop management systems, 111–
  112, 113, 382
integrated crop protection, 274–275, 276
integrated pest management (IPM), 155,
  163–165, 170–171, 187–188, 224
 farmer education, 165–170
intellectual property rights, 40–41, 150,
  219, 325, 362, 371, 377
intercropping, 109–110, 112, 178, 194,
  197–199, 235, 358
International Centre for Agricultural
  Research in Dry Areas, 325
International Crops Research Institute for
  the Semi-arid Tropics, 325
International Future Harvest, 111
International Plant Protection Convention,
  378
International Potato Center, 231, 233–234,
  237–238, 325
 and potato late blight, 157, 162–163
International Rice Research Institute, 166
International Service for National
  Agricultural Research, 364
International Service for the Acquisition of
  Agri-Biotech Applications, 364

International Undertaking on Plant Genetic
  Resources, 377
internet and information exchange, 171,
  382–383
invertebrates, 248
 and food source, 251
iodine deficiency, 54, 57
iron deficiency, 54, 57, 59–61
 and transgenic plants, 353
irrigation, 23, 27, 39, 42, 113, 119, 121
 and crop choice, 193
 crop transpiration efficiency, 122
 deficiency use, 130
 efficiency, 129–130, 131, 132, 139,
  374
 and green revolution, 102
 and pollution, 299
 rice yield increases, 286–287, 301
 and salinization, 84
island biogeography, 248
isoflavones, 55, **56,** 62

Japan
 edamame soybean, 318
 grain yield predictions, **36, 37,** 38
 population increase, **16**
Java grain yield growth decline, 38–39
jojoba (*Simmondsia chinensis*), 311, 318

karite (*Vitellaria paradoxa*) and Sahel cotton
  fields, 178
Kenya
 crop root research, 181
 maize and agroforestry research, 178–
  179, 180–181
kiwifruit (*Actinidia deliciosa*), 311, 313, 321,
  324
Korea grain yield growth decline, 38
kudzu, 216

labour restrictions and cropping
  management, 262–264
land cultivation loss of area, 2, 5
land degradation, 2, 5–6, **7,** 8–12, 23, 27,
  67, 103–105, 111–112
 in Andes, 234
 and grassland management, 73
land drainage, 245
land resources, 1–2, 3–4, 5, **7,** 48, 282
 and crop yield growth, 363, 365
 impact of infrastructure, 2–3
 marginal, 3
land scarcity, 103–105
land tenure, 23
 and land degradation, 10
 and sustainable farming, 75
land use, 383
 changes, 248
 intensification, 373

landraces, 208, 218, 358
landscape
    conservation, 66, 72
    heterogeneity, 243–257
landscape management, 253–256
    and industrial farming, 207–208
Latin America
    income growth, **17**
    population increase, **16**
leaf area index and photosynthesis, 291–293, 298, 300
leaf morphology and arid environments, 92
legumes, 317–318, 358–359, 375
    combined with cereals, **195**
    and crop diversification, 209
    crops and pasture rotation, 204, 206–207
    forage, 74
    grain, 352
    and nutrient management, 198–199
    world crop comparisons, **307, 308**
lentils (*Lens culinaris*), 202, 203–204, **205**
ley farming, 70, 77
    in marginal lands, 74
light interception and agroforestry, 180
livestock, 46–48
    genetic diversity, 219, 223
    industrialized production, 40
    production and resources, 48
low tillage agriculture, 244, 267, 357
lucerne, 129, **313**
    leaf photosynthesis, 292–293
lupins (*Lupinus* spp.), 201, 206–207, 318

maca (*Lepidium meyenii*), **234**, 240–241
Madagascar crop system change, 266–267
maize (*Zea mays*), 235, 320
    abiotic stress, 137, **138**, 140
    and agroforestry research, 178, 179
    assessment of increase requirement, 34, **35**
    Bt transgenic, 311, 313–314, 326, 351, 353, **354**, 357–358, 361, 365
    crop field weed infestation, 184
    crop rotation, 165
    crop selection for drought stress tolerance, 111
    domestication of species, 217, 315
    drought tolerance, **141**
    genome, 149
    grain yield predictions, **36, 37**
    and green manure, 112
    and intercrops, 198
    and iron and zinc deficiency, 60
    late blight disease, 324
    mimicry by weeds, 221
    in poor environments, 44
    root worm, 221–222
    rooting structure, 181
    selection for abiotic stress tolerance, 144–146

transgenic development, 316, 352
    water usage, 374
    yield and agroforestry, 180–181
    yield and soil organic matter, **104**
    yield during drought, 122
    yield potential, 39, 40
Malawi savanna agroforestry research, 179
Malaysia paddy field weed diversity, 221
mammals and field boundaries, 251
mannose selection markers, 331
marginal lands
    and household food security, 23
    and ley farming, 74
    and poverty, 30
    productivity and tree loppings, 108
marker-assisted selection (MAS), 146–147, 160–162
marker genes, 330–331, 330–332, **332,** 355
mashua (*Tropaeolum tuberosum*), **234,** 310
meadowfoam (*Limnanthes alba*), 311, 318, 324
meat and products, 47
    consumption, 18
    organic, 70
    world production, **308**
mechanization
    and crop production, 207, 209
    cropping systems and resources, 263
    and farm design, 244–245
    and intercrops, 198, 199, 201, 209
    and landscape patterning, 254
medic (*Medicago*) pasture, 204, **205**
medicinal plants, 322–323, 324
Mediterranean climate crop research, 124–126, 130
Mesembryanthemum genome, 87, 93–94
metapopulation theory, 248
methane gas, 67, 77
Mexico
    financial resources and crop system change, 266–267
    wheat yields, 283, 285, **286,** 287
micronutrient deficiencies, 54, 57–61
Middle East
    land resources, **3,** 6, **7**
    late blight disease, **238, 239**
    water resources, 84
milk
    and milk products, 73–74
    organic, 70
    products demand, 67
millet, 111, 262–263, 310, 324, 365
    pearl (*Pennisetum glaucum*), 317, 324
    in poor environments, 44
    semiarid crop and agroforestry, 180
    and tree intercropping, 182
    with tree shading, 178
modelling, 287–290, 376
    for crop management, 138, 139
    for crop systems, 262–265, 268, **269,** 270, 276
    for farm planning, 201, 206, 207, 210

millet *continued*
    for food security, 34–35, 41
    of spatial dynamics, 255–257
molecular biology, 325–326, 382
    and drought resistance, 123
    training, 364
monocultures, 192, 194, 196, 208, 372
    and crop rotation, 209–210
    and genetic diversity, 217, 223, 224–225
monogenic resistance to biotic stress, 158–160
Moroccan water shortages, 374
multilines vs. isolines, 199
mung bean (*Phaseolus aureus*), 318
    yield and soil organic matter, **104**

narcotic crops, 319–320
narrow-leaf lupin, 206–207
national food security, 12, 19
natural disasters
    and food aid, 28
    and household food security, 21, **22**, 23
natural resources, 41, 42
    and household food security, 21, **22**, 23
    preservation, 28–29, 34, 373–374
    and yield gain, 38
nature conservation, 373
nematodes, 185, 186
Netherlands intensive dairy farming, **69**
New Zealand
    crop diversification, 312–313
    potato cropping, 156
Nigeria groundwater use by crops and trees, 182
nitrate control, 69
nitrogen, 289, 295–296, **297**, 298
    fertilization, 125–126, **127**
    leaching, 252–253
    levels and grassland nutrition, 66
    supply and management, 43–44, 78, **353**
North Burkina millet growing, **263**
nut crops, 321–322
nutrient capture and yield, 294–295, 296
nutrient conservation, 29
nutrient deficiencies, 12, 40
    and maize abiotic stress, 137, **138**
nutrient depletion, 6, 9, 10, 103, 107
    and displacement to urban areas, 10, 12
nutrient management, 42–44, 45, 69, 197, 198. 199
nutrient replacement, 113
nutrient resources and rainfed lowland rice, 46
nutrition
    and diseases, 53–54
    education, 21, 58
    and food diversity, 26

    and phytochemicals, 53–62
    programmes, 28

oats (*Avena sativa*), 310, **314**
    multilines, 199
oca (*Oxalis tuberosa*), **234,** 310
    weevils, 170–171
oil industry and biofuel, 320–321
oilseeds, 318, 342
    and crop diversification, 209
    world crop comparisons, **307, 308**
    see also biofuels
organic farming, 70, 102, 114, 244, 372
ornamental plants, 323
osmotic stress tolerance, 87, 88, 89–91, 94, 121, 142, 335–336

Pacific area
    agricultural land shortage, 103–105
    land resources, **3**
Pakistan land resources, 3
palatine, 343–344
papaya transgenic varieties, 352, 353, **354**
parkland systems, **177,** 179
    and pest control, 186
pasture and crop rotation, 206–207
perennial pasture, 129, 132
perennials, 11, 375
    and diversification, 194
    for erosion control, 199
    and field boundaries, 249
    and resource conservation, 201–202
    vegetation and cropping systems, 128–129, 132
Peru
    Farmer Field Schools for pest management, 168, 170
    potato breeding, 239
    potato genetic erosion and conservation, 236, 237
    potato late blight, 157
    subsistence farming, 234–235
pest management, 29, 43
    and farmer education, 165–170
    and tree microclimates, 186–187
pest–predator relationships, 251–252
pesticides, 27
    botanical, 188
    plant extracts, 164
    poisoning and environmental damage, 164
pests and diseases, **47,** 382
    acquired resistance, 222, 224
    agroforestry, 182–185
    biological control, 167
    brown planthopper, 166–167
    and crop diversification, 196, 197, **197**
    crop losses, 156
    damage to trees, 184
    diversity in ancient agroecosystems, 220–221

genetic resistance to pesticides, 360, 365
late potato blight, 156-157, 162-163, 237-240
oca tuber weevils, 170–171
potato bacterial wilt, 168
resistance, **17,** 123, 357–358, 359, 364
resistance genes, 351
root worm, 221
and transgenic plants, **353**
pharmaceuticals, 68
cropping area, 4
phenolic compounds, 53, 54
phenology, 140
Philippines
paddy field arthropod diversity, 221
rainfed lowland rice, 46
rice production and pesticides, 166, 167
rice yields, 283, **284**
traditional plant knowledge, 218
phosphorus
and plant biodiversity, 72
utilization, 353, 363
photosynthesis, 337
and yield, 290–294, 300
phytase in transgenic plants, 353
phytic acid and iron deficiency, 59
phytochemicals and health, 53–62
phytooestrogens, 53, 54
phytosterols, 53, 54, 55, **56,** 57, 62
pigeon pea (*Cajanus cajan*), 111, 318
plant breeding, 30
for abiotic stress adaptation, 114
back-cross, 161
and biotechnology, 161–162
for biotic stress resistance, 157–158
conventional programmes, 144–146
for drought resistance, 121–123, 130
and increased phytochemicals, 62
and iron and zinc deficiency, 60
marker-assisted, 146–147, 160–162
research for nutritional value, 26
secondary traits, 145
selection for abiotic stress tolerance, 145–146, 147
selection for environment, 143–144
stressed vs. unstressed comparisons, 145
plant genetic resources management, 377
plant genomes and public databases, 378
political control of food production, 15, 19, **22,** 23, 24–25, 30, 243–244
pollution, 10, 27, 28, 29, 34, 68–69, 102, 103, 243, 273
management, 66–67, 68–69
polymer production, 334, 341–343
polyunsaturated fatty acids, 342
population biology, 248, 257
population dynamics, 216
population growth, 1, 10, 15, 16, 24, 33, 45, 73, 77, 101, 102, 370

and biodiversity, 236
and household food security, 21, **22,** 23
population-level genetic diversity strategies, 159
post harvest improvements, 334, 339–341
and transgenic plants, **353**
potatoes (*Solanum*), 315
Andean disease management, 167–170
bacterial wilt, 162, 168
Bt transgenic, 162, 338, 352, 353, **354,** 358
diversity conservation, 235–236
Farmer Field Schools for pest management, 168
genetic erosion in Andes, 236–237
genetic resources, 231, 233
late blight disease, 156–157, 162–163, 237–240
post harvest improvements, 339–340
virus Y and X, 237, 240
poverty, 25, 26, 27, 30, 45
and food security, 21, **22,** 23
precision agriculture, 42–43, 139, 244, 245, 270, 273
production cost reductions, 41
proteins production, 341
pseudograins, 310, 317

quantitative trait loci (QTLs), **82,** 83, 123, 146–147, 160–161, 163
quinoa (*Chenopodium quinoa*), 235, 317

radiation use efficiency, 44, 288–290
rainfall and grassland management, 73–75
rainfall patterns and crop growth, 124–126, 131–132
rainfall use efficiency, 178, 179
and agroforestry, 180–182
rainfed crops, 46, **47,** 193
rangeland management
in developed areas, 72–73
in dry areas, 74–75
rapeseed (*Brassica napus*), 198, 201, 202, 275, 318, 320, 324
herbicide resistant, 222
transgenic, 342, 352, 353, **354**
reactive oxygen species detoxification systems, 335–336
recreation, 2, 72, 74
regional food security, 44–46
reproductive growth and stress tolerance, 140–141
research funding, 325, 383
resistance genes, 157–158, 159, 238, 239–240, 336–337, 351
quantitatively inherited, 160–161
resistance management, 222
resource-poor agroecosystems, 44–46
resources
biota, 214

resources *continued*
    capture, 177–178
    conservation, 200, 201–202
    use management, 382
reverse genetics, 148
rhizosphere improvement, 109, 355
ribozyme technology, 329
rice (*Oryza sativa*), 93, 315
    assessment of increase requirement, 34,
        35
    brown planthopper, 166
    canopy photosynthesis, 290–294, 298,
        300
    drought stress tolerance, 140
    genetic erosion, 218
    genome, 149, 355, 366
    global production, 281–282
    grain yield predictions, **36, 37**
    hybrid varieties, 316
    irrigated systems and nitrogen, 295
    modified for iron content, 61
    and nitrogen fertilizer, 112
    osmotic stress tolerance, 142, 146
    pest management Farmer Field Schools,
        166–167
    prices, **20**
    production cost comparisons, 371
    productivity and tree loppings, 108,
        109
    quality, 282–283
    rainfed lowland rice, 46, **47**
    resistance breeding, 158, 159
    sink size, 296–298
    stress genomic research, 94
    thermal damage, 299, 300
    transgenic, 149, 159, 353
    and vitamin A deficiency, 58–59
    water deficit, 137
    water stress and yield reduction, 103
    weed diversity, 221
    yield barriers, 282–301
    yield modelling equations, 287–290
    yield potential, 39, 40, 44
    yield trends, 283–287, 300–301
Rio Convention, 243–244
riparian zones and nitrogen leaching, 253
risk management, 193–194, **195,** 203–204,
    209, 265
Rockefeller Foundation, 364
root
    distribution and drought, 141
    pruning, 188–189
    research and agroforestry, 179–180
    structure and function, 181–182
rubber, 308, 319

Sahel cotton and tree planting, 178
salicylic acid inducible genes, 333
salinity
    as abiotic stress, 81, **82,** 86
    stress genomic research, 94

salinization, 6, 9, 12, 104, **138,** 179, 199–
    200
    and crop diversity, **197**
    and deep drainage, 125, 127
    and perennial vegetation, 128–129
    plant sensing and signalling, 87–89,
        88, 89
    stress tolerance, **86**
    through irrigation, 119
    and transgenic crops, 363
savanna ecosystems, 176–177, 179
seed fairs, 236–237
seed market financial control, 371
seed technology
    hybrid production, 338–339
    and increased yield, 38
semiarid regions
    growth patterns and water supply,
        124–12, 132
    millet and agroforestry, 180
semiarid tropics
    and agroforestry, 176–177
    and alley cropping, 175
Senegal millet growing, **263**
shading microclimate, 179, 186–187
sheep
    in Syrian agriculture, 203–204
    in Western Australia, 206
shelter belts, 178
sink development, 337–338
skill development and crop diversification,
    201
slash and burn agriculture, 45, 110, 183
soil
    chemical changes, 373–374
    compaction, 6
    conservation, 29, 48, 103–104, 196,
        197, 201
    degradation, 27, 41, 102
    erosion, 6, 9, 199
    erosion and crop diversification, **197**
    erosion and retained biomass, 11–12
    microbial populations, 109
    quality, 44
soil fertility, 113
    and intercropping, 109–110
    and recycled manure, 47
    and unplanned biota, 226
soil organic matter, 11–12, 107, 108–109
    and crop yields, **104**
    and water retention, 103
sorghum, 316–317, 365
    osmotic stress tolerance, 142, 146
    in poor environments, 44
    with tree shading, 178
sorghum, sweet (*Sorghum bicolor*), 375
Soviet Union cereal exports, 19
soybean (*Glycine max*), 317–318, 324
    Bt herbicide resistant, 313–314
    and corn root worm, 221–222
    transgenic, 311, 324, 352, 353, **354,**
        361
    value-added processing, 315

species-area relationships, 215–216
species dispersal ability, 248–251
species diversity and grassland isolation, 251
species survival and genetic variation, 215–216
spelt (*Triticum spelta*), 310, 317
squash transgenic varieties, 352, 353, **354**
Sri Lanka genetic erosion of rice, 218
starch production, 341–342
staygreen, 142–143, 146
stress tolerance, 45
   and genomics, 139–143, 148–150
   and yield potential, 40
subsistence farming, 26, 45–46
   in Andes, 234–235
   and transgenic weeds, 364
   in tropical regions, 101–102
sugar beet biological pest control, 226
sugar crops, 315, 319, 343–344
   world crop comparisons, **307, 308**
sustainable agriculture, 5, 9, 106–107,
      111–112, 373, 375–376
   and agroforestry, 176, 177–178
   and biodiversity, 213, 224
   and crop diversification, 194, **195,** 196
   and the environment, 372, 373, 381
   and the green revolution, 102
   and industrial farming, 207–208, 210
   and land tenure, 75
   and landscape ecology, 255–256
   and landscape heterogeneity, 243–257
   and traditional agricultural methods,
      113–114
   and transgenic crops, 361–365
sustainable development
   agronomy and ecology blending, 256
sweet potatoes (*Ipomoea batatas*), 352
   genetic resources, 231, 233
   weevil, 170
Switzerland
   chestnut cultivars, 218
   ecological compensation areas, 225–226
Syria crop diversification, 203–204, **205**

Tanzania tree root structure research, 181
Taxol® (*Taxus brevifolia*), 311, 323, 324
temperature fluctuations, 81
temperature protectants, 142
Thailand
   grain yield predictions, **36, 37,** 38
   rainfed lowland rice, 46
thalwegs, 253, 255
timber, 68, 312, **313**
tobacco plants
   and nematodes, 185
   transgenic improvements, 340–341,
      344
tomato (*Lycopersicon esculentum*), **307,** 322,
   354
tourism, 67, 72, 74, 103
transgene expression, 331–334

transgenic crops, 41, 149–150, 159, 162,
      170–171, 219, 222, 311, 329–345, 351–366
   agronomic traits, **353**
   ecological aspects, 356, 357–360
   economic aspects, 360–361
   effect on non-target organisms, 359–360
   ethical aspects, 356, 361–362
   and food secutiry, 363
   future technology, 343
   industrial uses, **353**, 354
   male sterility, 324, 338, 355–356
   public concern over risks, 376
   quality traits, **353**
   rice with β-carotene, 58–59, 62, 283,
      316
   trait modification, 352, **353**
   world crop statistics, 352, **354, 355**
   see also genetically modified organisms
      (GMOs)
transpiration efficiency, 122, **123,** 178
transposons (jumping genes), 352
trees
   and annual intercrops, 198
   fallows, 183–186
   plantations, 4
   pruning, 181–182, 188–189
   roots compared with crop roots, 180–182
   strip planting, 201
   transgenic varieties, 353
trehalose, 343
tropical crop production stress, 101–111
tropical rainforest, 215
tropical soil fertility improvement, 139
tubers, world crop comparisons, **307, 308**
Turkey, zinc deficient soil, 59

ulluco (*Ullucus tuberosus*), **234,** 241,
   310
undernutrition, 21, 25–26
UNESCO Life Sciences Programme, 364
urbanization, 10, 17–18, 45, 236
Uruguay trade agreement, 27
USA
   crop rotation, 165
   crop yield improvement, 45
   crop yield improvements, **309**
   grain yield predictions, **36, 37,** 38
   Indiana crop diversification, 313–314
   insecticide use, 357–358
   limited crop range, 312
   maize yield increases, **284,** 285
   planting of kudzu, 216
   potato cropping, 156
   rice yields, 283, **284**
   soybean production, 318, 324
   speciality wheats, 317
   sugar beet biological pest control, 226
   transgenic crops, 219, 352
   vegetable crops, 322

USA *continued*
    water resources, 84

vegetables, 322, 325
    transgenic varieties, 353–354
    world crop comparisons, **307, 308**
vertical gene transfer, 359, 364
vetch (*Vicia sativa*), 203–204
viral disease resistance, 237, 240, 336, 337,
    353, 354, 360, 364
vitamin A deficiency, 57–58, 58–59
vitamin E, 344–345

wars and unrest
    and biodiversity, 236
    and food aid, 28
    and household food security, 21, **22,**
        23, 312
water
    deficit as abiotic stress, 84–85
    and economic choices, 103
    management and shortages, 28, 29
    quality, 27, 34, 253, 257
    retention and soil organic matter, 103,
        110
    retention and soil quality, 113
watermelons (*Citrullus lanatus*), 204, **205**
water resources, 23, 40, 41, 43, 44
    global, 119–132
    in tropical agriculture, 102–103
    world shortage, 374
water tables and salinization, 179, 199–200
water use efficiency plant breeding, 121–123
weather forecasting, 138, 150
weed control, 112
    and crop diversification, 196, 197, 198
    and herbicide-tolerant transgenic
        plants, 357–358
    and transgenic plants, **353**
weeds, 40, 46, **47,** 382
    and accidental gene transfer, 359, 360,
        364
    competition and agroforestry, 183–184
    crop mimic species, 221
    genetic resistance to herbicides, 222,
        224
    increase in diversity, 220–221
    native habitats, 219
    resistant to herbicides, 331
wheat yield productivity growth, 45
wheat (*Triticum aestivum*), 217, 312, **314**

canopy temperature, 146
carbon isotopic signature, 122, **123**
    and crop diversification, 202
    crop transpiration efficiency, 123
    economics, 200, 202
    fertilizer experimentation, **108**
    genetically modified, 316
    grain yield predictions, **36, 37**
    historical development, 315
    increase requirement, 34. **35**
    and iron and zinc deficiency, 60–61
    and legume crop rotation, 206–207
    nitrogen fertilization, 125–126, **127**
    osmotic stress tolerance, 142
    prices, **20**
    rainfall and crop yields, **131**
    resistance breeding, 158
    in Syrian agriculture, 203–204, **205**
    yield potential, 39
wild-crafted crop, 311, 313
wildlife conservation, 34, 47–48, 66
windbreaks, 186–187
world food security, 16–19
World Trade Organisation
    and cereal markets, 19
    Sanitary and Phytosanitary agreement,
        371–372, 378, 379

xylose selection markers, 331

yacon (*Smallanthus sonchifolius*), **234,** 241
yams
    iron and zinc deficiency, 60
    water stress and yield reduction, 103
yeast (*Saccharomyces cerevisiae*), 85, 86, 89–
        90
    osmotic stress tolerance, 87
    stress genomic research, 94
yield
    barriers, 281–301
    definitions, 282
    gain assessments, 34–39
    increase decline, 30
    limitation, 287, 288–293, 299, 300
yield potential, 42–44
    of grassland, 74
    prediction, 48
    and stress tolerance, 40

zinc deficiency, **47,** 54, 58, 59–61